Lecture Notes in Artificial Intelli

Edited by J.G. Carbonell and J. Siekmann

Subseries of Lecture Notes in Computer Science

Zbigniew W. Raś   Shusaku Tsumoto
Djamel Zighed (Eds.)

Mining Complex Data

ECML/PKDD 2007 Third International Workshop, MCD 2007
Warsaw, Poland, September 17-21, 2007
Revised Selected Papers

 Springer

Series Editors

Jaime G. Carbonell, Carnegie Mellon University, Pittsburgh, PA, USA
Jörg Siekmann, University of Saarland, Saarbrücken, Germany

Volume Editors

Zbigniew W. Raś
University of North Carolina
Department of Computer Science
Charlotte, NC 28223, USA
and
Polish-Japanese Institute of Information Technology
Dept. of Intelligent Systems
Koszykowa 86, 02-008 Warsaw, Poland
E-mail: ras@uncc.edu

Shusaku Tsumoto
Shimane Medical University
School of Medicine, Department of Medical Informatics
89-1 Enya-cho, Izumo 693-8501, Japan
E-mail: tsumoto@computer.org

Djamel Zighed
Université Lumière Lyon 2
5 avenue Pierre Mendès-France, 69676 Bron Cedex, France
E-mail: abdelkader.zighed@univ-lyon2.fr

Library of Congress Control Number: 2008926861

CR Subject Classification (1998): H.2.5, H.2.8, H.3.3

LNCS Sublibrary: SL 7 – Artificial Intelligence

ISSN   0302-9743
ISBN-10   3-540-68415-8 Springer Berlin Heidelberg New York
ISBN-13   978-3-540-68415-2 Springer Berlin Heidelberg New York

Springer is a part of Springer Science+Business Media

springer.com

© Springer-Verlag Berlin Heidelberg 2008

Typesetting: Camera-ready by author, data conversion by Scientific Publishing Services, Chennai, India
Printed on acid-free paper   SPIN: 12273379   06/3180   5 4 3 2 1 0

Preface

This volume contains 20 papers selected for presentation at the Third International Workshop on Mining Complex Data–*MCD 2007*–held in Warsaw, Poland, September 17–21, 2007. MCD is a workshop series that started in conjunction with the 5th IEEE International Conference on Data Mining (ICDM) in Houston, Texas, November 27–30, 2005. The second MCD workshop was held again in conjunction with the ICDM Conference in Hong Kong, December 18–22, 2006.

Data mining and knowledge discovery, as stated in their early definition, can today be considered as stable fields with numerous efficient methods and studies that have been proposed to extract knowledge from data. Nevertheless, the famous golden nugget is still challenging. Actually, the context evolved since the first definition of the KDD process, and knowledge now has to be extracted from data becoming more and more complex.

In the framework of data mining, many software solutions were developed for the extraction of knowledge from tabular data (which are typically obtained from relational databases). Methodological extensions were proposed to deal with data initially obtained from other sources, e.g., in the context of natural language (text mining) and image (image mining). KDD has thus evolved following a unimodal scheme instantiated according to the type of the underlying data (tabular data, text, images, etc.), which, at the end, always leads to working on the classical double entry tabular format.

However, in a large number of application domains, this unimodal approach appears to be too restrictive. Consider for instance a corpus of medical files. Each file can contain tabular data such as results of biological analyses, textual data coming from clinical reports, image data such as radiographies, echograms, or electrocardiograms. In a decision-making framework, treating each type of information separately has serious drawbacks. It appears therefore more and more necessary to consider these different data simultaneously, thereby encompassing all their complexity.

Hence, a natural question arises: how could one combine data of different nature and associate them with a same semantic unit, which is for instance the patient? On a methodological level, one could also wonder how to compare such complex units via similarity measures. The classical approach consists in aggregating partial dissimilarities computed on components of the same type. However, this approach tends to make superposed layers of information. It considers that the whole entity is the sum of its components. By analogy with the analysis of complex systems, it appears that knowledge discovery in complex data cannot simply consist of the concatenation of the partial information obtained from each part of the object. The aim, rather, would be to discover more global knowledge giving a meaning to the components and associating them with the semantic unit. This fundamental information cannot be extracted by the currently considered approaches and the available tools.

The new data mining strategies shall take into account the specificities of complex objects (units with which the complex data are associated). These specificities are summarized hereafter:

Different kind. The data associated to an object are of different types. Besides classical numerical, categorical or symbolic descriptors, text, image or audio/video data are often available.

Diversity of the sources. The data come from different sources. As shown in the context of medical files, the collected data can come from surveys filled in by doctors, textual reports, measures acquired from medical equipment, radiographies, echograms, etc.

Evolving and distributed. It often happens that the same object is described according to the same characteristics at different times or different places. For instance, a patient may often consult several doctors, each one of them producing specific information. These different data are associated with the same subject.

Linked to expert knowledge. Intelligent data mining should also take into account external information, also called expert knowledge, which could be taken into account by means of ontology. In the framework of oncology, for instance, the expert knowledge is organized under the form of decision trees and is made available under the form of best practice guides called standard option recommendations (SOR).

Dimensionality of the data. The association of different data sources at different moments multiplies the points of view and therefore the number of potential descriptors. The resulting high dimensionality is the cause of both algorithmic and methodological difficulties.

The difficulty of knowledge discovery in complex data lies in all these specificities.

We wish to express our gratitude to all members of the Program Committee and the Organizing Committee. Hakim Hacid (Chair of the Organizing Committee) did a terrific job of putting together and maintaining the home page for the workshop as well as helping us to prepare the workshop proceedings. Also, our thanks are due to Alfred Hofmann of Springer for his support.

December 2007

Zbigniew W. Raś
Shusaku Tsumoto
Djamel Zighed

Organization

MCD 2007 Workshop Committee

Workshop Chairs

Zbigniew W. Raś (University of North Carolina, Charlotte)
Shusaku Tsumoto (Shimane Medical University, Japan)
Djamel Zighed (University Lyon II, France)

Organizing Committee

Hakim Hacid (University Lyon II, France)(Chair)
Rory Lewis (University of North Carolina, Charlotte)
Xin Zhang (University of North Carolina, Charlotte)

Program Committee

Aijun An (York University, Canada)
Elisa Bertino (Purdue University, USA)
Ivan Bratko (University of Ljubljana, Slovenia)
Michelangelo Ceci (University of Bari, Italy)
Juan-Carlos Cubero (University of Granada, Spain)
Agnieszka Dardzińska (Białystok Technical University, Poland)
Tapio Elomaa (Tampere University of Technology, Finland)
Floriana Esposito (University of Bari, Italy)
Mirsad Hadzikadic (UNC-Charlotte, USA)
Howard Hamilton (University Regina, Canada)
Shoji Hirano (Shimane University, Japan)
Mieczyslaw Kłopotek (ICS PAS, Poland)
Bożena Kostek (Technical University of Gdańsk, Poland)
Nada Lavrac (Jozef Stefan Institute, Slovenia)
Tsau Young Lin (San Jose State University, USA)
Jiming Liu (University of Windsor, Canada)
Hiroshi Motoda (AFOSR/AOARD and Osaka University, Japan)
James Peters (University of Manitoba, Canada)
Jean-Marc Petit (LIRIS, INSA Lyon, France)
Vijay Raghavan (University of Louisiana, USA)
Jan Rauch (University of Economics, Prague, Czech Republic)
Henryk Rybiński (Warsaw University of Technology, Poland)
Dominik Ślezak (Infobright, Canada)

Roman Słowiński (Poznań University of Technology, Poland)
Jurek Stefanowski (Poznań University of Technology, Poland)
Alicja Wieczorkowska (PJIIT, Poland)
Xindong Wu (University of Vermont, USA)
Yiyu Yao (University Regina, Canada)
Ning Zhong (Maebashi Inst. of Tech., Japan)

Table of Contents

Session A1

Using Text Mining and Link Analysis for Software Mining 1
    Miha Grcar, Marko Grobelnik, and Dunja Mladenic

Generalization-Based Similarity for Conceptual Clustering 13
    S. Ferilli, T.M.A. Basile, N. Di Mauro, M. Biba, and F. Esposito

Trajectory Analysis of Laboratory Tests as Medical Complex Data
Mining 27
    Shoji Hirano and Shusaku Tsumoto

Session A2

Conceptual Clustering Applied to Ontologies: A Distance-Based
Evolutionary Approach 42
    Floriana Esposito, Nicola Fanizzi, and Claudia d'Amato

Feature Selection: Near Set Approach. 57
    James F. Peters and Sheela Ramanna

Evaluating Accuracies of a Trading Rule Mining Method Based on
Temporal Pattern Extraction 72
    Hidenao Abe, Satoru Hirabayashi, Miho Ohsaki, and
    Takahira Yamaguchi

Session A3

Discovering Word Meanings Based on Frequent Termsets. 82
    Henryk Rybinski, Marzena Kryszkiewicz, Grzegorz Protaziuk,
    Aleksandra Kontkiewicz, Katarzyna Marcinkowska, and
    Alexandre Delteil

Quality of Musical Instrument Sound Identification for Various Levels
of Accompanying Sounds 93
    Alicja Wieczorkowska and Elżbieta Kolczyńska

Discriminant Feature Analysis for Music Timbre Recognition and
Automatic Indexing 104
    Xin Zhang, Zbigniew W. Raś, and Agnieszka Dardzińska

Session A4

Contextual Adaptive Clustering of Web and Text Documents with
Personalization 116
   Krzysztof Ciesielski, Mieczysław A. Kłopotek, and
   Sławomir T. Wierzchoń

Improving Boosting by Exploiting Former Assumptions 131
   Emna Bahri, Nicolas Nicoloyannis, and Mondher Maddouri

Discovery of Frequent Graph Patterns that Consist of the Vertices with
the Complex Structures 143
   Tsubasa Yamamoto, Tomonobu Ozaki, and Takenao Ohkawa

Session B1

Finding Composite Episodes 157
   Ronnie Bathoorn and Arno Siebes

Ordinal Classification with Decision Rules........................ 169
   Krzysztof Dembczyński, Wojciech Kotłowski, and Roman Słowiński

Data Mining of Multi-categorized Data 182
   Akinori Abe, Norihiro Hagita, Michiko Furutani,
   Yoshiyuki Furutani, and Rumiko Matsuoka

ARAS: Action Rules Discovery Based on Agglomerative Strategy 196
   Zbigniew W. Raś, Elżbieta Wyrzykowska, and Hanna Wasyluk

Session B2

Learning to Order: A Relational Approach 209
   Donato Malerba and Michelangelo Ceci

Using Semantic Distance in a Content-Based Heterogeneous
Information Retrieval System..................................... 224
   Ahmad El Sayed, Hakim Hacid, and Djamel Zighed

Using Secondary Knowledge to Support Decision Tree Classification of
Retrospective Clinical Data 238
   Dympna O'Sullivan, William Elazmeh, Szymon Wilk, Ken Farion,
   Stan Matwin, Wojtek Michalowski, and Morvarid Sehatkar

POM Centric Multi-aspect Data Analysis for Investigating Human
Problem Solving Function 252
   Shinichi Motomura, Akinori Hara, Ning Zhong, and Shengfu Lu

Author Index 265

Using Text Mining and Link Analysis for Software Mining

Miha Grcar, Marko Grobelnik, and Dunja Mladenic

Jozef Stefan Institute, Dept. of Knowledge Technologies, Jamova 39,
1000 Ljubljana, Slovenia
`{miha.grcar,marko.grobelnik,dunja.mladenic}@ijs.si`

Abstract. Many data mining techniques are these days in use for ontology learning – text mining, Web mining, graph mining, link analysis, relational data mining, and so on. In the current state-of-the-art bundle there is a lack of "software mining" techniques. This term denotes the process of extracting knowledge out of source code. In this paper we approach the software mining task with a combination of text mining and link analysis techniques. We discuss how each instance (i.e. a programming construct such as a class or a method) can be converted into a feature vector that combines the information about how the instance is interlinked with other instances, and the information about its (textual) content. The so-obtained feature vectors serve as the basis for the construction of the domain ontology with OntoGen, an existing system for semi-automatic data-driven ontology construction.

Keywords: Software mining, text mining, link analysis, graph and network theory, feature vectors, ontologies, OntoGen, machine learning.

1 Introduction and Motivation

Many data mining (i.e. knowledge discovery) techniques are these days in use for ontology learning – text mining, Web mining, graph mining, network analysis, link analysis, relation data mining, stream mining, and so on [6]. In the current state-of-the-art bundle mining of software code and the associated documentations is not explicitly addressed. With the growing amounts of software, especially open-source software libraries, we argue that mining such data is worth considering as a new methodology. Thus we introduce the term "software mining" to refer to such methodology. The term denotes the process of extracting knowledge (i.e. useful information) out of data sources that typically accompany an open-source software library.

The motivation for software mining comes from the fact that the discovery of reusable software artifacts is just as important as the discovery of documents and multimedia contents. According to the recent Semantic Web trends, contents need to be semantically annotated with concepts from the domain ontology in order to be discoverable by intelligent agents. Because the legacy content repositories are relatively large, cheaper semi-automatic means for semantic annotation and domain ontology construction are preferred to the expensive manual labor. Furthermore, when dealing

Z.W. Raś, S. Tsumoto, and D. Zighed (Eds.): MCD 2007, LNAI 4944, pp. 1–12, 2008.
© Springer-Verlag Berlin Heidelberg 2008

with software artifacts it is possible to go beyond discovery and also support other user tasks such as composition, orchestration, and execution. The need for ontology-based systems has yield several research and development projects supported by EU that deal with this issue. One of these projects is TAO (http://www.tao-project.eu) which stands for Transitioning Applications to Ontologies. In this paper we present work in the context of software mining for the domain ontology construction. We illustrate the proposed approach on the software mining case study based on GATE [3], an open-source software library for natural-language processing written in Java programming language.

We interpret "software mining" as being a combination of methods for structure mining and for content mining. To be more specific, we approach the software mining task with the techniques used for text mining and link analysis. The GATE case study serves as a perfect example in this perspective. On concrete examples we discuss how each instance (i.e. a programming construct such as a class or a method) can be represented as a feature vector that combines the information about how the instance is interlinked with other instances, and the information about its (textual) content. The so-obtained feature vectors serve as the basis for the construction of the domain ontology with OntoGen [4], a system for semi-automatic, data-driven ontology construction, or by using traditional machine learning algorithms such as clustering, classification, regression, or active learning.

2   Related Work

When studying the literature we did not limit ourselves to ontology learning in the context of software artifacts – the reason for this is in the fact that the more general techniques also have the potential to be adapted for software mining.

Several knowledge discovery (mostly machine learning) techniques have been employed for ontology learning in the past. Unsupervised learning, classification, active learning, and feature space visualization form the core of OntoGen [4]. OntoGen employs text mining techniques to facilitate the construction of an ontology out of a set of textual documents. Text mining seems to be a popular approach to ontology learning because there are many textual sources available (one of the largest is the Web). Furthermore, text mining techniques are shown to produce relatively good results. In [8], the authors provide a lot of insight into the ontology learning in the context of the Text-To-Onto ontology learning architecture. The authors employ a multi-strategy learning approach and result combination (i.e. they combine outputs of several different algorithms) to produce a coherent ontology definition. In this same work a comprehensive survey of ontology learning approaches is presented.

Marta Sabou's thesis [13] provides valuable insights into ontology learning for Web Services. It summarizes ontology learning approaches, ontology learning tools, acquisition of software semantics, and describes – in detail – their framework for learning Web Service domain ontologies.

There are basically two approaches to building tools for software component discovery: the information retrieval approach and the knowledge-based approach. The first approach is based on the natural language documentation of the software components. With this approach no interpretation of the documentation is made – the information is extracted via statistical analyses of the words distribution. On the other hand, the knowledge-based approach relies on pre-encoded, manually provided information (the information is provided by a domain expert). Knowledge-based systems can be "smarter" than IR systems but they suffer from the scalability issue (extending the repository is not "cheap").

In [9], the authors present techniques for browsing amongst functionality related classes (rather than inheritance), and retrieving classes from object-oriented libraries. They chose the IR approach for which they believe is advantageous in terms of cost, scalability, and ease of posing queries. They extract information from the source code (a structured data source) and its associated documentation (an unstructured data source). First, the source code is parsed and the relations, such as derived-from or member-of, are extracted. They used a hierarchical clustering technique to form a browse hierarchy that reflected the degree of similarity between classes (the similarity is drawn from the class documentation rather than from the class structure). The similarity between two classes was inferred from the browse hierarchy with respect to the distance of the two classes from their common parent and the distance of their common parent from the root node.

In this paper we adopt some ideas from [9]. However, the purpose of our methodology is not to build browse hierarchies but rather to describe programming constructs with feature vectors that can be used for machine learning. In other words, the purpose of our methodology is to transform a source code repository into a feature space. The exploration of this feature space enables the domain experts to build a knowledge base in a "cheaper" semi-automatic interactive fashion.

3 Mining Content and Structure of Software Artifacts

In this section we present our approach and give an illustrative example of data pre-processing from documented source code using the GATE software library. In the context of the GATE case study the content is provided by the reference manual (textual descriptions of Java classes and methods), source code comments, programmer's guide, annotator's guide, user's guide, forum, and so on. The structure is provided implicitly from these same data sources since a Java class or method is often referenced from the context of another Java class or method (e.g. a Java class name is mentioned in the comment of another Java class). Additional structure can be harvested from the source code (e.g. a Java class contains a member method that returns an instance of another Java class), code snippets, and usage logs (e.g. one Java class is often instantiated immediately after another). In this paper we limit ourselves to the source code which also represents the reference manual (the so called *JavaDoc*) since the reference manual is generated automatically out of the source code comments by a documentation tool.

A software-based domain ontology should provide two views on the corresponding software library: the view on the data structures and the view on the functionality [13]. In GATE, these two views are represented with Java classes and their member methods – these are evident from the GATE source code. In our examples we limit ourselves to Java classes (i.e. we deal with the part of the domain ontology that covers the data structures of the system). This means that we will use the GATE Java classes as text mining instances (and also as graph vertices when dealing with the structure).

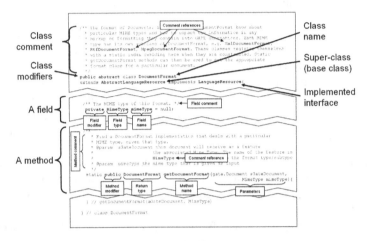

Fig. 1. Relevant parts of a typical Java class

Let us first take a look at a typical GATE Java class. It contains the following bits of information relevant for the understanding of this example (see also Fig. 1):

- **Class comment.** It should describe the purpose of the class. It is used by the documentation tool to generate the reference manual (i.e. JavaDoc).

  It is mainly a source of textual data but also provides structure – two classes are interlinked if the name of one class is mentioned in the comment of the other class.

- **Class name.** Each class is given a name that uniquely identifies the class. The name is usually a composed word that captures the meaning of the class.

  It is mainly a source of textual data but also provides structure – two classes are interlinked if they share a common substring in their names.

- **Field names and types.** Each class contains a set of member fields. Each field has a name (which is unique within the scope of the class) and a type. The type of a field corresponds to a Java class.

  Field names provide textual data. Field types mainly provide structure – two classes are interlinked if one class contains a field that instantiates the other class.

- **Field and method comments.** Fields and methods can also be commented. The comment should explain the purpose of the field or method.

  These comments are a source of textual data. They can also provide structure in the same sense as class comments do.

- **Method names and return types.** Each class contains a set of member methods. Each method has a name, a set of parameters, and a return type. The return type of a method corresponds to a Java class. Each parameter has a name and a type which corresponds to a Java class.

Methods can be treated similarly to fields with respect to taking their names and return types into account. Parameter types can be taken into account similarly to return types but there is a semantic difference between the two pieces of information. Parameter types denote classes that are "used/consumed" for processing while return types denote classes that are "produced" in the process.

- **Information about inheritance and interface implementation.** Each class inherits (fields and methods) from a base class. Furthermore, a class can implement one or more interfaces. An interface is merely a set of methods that need to be implemented in the derived class.

  The information about inheritance and interface implementation is a source of structural information.

3.1 Textual Content

Textual content is taken into account by assigning a textual document to each unit of the software code – in our illustrative example, to each GATE Java class. Suppose we focus on a particular arbitrary class – there are several ways to form the corresponding document.

It is important to include only those bits of text that are not misleading for the text mining algorithms. At this point the details of these text mining algorithms are pretty irrelevant provided that we can somehow evaluate the domain ontology that we build in the end.

Another thing to consider is how to include composed names of classes, fields, and methods into a document. We can insert each of these as:

- A omposed word (i.e. in its original form, e.g. "XmlDocumentFormat"),
- Sparate words (i.e. by inserting spaces, e.g. "Xml Document Format"), or
- Combination of both (e.g. "XmlDocumentFormat Xml Document Format").

The text-mining algorithms perceive two documents that have many words in common more similar that those that only share a few or no words. Breaking composed names into separate words therefore results in a greater similarity between documents that do not share full names but do share some parts of these names.

3.2 Determining the Structure

The basic units of the software code – in our case the Java classes – that we use as text-mining instances are interlinked in many ways. In this section we discuss how this structure which is often implicit can be determined from the source code.

As already mentioned, when dealing with the structure, we represent each class (i.e. each text mining instance) by a vertex in a graph. We can create several graphs – one for each type of associations between classes. This section describes several graphs that can be constructed out of object-oriented source code.

Comment Reference Graph. Every comment found in a class can reference another class by mentioning its name (for whatever the reason may be). In Fig. 1 we can see four such references, namely the class *DocumentFormat* references classes *XmlDocumentFormat*, *RtfDocumentFormat*, *MpegDocumentFormat*, and *MimeType* (denoted with "Comment reference" in the figure). A section of the comment reference graph for the GATE case study is shown in Fig. 2. The vertices represent GATE Java classes found in the "gate" subfolder of the GATE source code repository (we limited ourselves to a subfolder merely to reduce the number of vertices for the purpose of the illustrative visualization). An arc that connects two vertices is directed from the source vertex towards the target vertex (these two vertices represent the source and the target class, respectively). The weight of an arc (at least 1) denotes the number of times the name of the target class is mentioned in the comments of the source class. The higher the weight, the stronger is the association between the two classes. In the figure, the thickness of an arc is proportional to its weight.

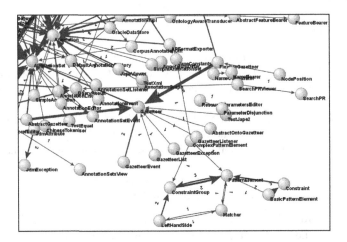

Fig. 2. A section of the GATE comment reference graph

Name Similarity Graph. A class usually represents a data structure and a set of methods related to it. Not every class is a data structure – it can merely be a set of (static) methods. The name of a class is usually a noun denoting either the data structure that the class represents (e.g. *Boolean*, *ArrayList*) or a "category" of the methods contained in the class (e.g. *System*, *Math*). If the name is composed (e.g. *ArrayList*) it is reasonable to assume that each of the words bears a piece of information about the class (e.g. an *ArrayList* is some kind of *List* with the properties of an *Array*). Therefore it is also reasonable to say that two classes that have more words in common are more similar to each other than two classes that have fewer words in common. According to this intuition we can construct the name similarity graph. This graph contains edges (i.e. undirected links) instead of arcs. Two vertices are linked when the two classes share at least one word. The strength of the link (i.e. the edge weight) can be computed by using the *Jaccard similarity* measure which is often used to measure the similarity of two sets of items (see http://en.wikipedia.org /wiki/Jaccard_index). The name similarity graph for the GATE case study is presented in Fig. 3. The vertices represent GATE Java classes found in the "gate" subfolder of the GATE source code repository. The Jaccard similarity measure

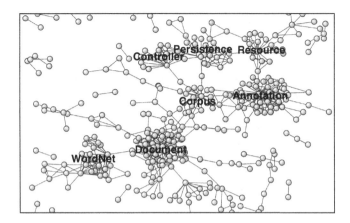

Fig. 3. The GATE name similarity graph. The most common substrings in names are shown for the most evident clusters.

was used to weight the edges. Edges with weights lower than 0.6 and vertices of degree 0 were removed to simplify the visualization. In Fig. 3 we have removed class names and weight values to clearly show the structure. The evident clustering of vertices is the result of the Kamada-Kawai graph drawing algorithm [14] employed by Pajek [1] which was used to create graph drawings in this paper. The Kamada-Kawai algorithm positions vertices that are highly interlinked closer together.

Type Reference Graph. Field types and method return types are a valuable source of structural information. A field type or a method return type can correspond to a class in the scope of the study (i.e. a class that is also found in the source code repository under consideration) – hence an arc can be drawn from the class to which the field or the method belongs towards the class represented by the type.

Inheritance and Interface Implementation Graph. Last but not least, structure can also be determined from the information about inheritance and interface implementation. This is the most obvious structural information in an object-oriented source code and is often used to arrange classes into the browsing taxonomy. In this graph, an arc that connects two vertices is directed from the vertex that represents a base class (or an interface) towards the vertex that represents a class that inherits from the base class (or implements the interface). The weight of an arc is always 1.

4   Transforming Content and Structure into Feature Vectors

Many data-mining algorithms work with feature vectors. This is true also for the algorithms employed by OntoGen and for the traditional machine learning algorithms such as clustering or classification. Therefore we need to convert the content (i.e. documents assigned to text-mining instances) and the structure (i.e. several graphs of interlinked vertices) into feature vectors. Potentially we also want to include other explicit features (e.g. in- and out-degree of a vertex).

4.1 Converting Content into Feature Vectors

To convert textual documents into feature vectors we resort to a well-known text mining approach. We first apply stemming[1] to all the words in the document collection (i.e. we normalize words by stripping them of their suffixes, e.g. *stripping* → *strip, suffixes* → *suffix*). We then search for *n-grams*, i.e. sequences of consecutive words of length *n* that occur in the document collection more than a certain amount of times [11]. Discovered *n-grams* are perceived just as all the other (single) words. After that, we convert documents into their bag-of-words representations. To weight words (and *n-grams*), we use the TF-IDF weighting scheme ([6], Section 1.3.2).

4.2 Converting Structure into Feature Vectors

Let us repeat that the structure is represented in the form of several graphs in which vertices correspond to text-mining instances. If we consider a particular graph, the task is to describe each vertex in the graph with a feature vector.

For this purpose we adopt the technique presented in [10]. First, we convert arcs (i.e. directed links) into edges (i.e. undirected links)[2]. The edges adopt weights from the corresponding arcs. If two vertices are directly connected with more than one arc, the resulting edge weight is computed by summing, maximizing, minimizing, or averaging the arc weights (we propose summing the weights as the default option). Then we represent a graph on *N* vertices as a *N×N* sparse matrix. The matrix is constructed so that the *X*th row gives information about vertex *X* and has nonzero components for the columns representing vertices from the neighborhood of vertex *X*. The neighborhood of a vertex is defined by its (restricted) domain. The *domain of a vertex* is the set of vertices that are path-connected to the vertex. More generally, a *restricted domain of a vertex* is a set of vertices that are path-connected to the vertex at a maximum distance of d_{max} steps [1]. The *X*th row thus has a nonzero value in the *X*th column (because vertex *X* has zero distance to itself) as well as nonzero values in all the other columns that represent vertices from the (restricted) domain of vertex *X*. A value in the matrix represents the importance of the vertex represented by the column for the description of the vertex represented by the row. In [10] the authors propose to compute the values as $1/2^d$, where *d* is the minimum path length between the two vertices (also termed the *geodesic* distance between two vertices) represented by the row and column.

We also need to include edge weights into account. The easiest way is to use the weights merely for thresholding. This means that we set a threshold and remove all the edges that have weights below this threshold. After that we construct the matrix which now indirectly includes the information about the weights (at least to a certain extent).

---

[1] We use the Porter stemmer for English (see http://de.wikipedia.org/wiki/Porter-Stemmer-Algorithmus).

[2] This is not a required step but it seems reasonable – a vertex is related to another vertex if they are interconnected regardless of the direction. In other words, if vertex *A references* vertex *B* then vertex *B is referenced* by vertex *A*.

The simple approach described above is based on more sophisticated approaches such as ScentTrails [12]. The idea is to metaphorically "waft" scent of a specific vertex in the direction of its out-links (links with higher weights conduct more scent than links with lower weights – the arc weights are thus taken into account explicitly). The scent is then iteratively spread throughout the graph. After that we can observe how much of the scent reached each of the other vertices. The amount of scent that reached a target vertex denotes the importance of the target vertex for the description of the source vertex.

The ScentTrails algorithm shows some similarities with the probabilistic framework: starting in a particular vertex and moving along the arcs we need to determine the probability of ending up in a particular target vertex within m steps. At each step we can select one of the available outgoing arcs with the probability proportional to the corresponding arc weight (assuming that the weight denotes the strength of the association between the two vertices). The equations for computing the probabilities are fairly easy to derive (see [7], Appendix C) but the time complexity of the computation is higher than that of ScentTrails and the first presented approach. The probabilistic framework is thus not feasible for large graphs.

4.3 Joining Different Representations into a Single Feature Vector

The next issue to solve is how to create a feature vector for a vertex that is present in several graphs at the same time (remember that the structure can be represented with more than one graph) and how to then also "append" the corresponding content feature vector. In general, this can be done in two different ways:

- **Horizontally.** This means that feature vectors of the same vertex from different graphs are first multiplied by factors α_i ($i = 1, ..., M$) and then concatenated into a feature vector with $M{\times}N$ components (M being the number of graphs and N the number of vertices). The content feature vector is multiplied by α_{M+1} and simply appended to the resulting structure feature vector.
- **Vertically.** This means that feature vectors of the same vertex from different graphs are first multiplied by factors α_i ($i = 1, ..., M$) and then summed together (component-wise) resulting in a feature vector with N components (N being the number of vertices). Note that the content feature vector cannot be summed together with the resulting structure feature vector since the features contained therein carry a different semantic meaning (not to mention that the two vectors are not of the same length). Therefore also in this case, the content feature vector is multiplied by α_{M+1} and appended to the resulting structure feature vector.

Fig. 4 illustrates these two approaches. A factor α_i ($i = 1, ..., M$) denotes the importance of information provided by graph i, relative to the other graphs. Factor α_{M+1}, on the other hand, denotes the importance of information provided by the content relative to the information provided by the structure. The easiest way to set the factors is to either include the graph or the content (i.e. $\alpha_i = 1$), or to exclude it (i.e. $\alpha_i = 0$). In general these factors can be quite arbitrary. Pieces of information with lower factors contribute less to the outcomes of similarity measures used in clustering algorithms

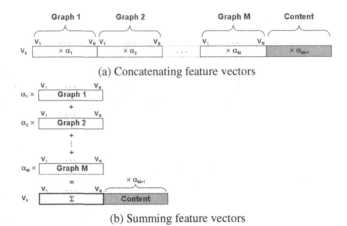

(a) Concatenating feature vectors

(b) Summing feature vectors

Fig. 4. The two different ways of joining several different representations of the same instance

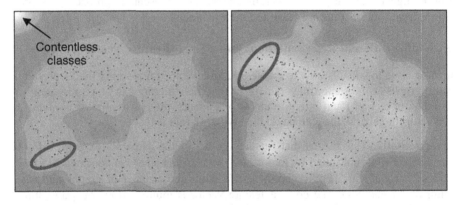

Fig. 5. Two different semantic spaces obtained by two different weighting settings

than those with higher factors. Furthermore, many classifiers are sensitive to this kind of weighting. For example, it has been shown in [2] that the SVM regression model is sensitive to how this kind of factors are set.

OntoGen includes a feature-space visualization tool called Document Atlas [5]. It is capable of visualizing high-dimensional feature space in two dimensions. The feature vectors are presented with two-dimensional points while the Euclidean distances between these points reflect cosine distances between feature vectors. It is not possible to perfectly preserve the distances from the high-dimensional space but even an approximation gives the user an idea of how the feature space looks like. Fig. 5 shows two such visualizations of the GATE case study data. In the left figure, only the class comments were taken into account (i.e. all the structural information was ignored and the documents assigned to the instances consisted merely of the corresponding class comments). In the right figure the information from the name similarity graph was added to the content information from the left figure. The content information was weighted twice higher than the structural information. d_{max} of the name similarity graph was set to 0.44.

The cluster marked in the left figure represents classes that provide functionality to consult WordNet (see http://wordnet.princeton.edu) to resolve synonymy[3]. The cluster containing this same functionality in the right figure is also marked. However, the cluster in the right figure contains more classes many of which were not commented thus were not assigned any content[4]. Contentless classes are stuck in the top left corner in the left figure because the feature-space visualization system did not know where to put them due to the lack of association with other classes. This missing association was introduced with the information from the name similarity graph. From the figures it is also possible to see that clusters are better defined in the right figure (note the dense areas represented with light color).

With these visualizations we merely want to demonstrate the difference in semantic spaces between two different settings. This is important because instances that are shown closer together are more likely to belong to the same cluster or category after applying clustering or classification. The weighting setting depends strongly on the context of the application of this methodology.

5 Conclusions

In this paper we presented a methodology for transforming a source code repository into a set of feature vectors, i.e. into a feature space. These feature vectors serve as the basis for the construction of the domain ontology with OntoGen, a system for semi-automatic data-driven ontology construction, or by using traditional machine learning algorithms such as clustering, classification, regression, or active learning. The presented methodology thus facilitates the transitioning of legacy software repositories into state-of-the-art ontology-based systems for discovery, composition, and potentially also execution of software artifacts.

This paper does not provide any evaluation of the presented methodology. Basically, the evaluation can be performed either by comparing the resulting ontologies with a golden-standard ontology (if such ontology exists) or, on the other hand, by employing them in practice. In the second scenario, we measure the efficiency of the users that are using these ontologies (directly or indirectly) in order to achieve certain goals. The aspects on the quality of the methods presented herein will be the focus of our future work.

We recently started developing an ontology-learning framework named LATINO which stands for Link-analysis and text-mining toolbox [7]. LATINO will be an open-source general purpose data mining platform providing (mostly) text mining, link analysis, machine learning, and data visualization capabilities.

Acknowledgments. This work was supported by the Slovenian Research Agency and the IST Programme of the European Community under TAO Transitioning Applications to Ontologies (IST-4-026460-STP) and PASCAL Network of Excellence (IST-2002-506778).

---

[3] The marked cluster in the left figure contains classes such as Word, VerbFrame, and Synset.
[4] The marked cluster in the right figure contains the same classes as the marked cluster in the left figure but also some contentless classes such as WordImpl, VerbFrameImpl, (Mutable)LexKBSynset(Impl), SynsetImpl, WordNetViewer, and IndexFileWordNetImpl.

References

1. Batagelj, V., Mrvar, A., de Nooy, W.: Exploratory Network Analysis with Pajek. Cambridge University Press, Cambridge (2004)
2. Brank, J., Leskovec, J.: The Download Estimation Task on KDD Cup 2003. In: ACM SIGKDD Explorations Newsletter, vol. 5(2), pp. 160–162. ACM Press, New York (2003)
3. Cunningham, H., Maynard, D., Bontcheva, K., Tablan, V.: GATE: A Framework and Graphical Development Environment for Robust NLP Tools and Applications. In: Proceedings of the 40th Anniversary Meeting of the Association for Computational Linguistics ACL 2002 (2002)
4. Fortuna, B., Grobelnik, M., Mladenic, D.: Semi-automatic Data-driven Ontology Construction System. In: Proceedings of the 9th International Multi-conference Information Society IS-2006, Ljubljana, Slovenia (2006)
5. Fortuna, B., Mladenic, D., Grobelnik, M.: Visualization of Text Document Corpus. Informatica 29, 497–502 (2005)
6. Grcar, M., Mladenic, D., Grobelnik, M., Bontcheva, K.: D2.1: Data Source Analysis and Method Selection. Project report IST-2004-026460 TAO, WP 2, D2.1 (2006)
7. Grcar, M., Mladenic, D., Grobelnik, M., Fortuna, B., Brank, J.: D2.2: Ontology Learning Implementation. Project report IST-2004-026460 TAO, WP 2, D2.2 (2006)
8. Maedche, A., Staab, S.: Discovering Conceptual Relations from Text. In: Proc. of ECAI 2000, pp. 321–325 (2001)
9. Helm, R., Maarek, Y.: Integrating Information Retrieval and Domain Specific Approaches for Browsing and Retrieval in Object-oriented Class Libraries. In: Proceedings of Object-oriented Programming Systems, Languages, and Applications, pp. 47–61. ACM Press, New York, USA (1991)
10. Mladenic, D., Grobelnik, M.: Visualizing Very Large Graphs Using Clustering Neighborhoods. In: Local Pattern Detection, Dagstuhl Castle, Germany, April 12–16, 2004 (2004)
11. Mladenic, D., Grobelnik, M.: Word Sequences as Features in Text Learning. In: Proceedings of the 17th Electrotechnical and Computer Science Conference ERK 1998, Ljubljana, Slovenia (1998)
12. Olston, C., Chi, H.E.: ScentTrails: Integrating Browsing and Searching on the Web. In: ACM Transactions on Computer-human Interaction TOCHI, vol. 10(3), pp. 177–197. ACM Press, New York (2003)
13. Sabou, M.: Building Web Service Ontologies. In: SIKS Dissertation Series No. 2004-4 (2006) ISBN 90-9018400-7
14. Kamada, T., Kawai, S.: An Algorithm for Drawing General Undirected Graphs. Information Processing Letters 31, 7–15 (1989)

Generalization-Based Similarity
for Conceptual Clustering

S. Ferilli, T.M.A. Basile, N. Di Mauro, M. Biba, and F. Esposito

Dipartimento di Informatica
Università di Bari
via E. Orabona, 4 - 70125 Bari, Italia
{ferilli,basile,ndm,biba,esposito}@di.uniba.it

Abstract. Knowledge extraction represents an important issue that concerns the ability to identify valid, potentially useful and understandable patterns from large data collections. Such a task becomes more difficult if the domain of application cannot be represented by means of an attribute-value representation. Thus, a more powerful representation language, such as First-Order Logic, is necessary. Due to the complexity of handling First-Order Logic formulæ, where the presence of relations causes various portions of one description to be possibly mapped in different ways onto another description, few works presenting techniques for comparing descriptions are available in the literature for this kind of representations. Nevertheless, the ability to assess similarity between first-order descriptions has many applications, ranging from description selection to flexible matching, from instance-based learning to clustering.

This paper tackles the case of Conceptual Clustering, where a new approach to similarity evaluation, based on both syntactic and semantic features, is exploited to support the task of grouping together similar items according to their relational description. After presenting a framework for Horn Clauses (including criteria, a function and composition techniques for similarity assessment), classical clustering algorithms are exploited to carry out the grouping task. Experimental results on real-world datasets prove the effectiveness of the proposal.

1 Introduction

The large amount of information available nowadays makes more difficult the task of extracting useful knowledge, i.e. valid, potentially useful and understandable patterns, from data collections. Such a task becomes more difficult if the collection requires a more powerful representation language than simple attribute-value vectors. First-order logic (FOL for short) is a powerful formalism, that is able to express relations between objects and hence can overcome the limitations shown by propositional or attribute-value representations. However, the presence of relations causes various portions of one description to be possibly mapped in different ways onto another description, which poses problems of computational effort when two descriptions have to be compared to each other.

Z.W. Raś, S. Tsumoto, and D. Zighed (Eds.): MCD 2007, LNAI 4944, pp. 13–26, 2008.

Specifically, an important subclass of FOL refers to sets of *Horn clauses*, i.e. logical formulæ of the form $l_1 \wedge \cdots \wedge l_n \Rightarrow l_0$ where the l_i's are *atoms*, usually represented in Prolog style as $l_0 :\text{-} l_1, \ldots, l_n$ to be interpreted as "l_0 (called *head* of the clause) is true, provided that l_1 and ... and l_n (called *body* of the clause) are all true". Without loss of generality [16], we will deal with the case of linked Datalog clauses.

The availability of techniques for the comparison between FOL (sub-)descriptions could have many applications: helping a subsumption procedure to converge quickly, guiding a generalization procedure by focussing on the components that are more similar and hence more likely to correspond to each other, implementing flexible matching, supporting instance-based classification techniques or conceptual clustering. Cluster analysis concerns the organization of a collection of unlabeled patterns into groups (clusters) of homogeneous elements based on their similarity. The similarity measure exploited to evaluate the distance between elements is responsible for the effectiveness of the clustering algorithms. Hence, the comparison techniques are generally defined in terms of a metric that must be carefully constructed if the clustering is to be relevant. In supervised clustering there is an associated output class value for each element and the efficacy of the metric exploited for the comparison of elements is evaluated according to the principle that elements belonging to the same class are clustered together as much as possible.

In the following sections, a similarity framework for first-order logic clauses will be presented. Then, Section 5 will deal with related work, and Section 6 will show how the proposed formula and criteria are able to effectively guide a clustering procedure for FOL descriptions. Lastly, Section 7 will conclude the paper and outline future work directions.

2   Similarity Formula

Intuitively, the evaluation of similarity between two items i' and i'' might be based both on the presence of common features, which should concur in a positive way to the similarity evaluation, and on the features of each item that are not owned by the other, which should concur negatively to the whole similarity value assigned to them [10]. Thus, plausible similarity parameters are:

n, the number of features owned by i' but not by i'' (*residual* of i' wrt i'');
l, the number of features owned both by i' and by i'';
m, the number of features owned by i'' but not by i' (*residual* of i'' wrt i').

A novel similarity function that expresses the degree of similarity between i' and i'' based on the above parameters, developed to overcome some limitations of other functions in the literature (e.g., Tverski's, Dice's and Jaccard's), is:

$$sf(i', i'') = sf(n, l, m) = 0.5 \frac{l+1}{l+n+2} + 0.5 \frac{l+1}{l+m+2} \tag{1}$$

It takes values in $]0, 1[$, to be interpreted as the degree of similarity between the two items. A complete overlapping of the two items tends to the limit of 1 as long as the number of common features grows. The full-similarity value 1 is never reached, and is reserved to the exact identification of items, i.e. $i' = i''$ (in the following, we assume $i' \neq i''$). Conversely, in case of no overlapping the function will tend to 0 as long as the number of non-shared features grows. This is consistent with the intuition that there is no limit to the number of different features owned by the two descriptions, which contribute to make them ever different. Since each of the two terms refers specifically to one of the two clauses under comparison, a weight could be introduced to give different importance to either of the two.

3   Similarity Criteria

The main contribution of this paper is in the exploitation of the formula in various combinations that can assign a similarity degree to the different clause constituents. In FOL formulæ, terms represent specific objects; unary predicates represent term properties and n-ary predicates express relationships. Hence, two levels of similarity between first-order descriptions can be defined: the *object* level, concerning similarities between terms in the descriptions, and the *structure* one, referring to how the nets of relationships in the descriptions overlap.

Example 1. Let us consider, as a running example throughout the paper, the following clause, representing a short description (with predicate names slightly changed for the sake of brevity), drawn from the real-world domain of scientific papers first-pages layout:

  observation(d) :-
num_pages(d,1), page_1(d,p1), page_w(p1,612.0), page_h(p1,792.0), last_page(p1),
frame(p1,f4), frame(p1,f2), frame(p1,f1), frame(p1,f6), frame(p1,f12), frame(p1,f10),
frame(p1,f3), frame(p1,f9),
t_text(f4), w_medium_large(f4), h_very_very_small(f4), center(f4), middle(f4),
t_text(f2), w_large(f2), h_small(f2), center(f2), upper(f2),
t_text(f1), w_large(f1), h_large(f1), center(f1), lower(f1),
t_text(f6), w_large(f6), h_very_small(f6), center(f6), middle(f6),
t_text(f12), w_medium(f12), h_very_very_small(f12), left(f12), middle(f12),
t_text(f10), w_large(f10), h_small(f10), center(f10), upper(f10),
t_text(f3), w_large(f3), h_very_small(f3), center(f3), upper(f3),
t_text(f9), w_large(f9), h_medium(f9), center(f9), middle(f9),
on_top(f4,f12), to_right(f4,f12), to_right(f6,f4), on_top(f4,f6), to_right(f1,f4),
on_top(f4,f1), to_right(f9,f4), on_top(f4,f9),
on_top(f10,f4), to_right(f10,f4), on_top(f2,f4), to_right(f2,f4), on_top(f3,f4),
to_right(f3,f4),
on_top(f2,f12), to_right(f2,f12),
on_top(f2,f6), valign_center(f2,f6),
on_top(f10,f2), valign_center(f2,f10),
on_top(f2,f1), valign_center(f2,f1),
on_top(f3,f2), valign_center(f2,f3),

on_top(f2,f9), valign_center(f2,f9),
on_top(f12,f1), to_right(f1,f12),
on_top(f6,f1), valign_center(f1,f6),
on_top(f10,f1), valign_center(f1,f10),
on_top(f3,f1), valign_center(f1,f3),
on_top(f9,f1), valign_center(f1,f9),
on_top(f6,f12), to_right(f6,f12),
on_top(f10,f6), valign_center(f6,f10),
on_top(f3,f6), valign_center(f6,f3),
on_top(f9,f6), valign_center(f6,f9),
on_top(f10,f12), to_right(f10,f12),
on_top(f3,f12), to_right(f3,f12),
on_top(f9,f12), to_right(f9,f12),
on_top(f3,f10), valign_center(f10,f3),
on_top(f10,f9), valign_center(f10,f9),
on_top(f3,f9), valign_center(f3,f9).

which reads as: "Observation d is made up of one page; page 1 is $p1$, which is also the last one and has width 612 pixels and height 792 pixels, and contains frames $f4$, $f2$, $f1$, $f6$, $f12$, $f10$, $f3$, $f9$. Frame $f4$ contains text, has width medium-large and height very very small and is placed in the center (horizontally) and in the midde (vertically) of the page; [...] frame $f9$ contains text, has large width and medium height and is placed in the center (horizontally) and in the midde (vertically) of the page. Frame $f4$ is on top of frames $f12$, $f6$, $f1$ and $f9$, the first one on its right and the others on its left; frames $f10$, $f2$ and $f3$ are in turn on top of frame $f4$, all of them on its right. [...]".

3.1   Object Similarity

Consider two clauses C' and C''. Call $A' = \{a'_1, \ldots, a'_n\}$ the set of terms in C', and $A'' = \{a''_1, \ldots, a''_m\}$ the set of terms in C''. When comparing a pair of objects $(a', a'') \in A' \times A''$, two kinds of object features can be distinguished: the properties they own as expressed by unary predicates (*characteristic features*), and the roles they play in n-ary predicates (*relational features*). More precisely, a *role* can be seen as a couple $R = (predicate, position)$ (written compactly as $R = predicate/arity.position$), since different positions actually refer to different roles played by the objects. For instance, a characteristic feature could be male(X), while relational features in a parent(X,Y) predicate are the 'parent' role ($parent/2.1$) the 'child' role ($parent/2.2$).

Two corresponding similarity values can be associated to a' and a'': a *characteristic similarity*,

$$\mathrm{sf}_c(a', a'') = \mathrm{sf}(n_c, l_c, m_c)$$

based on the set P' of properties related to a' and the set P'' of properties related to a'', for the following parameters:

$n_c = |P' \setminus P''|$ number of properties owned by a' in C' but not by a'' in C'' (*characteristic residual* of a' wrt a'');

$l_c = |P' \cap P''|$ number of common properties between a' in C' and a'' in C'';

$m_c = |P'' \setminus P'|$ number of properties owned by a'' in C'' but not by a' in C' (*characteristic residual* of a'' wrt a').

and a *relational similarity*,

$$\text{sf}_r(a', a'') = \text{sf}(n_r, l_r, m_r)$$

based on the *multisets* R' and R'' of roles played by a' and a'', respectively, for the following parameters:

$n_r = |R' \setminus R''|$ how many times a' plays in C' role(s) that a'' does not play in C'' (*relational residual* of a' wrt a'');

$l_r = |R' \cap R''|$ number of times that both a' in C' and a'' in C'' play the same role(s);

$m_r = |R'' \setminus R'|$ how many times a'' plays in C'' role(s) that a' does not play in C' (*relational residual* of a'' wrt a').

Overall, we can define the *object similarity* between two terms as

$$\text{sf}_o(a', a'') = \text{sf}_c(a', a'') + \text{sf}_r(a', a'')$$

Example 2. Referring to the clause concerning the document description, the set of properties of $f4$ is

$$\{t\_text, w\_medium\_large, h\_very\_very\_small, center, middle\}$$

while for $f9$ it is

$$\{t\_text, w\_large, h\_medium, center, middle\}$$

The multiset of roles of $f4$ (where $p \times n$ denotes that p has n occurrences in the multiset) is

$$\{frame/2.2, on\_top/2.1 \times 4, on\_top/2.2 \times 3, to\_right/2.1, to\_right/2.2 \times 6\}$$

while for $f9$ it is

$$\{frame/2.2, on\_top/2.1 \times 3, on\_top/2.2 \times 4, to\_right/2.1 \times 2, valign\_center/2.2 \times 5\}.$$

3.2   Structural Similarity

When checking for the structural similarity of two formulæ, many objects can be involved, and hence their mutual relationships represent a constraint on how each of them in the former formula can be mapped onto another in the latter. The structure of a formula is defined by the way in which n-ary *atoms* (predicates applied to a number of terms equal to their arity) are applied to the various objects to relate them. This is the most difficult part, since relations are specific to the first-order setting and are the cause of indeterminacy in mapping (parts of) a formula into (parts of) another one. In the following, we will call *compatible* two FOL (sub-)formulæ that can be mapped onto each other without yielding inconsistent term associations (i.e., a term in one formula cannot correspond to different terms in the other formula).

Given an n-ary literal, we define its *star* as the multiset of n-ary predicates corresponding to the literals linked to it by some common term (a predicate can appear in multiple instantiations among these literals). The *star similarity* between two compatible n-ary literals l' and l'' having stars S' and S'', respectively, can be computed for the following parameters:

$n_s = |S' \setminus S''|$ how many more relations l' has in C' than l'' has in C'' (*star residual* of l' wrt l'');

$l_s = |S' \cap S''|$ number of relations that both l' in C' and l'' in C'' have in common;

$m_s = |S'' \setminus S'|$ how many more relations l'' has in C'' than l' has in C' (*star residual* of l'' wrt l').

by taking into account also the object similarity values for all pairs of terms included in the association θ that map l' onto l'' of their arguments in corresponding positions:

$$\mathrm{sf}_s(l', l'') = \mathrm{sf}(n_s, l_s, m_s) + C^s(\{\mathrm{sf}_o(t', t'')\}_{t'/t'' \in \theta})$$

where C^s is a composition function (e.g., the average).

Example 3. In the document example, the star of $frame(p1, f4)$ is the multiset

$$\{page\_1/2, page\_w/2, page\_h/2, on\_top/2 \times 7, to\_right/2 \times 7\}$$

The star of $on\_top(f4, f9)$ is the multiset

$$\{frame/2 \times 2, on\_top/2 \times 11, to\_right/2 \times 8, valign\_center/2 \times 5\}$$

Then, Horn clauses can be represented as a graph in which atoms are the nodes, and edges connect two nodes *iff* they share some term, as described in the following. In particular, we will deal with *linked* clauses only (i.e. clauses whose associated graph is connected). Given a clause C, we define its *associated graph* G_C, where the edges to be represented form a Directed Acyclic Graph (DAG), *stratified* in such a way that the head is the only node at level 0 and each successive level is made up by nodes not yet reached by edges that have at least one term in common with nodes in the previous level. In particular, each node in the new level is linked by an incoming edge to each node in the previous level having among its arguments at least one term in common with it.

Example 4. Due to the graph representing the document example being too complex for being represented in the page, let us consider, as an additional example for the structural representation of a clause, the following toy clause:

$C : h(a) :\text{-} p(a, b), p(a, c), p(d, a), r(b, f), o(b, c), q(d, e), t(f, g),$
$\qquad \pi(a), \phi(a), \sigma(a), \tau(a), \sigma(b), \tau(b), \phi(b), \tau(d), \rho(d), \pi(f), \phi(f), \sigma(f).$

In the graph G_C, the head represents the 0-level of the stratification. Then directed edges may be introduced from $h(X)$ to $p(X, Y)$, $p(X, Z)$ and $p(W, X)$, which yields level 1 of the stratification. Now the next level can be built, adding directed edges from atoms in level 1 to the atoms not yet considered that share a variable with them: $r(Y, U)$ – end of an edge starting from $p(X, Y)$ –, $o(Y, Z)$ – end of edges starting from $p(X, Y)$ and $p(X, Z)$ – and $q(W, W)$ – end of an edge starting from $p(W, X)$. The third level of the graph includes the only remaining atom, $s(U, V)$ – having an incoming edge from $r(Y, U)$.

Now, all possible paths starting from the head and reaching *leaf* nodes are univoquely determined, which reduces the amount of indeterminacy in the comparison. Given two clauses C' and C'', we define the *intersection* between two paths $p' = <l'_1, \dots, l'_{n'}>$ in $G_{C'}$ and $p'' = <l''_1, \dots, l''_{n''}>$ in $G_{C''}$ as the pair of longest compatible initial subsequences of p' and p'':

$$p' \cap p'' = (p_1, p_2) = (<l'_1, \dots, l'_k>, <l''_1, \dots, l''_k>) \text{ s.t.}$$
$$\forall i = 1, \dots, k : l'_1, \dots, l'_i \text{ compatible with } l''_1, \dots, l''_i \wedge$$
$$(k = n' \vee k = n'' \vee l'_1, \dots, l'_{k+1} \text{ incompatible with } l''_1, \dots, l''_{k+1})$$

and the two residuals as the incompatible trailing parts:

$$p' \setminus p'' = <l'_{k+1}, \dots, l'_{n'}> \qquad p'' \setminus p' = <l''_{k+1}, \dots, l''_{n''}>)$$

Hence, the *path similarity* between p' and p'', $\mathrm{sf}_s(p', p'')$, can be computed by applying (1) to the following parameters:

$n_p = |p' \setminus p''| = n' - k$ is the length of the trail incompatible sequence of p' wrt p'' (*path residual* of p' wrt p'');

$l_p = |p_1| = |p_2| = k$ is the length of the maximum compatible initial sequence of p' and p'';

$m_p = |p'' \setminus p'| = n'' - k$ is the length of the trail incompatible sequence of p'' wrt p' (*path residual* of p'' wrt p').

by taking into account also the star similarity values for all pairs of literals associated by the initial compatible sequences:

$$\mathrm{sf}_p(p', p'') = \mathrm{sf}(n_p, l_p, m_p) + C^p(\{\mathrm{sf}_s(l'_i, l''_i)\}_{i=1,\dots,k})$$

where C^p is a composition function (e.g., the average).

Example 5. The paths in C (ignoring the head that, being unique, can be univoquely matched) are

$$\{<p(a, b), r(b, f), t(f, g)>, <p(a, b), o(b, c)>, <p(a, c), o(b, c)>,$$
$$<p(d, a), q(d, e)>\}.$$

Some paths in the document example are

$$<num\_pages(d, 1)>, <page\_1(d, p1), page\_w(p1, 612.0)>,$$
$$<page\_1(d, p1), page\_h(p1, 792.0)>, <page\_1(d, p1), last\_page(p1)>,$$
$$<page\_1(d, p1), frame(p1, f4), on\_top(f4, f9)>,$$

$< page\_1(d, p1), frame(p1, f9), on\_top(f4, f9) >,$
$< page\_1(d, p1), frame(p1, f9), valign\_center(f10, f9) >,$
...

Note that no single criterion is by itself neatly discriminant, but their cooperation succeeds in assigning sensible similarity values to the various kinds of components, and in distributing on each kind of component a proper portion of the overall similarity, so that the difference becomes ever clearer as long as they are composed one ontop the previous ones.

4   Clause Similarity

Now, similarity between two (tuples of) terms reported in the head predicates of two clauses, according to their description reported in the respective bodies, can be computed based on their generalization. In particular, one would like to exploit their *least general generalization*, i.e. the most specific model for the given pair of descriptions. Unfortunately, such a generalization is not easy to find: either classical θ-subsumption is used as a generalization model, and then one can compute Plotkin's least general generalization [13], at the expenses of some undesirable side-effects concerning the need of computing its reduced equivalent (and also of some counter-intuitive aspects of the result), or, as most ILP learners do, one requires the generalization to be a subset of the clauses to be generalized. In the latter option, that we choose for the rest of the work, the θ_{OI} generalization model [5], based on the Object Identity assumption, represents a supporting framework with solid theoretical foundations to be exploited.

Given two clauses C' and C'', call $C = \{l_1, \ldots, l_k\}$ their least general generalization, and consider the substitutions θ' and θ'' such that $\forall i = 1, \ldots, k :$ $l_i\theta' = l'_i \in C'$ and $l_i\theta'' = l''_i \in C''$, respectively. Thus, a formula for assessing the overall similarity between C' and C'', called *formulæ similitudo* and denoted fs, can be computed according to the amounts of common and different literals:

$n = |C'| - |C|$ how many literals in C' are not covered by its least general generalization with respect to C'' (*clause residual* of C' wrt C'');

$l = |C| = k$ maximal number of literals that can be put in correspondence between C' and C'' according to their least general generalization;

$m = |C''| - |C)|$ how many literals in C'' are not covered by its least general generalization with respect to C' (*clause residual* of C'' wrt C').

and of common and different objects:

$n_o = |terms(C')| - |terms(C)|$ how many terms in C' are not associated by its least general generalization to terms in C'' (*object residual* of C' wrt C'');

$l_o = |terms(C)|$ maximal number of terms that can be put in correspondence in C' and C'' as associated by their least general generalization;

$m_o = |terms(C'')| - |terms(C))|$ how many terms in C'' are not associated by its least general generalization to terms in C' (*object residual* of C'' wrt C').

by taking into account also the star similarity values for all pairs of literals associated by the least general generalization:

$$\text{fs}(C', C'') = \text{sf}(n, l, m) \cdot \text{sf}(n_o, l_o, m_o) + C^c(\{\text{sf}_s(l'_i, l''_i)\}_{i=1,\dots,k})$$

where C^c is a composition function (e.g., the average). This function evaluates the similarity of two clauses according to the composite similarity of a maximal subset of their literals that can be put in correspondence (which includes both structural and object similarity), smoothed by adding the overall similarity in the number of overlapping and different literals and objects between the two (whose weight in the final evaluation should not overwhelm the similarity coming from the detailed comparisons, hence the multiplication).

In particular, the similarity formula itself can be exploited for computing the generalization. The path intersections are considered by decreasing similarity, adding to the partial generalization generated thus far the common literals of each pair whenever they are compatible [6]. The proposed similarity framework proves actually able to lead towards the identification of the proper sub-parts to be put in correspondence in the two descriptions under comparison, as shown indirectly by the portion of literals in the clauses to be generalized that is preserved by the generalization. More formally, the compression factor (computed as the ratio between the length of the generalization and that of the shortest clause to be generalized) should be as high as possible. Interestingly, on the document dataset (see section 6 for details) the similarity-driven generalization preserved on average more than 90% literals of the shortest clause, with a maximum of 99,48% (193 literals out of 194, against an example of 247) and just 0,006 variance. As a consequence, one woud expect that the produced generalizations are least general ones or nearly so. Noteworthly, using the similarity function on the document labelling task leads to runtime savings that range from $1/3$ up to $1/2$, in the order of hours.

5   Related Works

Few works faced the definition of similarity or distance measures for first-order descriptions. [4] proposes a distance measure based on probability theory applied to the formula components. Compared to that, our function does not require the assumptions and simplifying hypotheses to ease the probability handling, and no *a-priori* knowledge of the representation language is required. It does not require the user to set weights on the predicates' importance, and is not based on the presence of 'mandatory' relations, like for the $G1$ subclause in [4]. *KGB* [1] uses a similarity function, parameterized by the user, to guide generalization; our approach is more straightforward, and can be easily extended to handle negative information in the clauses. In *RIBL* [3] object similarity depends on the similarity of their attributes' values and, recursively, on the similarity of the objects related to them, which poses the problem of indeterminacy. [17] presents an approach for the induction of a distance on FOL examples, that exploits the truth values of whether each clause covers the example or not as features for a distance on the space $\{0,1\}^k$ between the examples. [12] organizes terms in an importance-related hierarchy, and proposes a distance between terms based on interpretations and a level mapping function that maps every simple expression on a natural number. [14] presents a distance function between atoms based on the difference with their lgg, and uses it to compute distances between clauses. It consists of a pair where the second component allows to differentiate cases where the first component cannot.

As pointed out, we focus on the identification and exploitation of similarity measures for first-order descriptions in the clustering task. Many research efforts on data representation, elements' similarity and grouping strategies have produced several successful clustering methods (see [9] for a survey). The classical strategies can be divided in bottom-up and top-down. In the former, each element of the dataset is considered as a cluster. Successively, the algorithm tries to group the clusters that are more similar according to the similarity measure. This step is performed until the number of clusters the user requires as a final result is reached, or the minimal similarity value among clusters is greater than a given threshold. In the latter approach, known as hierarchical clustering, at the beginning all the elements of the dataset form a unique cluster. Successively, the cluster is partitioned into clusters made up of elements that are more similar according to the similarity measure. This step is performed until the number of clusters required by the user as a final result is reached. A further classification is based on whether an element can be assigned (NotExclusive or Fuzzy Clustering) or not (Exclusive or Hard Clustering) to more than one cluster. Also the strategy exploited to partition the space is a criterion used to classify the clustering techniques: in Partitive Clustering a representative point (centroid, medoid, etc.) of the cluster in the space is chosen; Hierarchical Clustering produces a nested series of partitions by merging (Hierarchical Agglomerative) or splitting (Hierarchical Divisive) clusters, Density-based Clustering considers the density of the elements around a fixed point.

Closely related to data clustering is Conceptual Clustering, a Machine Learning paradigm for unsupervised classification which aims at generating a concept description for each generated class. In conceptual clustering both the inherent structure of the data and the description language, available to the learner, drive cluster formation. Thus, a concept (regularity) in the data could not be learned by the system if the description language is not powerful enough to describe that particular concept (regularity). This problem arises when the elements simultaneously describe several objects whose relational structures change from one element to the other. First-Order Logic representations allow to overcome these problems. However, most of the clustering algorithms and systems work on attribute-value representation (e.g., CLUSTER/2 [11], CLASSIT [8], COBWEB [7]). Other systems such as LABYRINTH [18] can deal with structured objects exploiting a representation that is not powerful enough to express the dataset in a lot of domains. There are few systems that cluster examples represented in FOL (e.g., AUTOCLASS-like [15], KBG [1]), some of which still rely on propositional distance measures (e.g., TIC [2]).

6   Experiments on Clustering

The proposed similarity framework was tested on the conceptual clustering task, where a set of items must be grouped into homogeneous classes according to the similarity between their first-order logic description. In particular, we adopted the classical K-means clustering technique. However, since first-order logic formulæ do not induce an euclidean space, it was not possible to identify/build a *centroid* prototype for the various clusters according to which the next distribution in the loop would be performed. For this reason, we based the distribution on the concept of *medoid* prototypes, where a medoid is defined as the observation that actually belongs to a cluster and that has the minimum average distance from all the other members of the cluster. As to the stop criterion, it was set as the moment in which a new iteration outputs a partition already seen in previous iterations. Note that it is different than performing the same check on the set of prototypes, since different prototypes could yield the same partition, while there cannot be several different sets of prototypes for one given partition. In particular, it can happen that the last partition is the same as the last-but-one, in which case a fixed point is reached and hence a single solution has been found and has to be evaluated. Conversely, when the last partition equals a previous partition, but not the last-but-one one, a loop is identified, and one cannot focus on a single minimum to be evaluated.

Experiments on Conceptual Clustering were run on a real-world dataset[1] containing 353 descriptions of scientific papers first page layout, belonging to 4 different classes: Elsevier journals, Springer-Verlag Lecture Notes series (SVLN), Journal of Machine Learning Research (JMLR) and Machine Learning Journal (MLJ). The complexity of such a dataset is considerable, and concerns several aspects of the dataset: the journals layout styles are quite similar, so that it is not easy to grasp the difference when trying to group them in distinct classes; moreover, the 353 documents are described with a total of 67920 literals, for an average of more than 192 literals per description (some descriptions are made up of more than 400 literals); last, the description is heavily based on a *part_of* relation, expressed by the *frame* predicate, that increases indeterminacy.

Since the class of each document in the dataset is known, we performed a supervised clustering: after hiding the correct class to the clustering procedure, we provided it with the 'anonymous' dataset, asking for a partition of 4 clusters. Then, we compared each outcoming cluster with each class, and assigned it to the best-matching class according to precision and recall. In practice, we found that for each cluster the precision-recall values were neatly high for one class, and considerably low for all the others; moreover, each cluster had a different best-matching class, so that the association and consequent evaluation became straightforward.

The clustering procedure was run first on 40 documents randomly selected from the dataset, then on 177 documents and lastly on the whole dataset, in order to evaluate its performance behaviour when takling increasingly large data.

---

[1] http://lacam.di.uniba.it:8000/systems/inthelex/index.htm#datasets

Table 1. Experimental results

Instances	Cluster	Class	Intersection	Prec (%)	Rec (%)	Total Overlapping
40	8	Elsevier (4)	4	50	100	35
	6	SVLN (6)	5	83,33	83,33	
	8	JMLR (8)	8	100	100	
	18	MLJ (22)	18	100	81,82	
177	30	Elsevier (22)	22	73,33	100	164
	36	SVLN (38)	35	97,22	92,11	
	48	JMLR (45)	45	93,75	100	
	63	MLJ (72)	62	98,41	86,11	
353	65	Elsevier (52)	52	80	100	326
	65	SVLN (75)	64	98,46	85,33	
	105	JMLR (95)	95	90,48	100	
	118	MLJ (131)	115	97,46	87,79	

Table 2. Experimental results statistics

Instances	Runtime	Comparisons	Avg Runtime (sec)	Prec (%)	Rec (%)	Pur (%)
40	25'24"	780	1,95	83,33	91,33	87,5
177	9h 34' 45"	15576	2,21	90,68	94,56	92,66
353	39h 12' 07"	62128	2,27	91,60	93,28	92,35

Results are reported in Table 1: for each dataset size it reports the number of instances in each cluster and in the corresponding class, the number of matching instances between the two and the consequent precision (Prec) and recall (Rec) values, along with the overall number of correctly split documents in the dataset.

Compound statistics, shown in Table 2, report the average precision and recall for each dataset size, along with the overall accuracy, plus some information about runtime and number of description comparisons to be carried out. The overall results show that the proposed method is highly effective since it is able to autonomously recognize the original classes with precision, recall and purity (Pur) well above 80% and, for larger datasets, always above 90%. This is very encouraging, especially in the perspective of the representation-related difficulties (the lower performance on the reduced dataset can probably be explained with the lack of sufficient information for properly discriminating the clusters, and suggests further investigation). Runtime refers almost completely to the computation of the similarity between all couples of observations: computing each similarity takes on average about 2sec, which can be a reasonable time considering the descriptions complexity and the fact that the prototype has no optimization in this preliminary version. Also the semantic perspective is quite satisfactory: an insight of the clustering outcomes shows that errors are made on very ambiguous documents (the four classes have a very similar layout style), while the induced cluster descriptions highlight interesting and characterizing layout clues. Preliminary comparisons on the 177 dataset with other classical measures report an improvement with respect to both Jaccard's, Tverski's and

Dice's measures up to +5,48% for precision, up to + 8,05% for recall and up to + 2,83% for purity.

7   Conclusions

Knowledge extraction concerns the ability to identify valid, potentially useful and understandable patterns from large data collections. Such a task becomes more difficult if the domain of application requires a First-Order Logic representation language, due to the problem of indeterminacy in mapping portions of descriptions onto each other. Nevertheless, the ability to assess similarity between first-order descriptions has many applications, ranging from description selection to flexible matching, from instance-based learning to clustering.

This paper deals with Conceptual Clustering, and proposes a framework for Horn Clauses similarity assessment. Experimental results on real-world datasets prove that, endowing classical clustering algorithms with this framework, considerable effectiveness can be reached. Future work will concern fine-tuning of the similarity computation methodology, and a more extensive experimentation.

References

[1] Bisson, G.: Conceptual clustering in a first order logic representation. In: ECAI 1992: Proceedings of the 10th European conference on Artificial intelligence, pp. 458–462. John Wiley & Sons Inc., Chichester (1992)

[2] Blockeel, H., De Raedt, L., Ramon, J.: Top-down induction of clustering trees. In: Shavlik, J. (ed.) Proceedings of the 15th International Conference on Machine Learning, pp. 55–63. Morgan Kaufmann, San Francisco (1998)

[3] Emde, W., Wettschereck, D.: Relational instance based learning. In: Saitta, L. (ed.) Proc. of ICML 1996, pp. 122–130 (1996)

[4] Esposito, F., Malerba, D., Semeraro, G.: Classification in noisy environments using a distance measure between structural symbolic descriptions. IEEE Transactions on PAMI 14(3), 390–402 (1992)

[5] Esposito, F., Fanizzi, N., Ferilli, S., Semeraro, G.: A generalization model based on oi-implication for ideal theory refinement. Fundam. Inform. 47(1-2), 15–33 (2001)

[6] Ferilli, S., Basile, T.M.A., Di Mauro, N., Biba, M., Esposito, F.: Similarity-guided clause generalization. In: Basili, R., Pazienza, M.T. (eds.) AI*IA 2007. LNCS (LNAI), vol. 4733, pp. 278–289. Springer, Heidelberg (2007)

[7] Fisher, D.H.: Knowledge acquisition via incremental conceptual clustering. Machine Learning 2(2), 139–172 (1987)

[8] Gennari, J.H., Langley, P., Fisher, D.: Models of incremental concept formation. Artificial Intelligence 40(1-3), 11–61 (1989)

[9] Jain, A.K., Murty, M.N., Flynn, P.J.: Data clustering: a review. ACM Computing Surveys 31(3), 264–323 (1999)

[10] Lin, D.: An information-theoretic definition of similarity. In: Proc. 15th International Conf. on Machine Learning, pp. 296–304. Morgan Kaufmann, San Francisco, CA (1998)

[11] Michalski, R.S., Stepp, R.E.: Learning from observation: Conceptual clustering. In: Michalski, R.S., Carbonell, J.G., Mitchell, T.M. (eds.) Machine Learning: An Artificial Intelligence Approach, pp. 331–363. Springer, Berlin (1984)

[12] Nienhuys-Cheng, S.: Distances and limits on herbrand interpretations. In: Page, D.L. (ed.) ILP 1998. LNCS, vol. 1446, pp. 250–260. Springer, Heidelberg (1998)

[13] Plotkin, G.D.: A note on inductive generalization. Machine Intelligence 5, 153–163 (1970)

[14] Ramon, J.: Clustering and instance based learning in first order logic. PhD thesis, Dept. of Computer Science, K.U.Leuven, Belgium (2002)

[15] Ramon, J., Dehaspe, L.: Upgrading bayesian clustering to first order logic. In: Proceedings of the 9th Belgian-Dutch Conference on Machine Learning, pp. 77–84, Department of Computer Science, K.U. Leuven (1999)

[16] Rouveirol, C.: Extensions of inversion of resolution applied to theory completion. In: Inductive Logic Programming, pp. 64–90. Academic Press, London (1992)

[17] Sebag, M.: Distance induction in first order logic. In: Džeroski, S., Lavrač, N. (eds.) ILP 1997. LNCS, vol. 1297, pp. 264–272. Springer, Heidelberg (1997)

[18] Thompson, K., Langley, P.: Incremental concept formation with composite objects. In: Proceedings of the sixth international workshop on Machine learning, pp. 371–374. Morgan Kaufmann Publishers Inc., San Francisco (1989)

Trajectory Analysis of Laboratory Tests as Medical Complex Data Mining

Shoji Hirano and Shusaku Tsumoto

Department of Medical Informatics, Shimane University, School of Medicine
89-1 Enya-cho, Izumo, Shimane 693-8501, Japan
hirano@ieee.org, tsumoto@computer.org

Abstract. Finding temporally covariant variables is very important for clinical practice because we are able to obtain the measurements of some examinations very easily, while it takes a long time for us to measure other ones. Also, unexpected covariant patterns give us new knowledge for temporal evolution of chronic diseases. This paper focuses on clustering of trajectories of temporal sequences of two laboratory examinations. First, we map a set of time series containing different types of laboratory tests into directed trajectories representing temporal change in patients' status. Then the trajectories for individual patients are compared in multiscale and grouped into similar cases by using clustering methods. Experimental results on the chronic hepatitis data demonstrated that the method could find the groups of trajectories which reflects temporal covariance of platelet, albumin and choline esterase.

1 Introduction

Hosptial information system (HIS) collects all the data from all the branches of departments in a hospital, including laboratory tests,physiological tests, electronic patient records. Thus, HIS can be viewed as a large heterogenous database, which stores chronological changes in patients' status. Recent advances not only in informaiton technology, but also other developments in devices enable us to collect huge amount of temporal data automatically, one of whose advantage is that we are able not only to analyze the data within one patient, but also the data in a cross-sectoral manner. It may reveal a underlying mechanism in temporal evolution of (chronic) diseases with some degree of evidence, which can be used to predict or estimate a new case in the future. Especially, finding temporally covariant variables is very important for clinical practice because we are able to obtain the measurements of some examinations very easily, while it takes a long time for us to measure other ones. Also, unexpected covariant patterns give us new knowledge for temporal evolution of chronic diseases. However, despite of its importance, large-scale analysis of time-series medical databases has rarely been reported due to the following problems: (1) sampling intervals and lengths of data can be both irregular, as they depend on the condition of each patient. (2) a time series can include various types of events such as acute changes and chronic changes. When comparing the time series, one is required to appropriately determine the correspondence of data points to be compared taking into account the above issues. Additionally, the dimensionality of data can be usually high due to the variety of medical examinations. These fearures prevent us from using conventional time series analysis methods.

Z.W. Raś, S. Tsumoto, and D. Zighed (Eds.): MCD 2007, LNAI 4944, pp. 27–41, 2008.

This paper presents a novel cluster analysis method for multivariate time-series data on medical laboratory tests. Our method represents time series of test results as trajectories in multidimensional space, and compares their structural similarity by using the multiscale comparison technique [1]. It enables us to find the part-to-part correspondences between two trajectories, taking into account the relationships between different tests. The resultant dissimilarity can be further used as input for clustering algorithms for finding the groups of similar cases. In the experiments we demonstrate the usefulness of our approach through the grouping tasks of artificially generated digit stroke trajectories and medical test trajectories on chronic hepatitis patients.

The remainder of this paper is organized as follows. In Section 2 we describe the methodoology, including preprocessing of the data. In Section 3 we show experimental results on a synthetic data (digit strokes) and chronic hepatitis data (albumin-platelet trajectories and cholinesterase-platelet trajectories). Finally, Section 5 is a conclusion of this paper.

2   Methods

2.1   Overview

Figure 1 shows an overview of the whole process of clustering of trajectories. First, we apply preprocessing of a raw temporal sequence for each variable (Subsection 2.2). Secondly, a trajectory of laboratory tests is calculated for each patient, segmentation technique is applied to each sequence for generation of a segmentation hiearchy (Subsection 2.3). Third, we trace segemented sequences and search for matching between two sequences in a hiearchical way (Subsection 2.4). Then, dissimilarities are calculated for matched sequences (Subsection 2.5 and 2.6). Finally, we apply clustering to the dissimilarities obtained (Subsection 2.7).

2.2   Preprocessing

Time-series examination data is often represented as a tuple of examination date and results. Interval of examinations is usually irregular, as it depends on the condition of a patient. However, in the process of multiscale matching, it is neccessary to represent time-series as a set of data points with a constant interval in order to represent the time span by the number of data points. Therefore, we employed linear interpolation and constructed new equi-interval data.

2.3   Multiscale Description of Trajectories by the Modified Bessel Function

Let us consider examination data for one person, consisting of I different time-series examinations. Let us denote the time series of i-th examination by $ex_i(t)$, where $i \in I$. Then the trajectory of examination results, $c(t)$ is denoted by

$$c(t) = \{ex_1(t), ex_2(t), \ldots, ex_I(t)\}$$

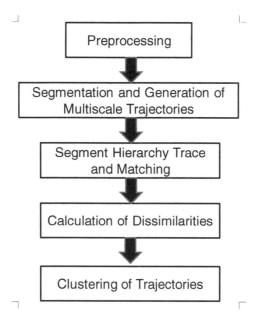

Fig. 1. Overview of Trajectory Clustering

Next, let us denote an observation scale by σ and denote a Gaussian function with scale parameter σ^2 by $g(t, \sigma)$. Then the time-series of the i-th examination at scale σ, $EX_i(t, \sigma)$ is derived by convoluting $ex_i(t)$ with $g(t, \sigma)$ as follows.

$$EX_i(t, \sigma) = ex_i(t) \otimes g(t, \sigma) = \int_{-\infty}^{+\infty} \frac{ex_i(u)}{\sigma \sqrt{2\pi}} e^{\frac{-(t-u)^2}{2\sigma^2}} du$$

Applying the above convolution to all examinations, we obtain the trajectory of examination results at scale σ, $C(t, \sigma)$, as

$$C(t, \sigma) = \{EX_1(t, \sigma), EX_2(t, \sigma), \ldots, EX_I(t, \sigma)\}$$

By changing the scale factor σ, we can represent the trajectory of examination results at various observation scales. Figure 2 illustrates an example of multiscale representation of trajectories where $I = 2$. Increase of σ induces the decrease of convolution weights for neighbors. Therefore, more flat trajectories with less inflection points will be observed at higher scales.

Curvature of the trajectory at time point t is defined by, for $I = 2$,

$$K(t, \sigma) = \frac{EX_1' EX_2'' + EX_1'' EX_2'}{(EX_1'^2 + EX_2'^2)^{3/2}}$$

where EX_i' and EX_i'' denotes the first- and second-order derivatives of $EX_i(t, \sigma)$ respectively. The m-th order derivative of $EX_i(t, \sigma)$, $EX_i^{(m)}(t, \sigma)$, is defined by

$$EX_i^{(m)}(t, \sigma) = \frac{\partial^m EX_i(t, \sigma)}{\partial t^m} = ex_i(t) \otimes g^{(m)}(t, \sigma)$$

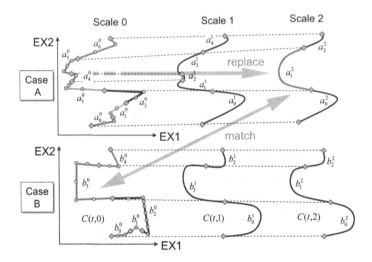

Fig. 2. Multiscale representation and matching scheme

It should be noted that many of the real-world time-series data, including medical data, can be discrete in time domain. Thus, a sampled Gaussian kernel is generally used for calculation of $EX_i(t, \sigma)$, changing an integral to summation. However, Lindeberg [2] pointed out that, a sampled Gaussian may lose some of the properties that a continuous Gaussian has, for example, non-creation of local extrema with the increase of scale. Additionally, in a sampled Gaussian kernel, the center value can be relatively large and imbalanced when the scale is very small. Ref. [2] suggests the use of kernel based on the modified Bessel function, as it is derived by incorporating the discrete property. Since this influences the description ability about detailed structure of trajectories, we employed the Lindeberg's kernel and derive $EX_i(t, \sigma)$ as follows.

$$EX_i(t, \sigma) = \sum_{n=-\infty}^{\infty} e^{-\sigma} I_n(\sigma) ex_i(t - n)$$

where $I_n(\sigma)$ denotes the modified Bessel function of order n. The first- and second-order derivatives of $EX_i(t, \sigma)$ are obtained as follows.

$$EX_i'(t, \sigma) = \sum_{n=-\infty}^{\infty} -\frac{n}{\sigma} e^{-\sigma} I_n(\sigma) ex_i(t - n)$$

$$EX_i''(t, \sigma) = \sum_{n=-\infty}^{\infty} \frac{1}{\sigma} (\frac{n^2}{\sigma} - 1) e^{-\sigma} I_n(\sigma) ex_i(t - n).$$

2.4 Segment Hierarchy Trace and Matching

For each trajectory represented by multiscale description, we find the places of inflection points according to the sign of curvature. Then we divide each trajectory into a set of convex/concave segments, where both ends of a segment correspond to adjacent inflection points. Let A be a trajectory at scale k composed of $M^{(k)}$ segments. Then A is represented by $\mathbf{A}^{(k)} = \{a_i^{(k)} \mid i = 1, 2, \cdots, M^{(k)}\}$, where $a_i^{(k)}$ denotes i-th segment at scale k. Similarly, another trajectory B at scale h is represented by $\mathbf{B}^{(h)} = \{b_j^{(h)} \mid j = 1, 2, \cdots, N^{(h)}\}$.

Next, we chase the cross-scale correspondence of inflection points from top scales to bottom scale. It defines the hierarchy of segments and enables us to guarantee the connectivity of segments represented at different scales. Details of the algorithm for checking segment hierarchy is available on ref. [1]. In order to apply the algorithm for closed curve to open trajectory, we modified it to allow replacement of odd number of segments at sequence ends, since cyclic property of a set of inflection points can be lost.

The main procedure of multiscale matching is to search the best set of segment pairs that satisfies both of the following conditions:

1. Complete Match: By concatenating all segments, the original trajectory must be completely formed without any gaps or overlaps.
2. Minimal Difference: The sum of segment dissimilarities over all segment pairs should be minimized.

The search is performed throughout all scales. For example, in Figure 2, three contiguous segments $a_3^{(0)} - a_5^{(0)}$ at the lowest scale of case A can be integrated into one segment $a_1^{(2)}$ at upper scale 2, and the replaced segment well matches to one segment $b_3^{(0)}$ of case B at the lowest scale. Thus the set of the three segments $a_3^{(0)} - a_5^{(0)}$ and one segment $b_3^{(0)}$ will be considered as a candidate for corresponding segments. On the other hand, segments such as $a_6^{(0)}$ and $b_4^{(0)}$ are similar even at the bottom scale without any replacement. Therefore they will be also a candidate for corresponding segments. In this way, if segments exhibit short-term similarity, they are matched at a lower scale, and if they present long-term similarity, they are matched at a higher scale.

2.5 Local Segment Difference

In order to evaluate the structural (dis-)similarity of segments, we first describe the structural feature of a segment by using shape parameters defined below.

1. Gradient at starting point: $g(a_m^{(k)})$
2. Rotation angle: $\theta(a_m^{(k)})$
3. Velocity: $v(a_m^{(k)})$

Figure 3 illustrates these parameters. Gradient represents the direction of the trajectory at the beginning of the segment. Rotation angle represents the amount of change of direction along the segment. Velocity represents the speed of change in the segment, which is calculated by dividing segment length by the number of points in the segment.

Next, we define the local dissimilarity of two segments, $a_m^{(k)}$ and $b_n^{(h)}$, as follows.

$$d(a_m^{(k)}, b_n^{(h)}) = \sqrt{\left(g(a_m^{(k)}) - g(b_n^{(h)})\right)^2 + \left(\theta(a_m^{(k)}) - \theta(b_n^{(h)})\right)^2}$$
$$+ \left|v(a_m^{(k)}) - v(b_n^{(h)})\right| + \gamma \left\{cost(a_m^{(k)}) + cost(b_n^{(h)})\right\}$$

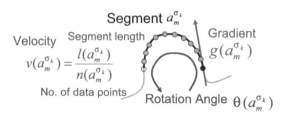

Fig. 3. Segment Parameters

where $cost()$ denotes a cost function used for suppressing excessive replacement of segments, and γ is the weight of costs. We define the cost function using local segment dissimilarity as follows. For a segment $a_m^{(k)}$ that replaces p segments $a_r^{(0)} - a_{r+p-1}^{(0)}$ at the bottom scale,

$$cost(a_m^{(k)}) = \sum_{q=r}^{r+p-1} d(a_q^{(0)}, a_{q+1}^{(0)}).$$

2.6 Sequence Dissimilarity

After determining the best set of segment pairs, we newly calculate value-based dissimilarity for each pair of matched segments. The local segment dissimilarity defined in the previous section reflects the structural difference of segments, but does not reflect the difference of original sequence values; therefore, we calculate the value-based dissimilarity that can be further used as a metric for proximity in clustering.

Suppose we obtained L pairs of matched segments after multiscale matching of trajectories A and B. The value-based dissimilarity between A and B, $D_{val}(A, B)$, is defined as follows.

$$D_{val}(A, B) = \sum_{l=1}^{L} d_{val}(\alpha_l, \beta_l)$$

where α_l denotes a set of contiguous segments of A at the lowest scale that constitutes the l-th matched segment pair $(l \in L)$, and β_l denotes that of B. For example, suppose that segments $a_3^{(0)} \sim a_5^{(0)}$ of A and segment $b_3^{(0)}$ of B in Figure 2 constitute the l-th matched pair. Then, $\alpha_l = a_3^{(0)} \sim a_5^{(0)}$ and $\beta_l = b_3^{(0)}$, respectively. $d_{val}(\alpha_l, \beta_l)$ is the difference between α_l and β_l in terms of data values at the peak and both ends of the segments. For the i-th examination $(i \in I)$, $d_{val_i}(\alpha_l, \beta_l)$ is defined as

$$d_{val_i}(\alpha_l, \beta_l) = peak_i(\alpha_l) - peak_i(\beta_l)$$
$$+ \frac{1}{2}\{left_i(\alpha_l) - left_i(\beta_l)\} + \frac{1}{2}\{right_i(\alpha_l) - right_i(\beta_l)\}$$

where $peak_i(\alpha_l)$, $left_i(\alpha_l)$, and $right_i(\alpha_l)$ denote data values of the i-th examination at the peak, left end and right end of segment α_l, respectively. If α_l or β_l is composed of plural segments, the centroid of the peak points of those segments is used as the peak of α_l. Finally, d_{val_i} is integrated over all examinations as follows.

$$d_{val}(\alpha_l, \beta_l) = \frac{1}{I}\sqrt{\sum_i d_{val_i}(\alpha_l, \beta_l)}.$$

2.7 Clustering

For clustering, we employ two methods: agglomerative hierarchical clustering (AHC) [3] and rough set-based clustering (RC) [4]. The sequence comparison part performs pairwise comparison for all possible pairs of time series, and then produces a dissimilarity matrix. The clustering part performs grouping of trajectories according to the given dissimilarity matrix.

3   Experimental Results

We applied our method to the chronic hepatitis dataset which was a common dataset in ECML/PKDD discovery challenge 2002-2004 [5]. The dataset contained time series laboratory examinations data collected from 771 patients of chronic hepatitis B and C. In this work, we focused on analyzing the temporal relationships between platelet count (PLT), albumin (ALB) and cholinesterase (CHE), that were generally used to examine the status of liver function. Our goals were set to: (1) find groups of trajectories that exhibit interesting patterns, and (2) analyze the relationships between these patterns and the stage of liver fibrosis.

We selected a total of 488 cases which had valid examination results for all of PLT, ALB, CHE and liver biopsy. Constitution of the subjects classified by virus types and administration of interferon (IFN) was as follows. Type B: 193 cases, Type C with IFN: 296 cases, Type C without IFN: 99 cases. In the following sections, we mainly describe the results about Type C without IFN cases, which contained the natural courses of Type C viral hepatitis.

Experiments were conducted as follows. This procedure was applied separately for ALB-PLT, CHE-PLT and ALB-CHE trajectories.

1. Select a pair of cases (patients) and calculate the dissimilarity by using the proposed method. Apply this procedure for all pairs of cases, and construct a dissimilarity matrix.
2. Create a dendrogram by using conventional hierarchical clustering [3] and the dissimilarity matrix. Then perform cluster analysis.

Parameters for multiscale matching were empirically determined as follows: starting scale = 0.5, scale interval = 0,5, number of scales = 100, weight for segment replacement cost = 1.0. We used group average as a linkage criterion for hierarchical clustering. The experiments were performed on a small PC cluster consisted of 8 DELL PowerEdge 1750 (Intel Xeon 2.4GHz 2way) workstations. It took about three minutes to make the dissimilarity matrix for all cases.

Results on ALB-PLT Trajectories. Figure 4 shows the dendrogram generated from the dataset on Type C without IFN cases. The dendrogram suggested splitting of the data into two or three clusters; however, in order to carefully examine the data structure, we

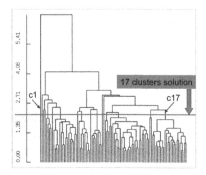

Fig. 4. Dendrogram for ALB-PLT trajectories in Type C without IFN dataset

Table 1. Cluster constitutions of ALB-PLT trajectories, stratified by fibrotic stages. Small clusters of $N < 2$ were omitted.

Cluster	# of Cases / Fibrotic stage				Total
	F0,F1	F2	F3	F4	
5	0	1	0	3	4
7	3	2	2	9	16
9	6	2	0	0	8
11	7	0	0	0	7
14	2	1	0	0	3
15	17	2	7	1	27
16	1	0	1	0	2
17	20	2	1	0	23

avoided excessive merge of clusters and determined to split it into 17 clusters where dissimilarity increased relatively largely at early stage. For each of the 8 clusters that contained ≥ 2 cases, we classified cases according to the fibrotic stage. Table 1 shows the summary. The leftmost column shows cluster number. The next column shows the number of cases whose fibrotic stages were F0 or F1. The subsequent three columns show the number of F2, F3, and F4 cases respectively. The rightmost column shows the total number of cases in each cluster.

From Table 1, it could be recognized that the clusters can be globally classified into one of the two categories: one containing progressed cases of liver fibrosis (clusters 5 and 7) and another containing un-progressed cases (clusters 9, 11, 14, 15, 16 and 17). This can be confirmed from the dendrogram in Figure 4, where these two types of clusters appeared at the second devision from the root. This implied that the difference about ALB and PLT might be related to the fibrotic stages.

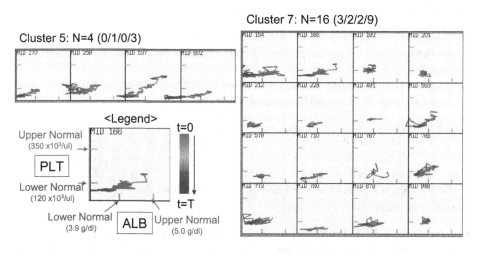

Fig. 5. Trajectories in Cluster 5   **Fig. 6.** Trajectories in Cluster 7

In order to recognize the detailed characteristics of 8 clusters, we observed the feature of grouped trajectories. Figures 5-8 show the examples of grouped ALB-PLT trajectories. Each quadrate region contains a trajectory of ALB-PLT values for a patient. If the number of cases in a cluster was larger than 16, the first 16 cases w.r.t. ID number were selected for visualization. The bottom part of Figure 5 provides the legend. The horizontal axis represents ALB value, and the vertical axis represents PLT value. Lower end of the normal range (ALB:$3.9g/dl$, PLT:$120 \times 10^3/ul$) and Upper end of the normal range (ALB:$5.0g/dl$, PLT:$350 \times 10^3/ul$) were marked with blue and red short lines on each axis respectively. Time phase on each trajectory was represented by color phase: red represents the start of examination, and it changes toward blue as time proceeds.

Figure 5 shows cases grouped into cluster 5 which contained remarkably many F4 cases (3/4). The skewed trajectory of ALT and PLT clearly demonstrated that both values decreased from the normal range to the lower range as time proceeded, due to the dysfunction of the liver. Cluster 7, shown in Figure 6, also contained similarly large number of progressed cases (F4:9/16, F3:2/16) and exhibited the similar characteristics, though it was relatively weaker than in cluster 5.

On the contrary, clusters that contained many un-progressed cases exhibited different characteristics. Figure 7 shows the trajectories grouped into cluster 17, where the number of F0/F1 cases was large (20/23). Most of the trajectories moved within the normal range, and no clear feature about time-direction dependency was observed. Figure 8 (top) shows the trajectories in cluster 11, where all of 7 cases were F0/F1. They moved within the normal range, but the PLT range was higher than in cluster 17.

Figure 8 (bottom) shows the trajectories in cluster 14, where trajectories exhibited skewed shapes similarly to cluster 5. But this cluster consisted of F0/F1 and F2 cases, whereas cluster 5 contained mainly progressed cases. The reason why these cases were separated into different clusters should be investigated further, but it seemed that the difference of progress speed of liver fibrosis, represented as a velocity term, might be a candidate cause.

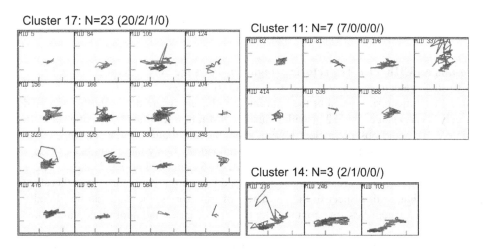

Fig. 7. Trajectories in Cluster 17   **Fig. 8.** Trajectories in Cluster 11 and 14

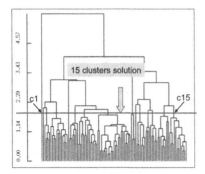

Fig. 9. Dendrogram for Type C without IFN dataset (CHE-PLT trajectories)

Table 2. Cluster constitutions of CHE-PLT trajectories, stratified by fibrotic stages. Small clusters of $N < 2$ were omitted.

Cluster	# of Cases / Fibrotic stage				Total
	F0,F1	F2	F3	F4	
3	0	0	1	3	4
4	2	1	2	7	12
6	3	0	1	2	6
7	5	2	3	3	13
8	9	8	4	2	23
9	1	2	0	0	3
11	4	2	0	0	6
12	2	0	1	0	3
13	5	0	0	0	5
14	8	0	0	0	8
15	12	0	0	0	12

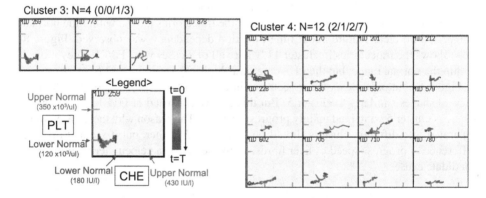

Fig. 10. Trajectories in Cluster 3   **Fig. 11.** Trajectories in Cluster 4

Results on CHE-PLT Trajectories. Figure 9 shows the dendrogram generated from CHE-PLT trajectories of 99 Type C without IFN cases. Similarly to the case of ALB-PLT trajectories, we split the data into 15 clusters where dissimilarity increased largely at early stage. Table 2 provides cluster constitution stratified by fibrotic stage. In Table 2, we could observe a clear feature about the distribution of fibrotic stages over clusters. Clusters such as 3, 4, 6, 7 and 8 contained relatively large number of F3/F4 cases, whereas clusters such as 9, 11, 12, 13, 14, 15 contained no F3/F4 cases. These two types of clusters were divided at the second branch on the dendrogram; therefore it implied that, with respect to the similarity of trajectories, the data can be globally split into two categories, one contains the progressed cases and another contained unprogressed cases.

Now let us examine the features of trajectories grouped into each cluster. Figure 10 shows CHE-PLT trajectories grouped into cluster 3. The bottom part of the figure

Cluster 15: N=12 (12/0/0/0)

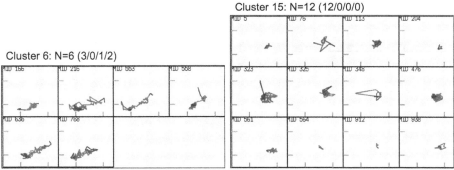

Cluster 6: N=6 (3/0/1/2)

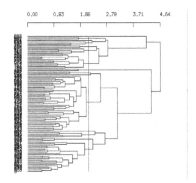

Fig. 12. Trajectories in Cluster 6   **Fig. 13.** Trajectories in Cluster 15

Fig. 14. Dendrogram for ALB-CHE trajectories in Type C without IFN dataset

Table 3. Cluster constitutions of ALB-CHE trajectories, stratified by fibrotic stages. Small clusters of $N < 2$ were omitted.

Cluster	# of Cases / Fibrotic stage				Total
	F0,F1	F2	F3	F4	
2	0	0	0	2	2
4	0	0	0	3	3
5	0	0	1	1	2
6	0	0	0	4	4
7	3	1	2	5	11
8	1	1	0	0	2
9	2	0	0	0	2
11	22	9	8	1	40
13	2	2	0	0	4
14	3	0	0	0	3
15	19	2	0	1	22

provides the legend. The horizontal axis corresponds to CHE, and the vertical axis corresponds to PLT. This cluster contained four cases: one F3 and three F4. The trajectories settled around the lower bounds of the normal range for PLT ($120 \times 10^3/ul$), and below the lower bounds of CHE (180 IU/l), with global direction toward lower values. This meant that, in these cases, CHE deviated from normal range earlier than PLT.

Figure 11 shows trajectories grouped into cluster 4, which contained nine F3/F4 cases and three other cases. Trajectories in this cluster exhibited interesting characteristics. First, they had very clear descending shapes; in contrast to trajectories in other clusters in which trajectories changed directions frequently and largely, they moved toward the left corner with little directional changes. Second, most of the trajectories settled below the normal bound of PLT whereas their CHE values ranged within normal range at early phase. This meant that, in these cases, CHE deviated from normal range later than PLT.

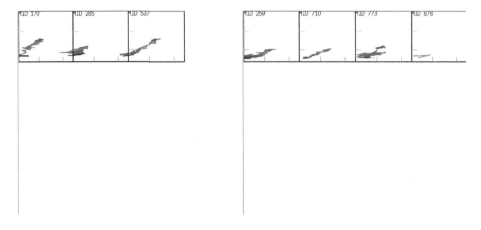

Fig. 15. Trajectories in Cluster 4   **Fig. 16.** Trajectories in Cluster 6

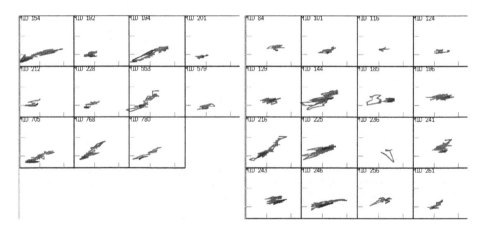

Fig. 17. Trajectories in Cluster 7   **Fig. 18.** Trajectories in Cluster 11

Figure 12 shows trajectories grouped into cluster 6, which contained three F3/F4 cases and three other cases. Trajectories in this cluster exhibited descending shapes similarly to the cases in cluster 4. The average levels of PLT were higher than those in cluster 4, and did not largely deviated from the normal range. CHE remained within the normal range for most of the observations.

Figure 13 shows trajectories grouped into cluster 15, which contained twelve F0/F1 cases and no other cases. In contrast to the high stage cases mentioned above, trajectories settled within the normal ranges for both CHE and PLT and did not exhibit any remarkable features about their directions.

These results suggested the followings about the CHE-PLT trajectories on type C without IFN cases used in this experiment: (1) They could be globally divided into two categories, one containing high-stage cases and another containing low-stage cases,

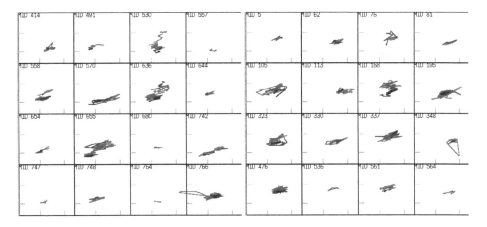

Fig. 19. Trajectories in Cluster 11(2)   **Fig. 20.** Trajectories in Cluster 15

(2) trajectories in some high-stage clusters exhibited very clear descending shapes. (3) in a group containing descending trajectories, PLT deviated from normal range faster than CHE, however, in another group containing descending trajectories, PLT deviated from normal range later than CHE.

Results on ALB-CHE Trajectories. Figure 14 shows the dendrogram generated from the dataset on Type C without IFN cases. The dendrogram suggested splitting of the data into two or three clusters; however, in order to carefully examine the data structure, we avoided excessive merge of clusters and determined to split it into 15 clusters where dissimilarity increased relatively largely at early stage. For each of the 8 clusters that contained ≥ 2 cases, we classified cases according to the fibrotic stage. Table 1 shows the summary. The leftmost column shows cluster number. The next column shows the number of cases whose fibrotic stages were F0 or F1. The subsequent three columns show the number of F2, F3, and F4 cases respectively. The rightmost column shows the total number of cases in each cluster.

From Table 3, it could be recognized that the clusters can be globally classified into one of the two categories: one containing progressed cases of liver fibrosis (clusters 2, 4, 5, 6and 7) and another containing un-progressed cases (clusters 8, 9, 11, 13, 14 and 15). This can be confirmed from the dendrogram in Figure 14, where these two types of clusters appeared at the second devision from the root. This implied that the difference about ALB and PLT might be related to the fibrotic stages.

In order to recognize the detailed characteristics of 8 clusters, we observed the feature of grouped trajectories. Figures 5–8 show the examples of grouped ALB-PLT trajectories. Each quadrate region contains a trajectory of ALB-PLT values for a patient. If the number of cases in a cluster was larger than 16, the first 16 cases w.r.t. ID number were selected for visualization. The bottom part of Figure 5 provides the legend. The horizontal axis represents ALB value, and the vertical axis represents CHE value. Time phase on each trajectory was represented by color phase: red represents the start of examination, and it changes toward blue as time proceeds.

Figure 15 and 16 shows cases grouped into cluster 4 and 6 which contained only F4 cases (3 and 4). The skewed trajectory of ALT and CHE clearly demonstrated that both values decreased from the normal range to the lower range as time proceeded, due to the dysfunction of the liver. Cluster 7, shown in Figure 17, also contained similarly large number of progressed cases (F1: 3, F2: 1, F3: 2, F4: 5) and exhibited the similar characteristics, though it was relatively weaker than in cluster 4 and 6.

On the contrary, clusters that contained many un-progressed cases exhibited different characteristics. Figure 18 and 19 show the trajectories grouped into cluster 11, where the number of F0/F1 cases was large (31/40). Most of the trajectories moved within the normal range, but some decreasing cases were included in this cluster, and no clear feature about time-direction dependency was observed. Figure 20 shows the trajectories in cluster 15, where 19 of 22 cases were F0/F1 and the sequences moved within the normal range.

In summary, the degree of covariance between ALB and CHE is higher than those between ALB and PLT or CHE and PLT. Samples are better split into F4-dominant cases and F0/F1-dominant cases.

4   Discussion

Table 4 compares the characteristics of clustering results. As shown in the table, it seems that a combination of ALB and CHE generates a slightly better results than other pairs with respect to the degree of seperation of fibrotic stages.

Table 5 shows the contingency table between CHE-PLT and ALB-PLT whose examples belong to Cluster No.7 in ALB-CHE. Compared with the results in Table 2 and Table 1, these cases covers impure clusters in CHE-PLT and ALB-PLT. This observation shows that this cluster should be carefully examined by additional information, since the cluster includes several F0/F1 cases whose PLT is decreasing.

Table 4. Comparison of Clustering Results

Pair	#Clusters	$\#Examples$	Most Impurity Clusters				
		> 1	#	F1	F2	F3	F4
ALB-PLT	17	8	16	3	2	2	9
CHE-PLT	15	11	13	5	2	3	3
ALB-CHE	15	11	11	3	1	2	5

Table 5. Contingency Table of Cluster No.7 in ALB-CHE

		ALB-PLT			
		No.7	No.14	No.15	Total
CHE-PLT	No. 4	7	1	0	8
	No. 6	2	0	0	2
	No. 7	0	0	1	1
	Total	9	1	1	11

5 Conclusions

In this paper we propose a trajectory clustering method as multivariate temporal data mining and shows its application to data on chronic hepatits. Our method consists of a two-stage approach. Firstly, it compares two trajectories based on their structural similarity and determines the best correspondence of partial trajectories. Next, it calculates the value-based dissimilarity for the all pairs of matched segments and outputs the total sum as dissimilarity of the two trajectories.

Clustering experiments on the chronic hepatitis dataset yielded several interesting results. First, the clusters constructed with respect to the similarity of trajectories well matched with the distribution of fibrotic stages, especially with the distribution of high-stage cases and low-stage cases, for ALB-PLT, CHE-PLT and ALB-CHE trajectories. Among three combinations, ALB-CHE shows the highest degree of covariance, which means that CHE can be used to evaluate the trends of ALB.

Our next step is to extend bivariate trajectory analysis into multivariate one. From the viewpoint of medical application, our challenging issue will be to find a variable whose chronological trend is fitted to PLT.

References

1. Ueda, N., Suzuki, S.: A Matching Algorithm of Deformed Planar Curves Using Multiscale Convex/Concave Structures. IEICE Transactions on Information and Systems J73-D-II(7), 992–1000 (1990)
2. Lindeberg, T.: Scale-Space for Discrete Signals. IEEE Trans. PAMI 12(3), 234–254 (1990)
3. Everitt, B.S., Landau, S., Leese, M.: Cluster Analysis, 4th edn. Arnold Publishers (2001)
4. Hirano, S., Tsumoto, S.: An Indiscernibility-Based Clustering Method with Iterative Refinement of Equivalence Relations - Rough Clustering -. Journal of Advanced Computational Intelligence and Intelligent Informatics 7(2), 169–177 (2003)
5. URL: http://lisp.vse.cz/challenge/
6. Tsumoto, S., Hirano, S., Takabayashi, K.: Development of the Active Mining System in Medicine Based on Rough Sets. Journal of Japan Society of Artificial Intelligence 20(2), 203–210 (2005)

Conceptual Clustering Applied to Ontologies
A Distance-Based Evolutionary Approach

Floriana Esposito, Nicola Fanizzi, and Claudia d'Amato

LACAM – Dipartimento di Informatica, Università degli studi di Bari
Campus Universitario, Via Orabona 4 – 70125 Bari, Italy
{esposito,fanizzi,claudia.damato}@di.uniba.it

Abstract. A clustering method is presented which can be applied to semantically annotated resources in the context of ontological knowledge bases. This method can be used to discover emerging groupings of resources expressed in the standard ontology languages. The method exploits a language-independent semi-distance measure over the space of resources, that is based on their semantics w.r.t. a number of dimensions corresponding to a committee of discriminating features represented by concept descriptions. A maximally discriminating group of features can be constructed through a feature construction method based on genetic programming. The evolutionary clustering algorithm proposed is based on the notion of medoids applied to relational representations. It is able to induce a set of clusters by means of a fitness function based on a discernibility criterion. An experimentation with some ontologies proves the feasibility of our method.

1 Introduction

In the perspective of the Semantic Web [2] knowledge bases will contain rich data and meta-data described with complex representations. This requires re-thinking the current data mining approaches to cope with the challenge of the new representation and semantics. In this work, unsupervised learning is tackled in the context of the standard concept languages used for representing ontologies which are based on *Description Logics* (henceforth DLs) [1]. In particular, we focus on the problem of *conceptual clustering* [25] for semantically annotated resources.

The benefits of clustering in the context of semantically annotated knowledge bases are manifold. Clustering enables the definition of new emerging categories (*concept formation*) on the grounds of the primitive concepts asserted in a knowledge base [9]; supervised methods can exploit these clusters to induce new concept definitions or to refining existing ones *ontology evolution*; intensionally defined groupings may speed-up the task of *discovery* and search in general.

Essentially, many existing clustering methods are based on the application of similarity (or density) measures defined over a fixed set of attributes of the domain objects. Classes of objects are taken as collections that exhibit low interclass similarity (density) and high intraclass similarity (density). Thus, clustering methods have aimed at defining groups of objects through conjunctive descriptions based on selected attributes [25].

Often these methods cannot into account any form of *prior knowledge* at a conceptual level encoding some semantic relationships. This hinders the interpretation of the

Z.W. Raś, S. Tsumoto, and D. Zighed (Eds.): MCD 2007, LNAI 4944, pp. 42–56, 2008.

outcomes of these methods which is crucial in the Semantic Web perspective in which the expressiveness of the language adopted for describing objects and clusters is extremely important. Specific logic-based approaches, intended for terminological representations [1], have have been proposed as language-dependent methods [16, 9]. These methods have been criticized for suffering from noise in the data. This motivates our investigation on similarity-based clustering approaches which can be more noise-tolerant, and as language-independent as possible. Specifically we propose a multi-relational extension of effective clustering techniques intended for grouping similar resources w.r.t. a semantic dissimilarity measure, which is tailored for the standard representations of Semantic Web context.

From a technical viewpoint, adapting existing algorithms to work on complex representations, requires semantic measures that are suitable for such concept languages. Recently, dissimilarity measures for specific DLs have been proposed [5]. Although they turned out to be quite effective for the inductive tasks, they were still partly based on structural criteria which makes them fail to fully capture the underlying semantics and hardly scale to any standard ontology language. As pointed out in a seminal paper on similarity measures for DLs [4], most of the existing measures focus on the similarity of atomic concepts within hierarchies or simple ontologies. Moreover, they have been conceived for assessing *concept* similarity, whereas, for other tasks, a notion of similarity between *individuals* is required.

Therefore, we have devised a family of dissimilarity measures for semantically annotated resources, which can overcome the mentioned limitations [8]. Following the criterion of semantic discernibility of individuals, these measures are suitable for a wide range of concept languages since they are merely based on the discernibility of the input individuals with respect to a fixed committee of features represented by concept definitions. As such the new measures are not absolute, yet they depend on the knowledge base they are applied to. Thus, also the choice of the optimal feature sets deserves a preliminary feature construction phase, which may be performed by means of a randomized search procedure based on *genetic programming*, whose operators are borrowed from recent works on ontology evolution [13].

The clustering algorithm that we propose adopts an evolutionary learning approach for adapting classic distance-based clustering approaches, such as the K-MEANS [14]. In our setting, instead of the notion of *centroid* that characterizes algorithms originally developed for numeric or ordinal features, we recur to the notion of *medoids* [15] as central individuals in a cluster. The clustering problem is solved by considering populations made up of strings of medoids with different lengths. The medoids are computed according to the semantic measure induced with the methodology introduced above. On each generation, the strings in the current population are evolved by mutation and cross-over operators, which are also able to change the number of medoids. Thus, this algorithm is also able to autonomously suggest a promising number of clusters.

The paper is organized as follows. Sect. 2 presents the basics of the representation and the similarity measure adopted in the clustering algorithm. This algorithm is illustrated and discussed in Sect. 3. Related methods and distance measures are recalled in Sect. 4 then an experimental session applying the method on real ontologies is reported in Sect. 5. Conclusions and extensions are finally examined in Sect. 6.

2   Semantic Distance Measures

One of the advantages of our method is that it does not rely on a particular language for semantic annotations. Hence, in the following, we assume that resources, concepts and their relationship may be defined in terms of a generic ontology language that may be mapped to some DL language with the standard open-world semantics (see the handbook [1] for a thorough reference).

In this context, a *knowledge base* $\mathcal{K} = \langle \mathcal{T}, \mathcal{A} \rangle$ is made up of a *TBox* $\mathcal{T}$ and an *ABox* $\mathcal{A}$. $\mathcal{T}$ is a set of concept definitions. $\mathcal{A}$ contains assertions (ground facts) concerning individuals. The set of the individuals occurring in $\mathcal{A}$ will be denoted with $\mathsf{Ind}(\mathcal{A})$. The *unique names assumption* can be made for such individuals: each is assumed to be identified by its own URI.

As regards the inference services, like all other instance-based methods, our procedure may require performing *instance-checking*, which amounts to determining whether an individual, say a, belongs to a concept extension, i.e. whether $C(a)$ holds for a certain concept C.

2.1   A Semantic Semi-distance for Individuals

Moreover, for our purposes, we need a function for measuring the similarity of individuals rather than concepts. It can be observed that individuals do not have a syntactic structure that can be compared. This has led to lifting them to the concept description level before comparing them (recurring to the approximation of the *most specific concept* of an individual w.r.t. the ABox).

We have developed new measures whose definition totally depends on semantic aspects of the individuals in the knowledge base [8]. On a semantic level, similar individuals should behave similarly with respect to the same concepts. We introduce a novel measure for assessing the similarity of individuals in a knowledge base, which is based on the idea of comparing their semantics along a number of dimensions represented by a committee of concept descriptions. Following the ideas borrowed from ILP [24] and *multi-dimensional scaling*, we propose the definition of totally semantic distance measures for individuals in the context of a knowledge base.

The rationale of the new measure is to compare them on the grounds of their behavior w.r.t. a given set of hypotheses, that is a collection of concept descriptions, say $\mathsf{F} = \{F_1, F_2, \ldots, F_m\}$, which stands as a group of discriminating *features* expressed in the language taken into account.

In its simple formulation, a family of distance functions for individuals inspired to Minkowski's distances can be defined as follows:

Definition 2.1 (dissimilarity measures). *Let* $\mathcal{K} = \langle \mathcal{T}, \mathcal{A} \rangle$ *be a knowledge base. Given a set of concept descriptions* $\mathsf{F} = \{F_1, F_2, \ldots, F_m\}$, *a family of functions*

$$d_p^\mathsf{F} : \mathsf{Ind}(\mathcal{A}) \times \mathsf{Ind}(\mathcal{A}) \mapsto [0, 1]$$

defined as follows:
$\forall a, b \in \mathsf{Ind}(\mathcal{A})$

$$d_p^\mathsf{F}(a, b) := \frac{1}{m} \left(\sum_{i=1}^{m} \mid \pi_i(a) - \pi_i(b) \mid^p \right)^{1/p}$$

where $p > 0$ and $\forall i \in \{1, \ldots, m\}$ the projection function π_i *is defined by:*
$\forall c \in \mathsf{Ind}(\mathcal{A})$

$$\pi_i(c) = \begin{cases} 1 & \mathcal{K} \models F_i(c) \\ 0 & \mathcal{K} \models \neg F_i(c) \\ 1/2 & otherwise \end{cases} \qquad (1)$$

The case of $\pi_i(c) = 1/2$ corresponds to the case when a reasoner cannot give the truth value for a certain membership query. This is due to the *Open World Assumption* (OWA) normally made in the descriptive semantics [1].

It can be proved that these functions have almost all standard properties of distances [8]:

Proposition 2.1 (semi-distance). *For a fixed feature set* F *and* $p > 0$ *the function* d_p^{F} *is a semi-distance.*

It cannot be proved that $d_p(a, b) = 0$ iff $a = b$. This is the case of *indiscernible* individuals with respect to the given set of hypotheses F.

Compared to other proposed distance (or dissimilarity) measures [4], the presented function does not depend on the constructors of a specific language, rather it requires only retrieval or instance-checking service used for deciding whether an individual is asserted in the knowledge base to belong to a concept extension (or, alternatively, if this could be derived as a logical consequence).

Note that the π_i functions ($\forall i = 1, \ldots, m$) for the training instances, that contribute to determine the measure with respect to new ones, can be computed in advance thus determining a speed-up in the actual computation of the measure. This is very important for the measure integration in algorithms which massively use this distance, such as all instance-based methods.

The underlying idea for the measure is that similar individuals should exhibit the same behavior w.r.t. the concepts in F. Here, we make the assumption that the feature-set F represents a sufficient number of (possibly redundant) features that are able to discriminate really different individuals.

2.2 Committee Optimization

The choice of the concepts to be included in the committee – *feature selection* – may be crucial. Experimentally, it was observed that good results could be obtained by using the very set of both primitive and defined concepts found in the ontology. However, some ontologies define very large sets of concepts which make the task unfeasible. Thus, we have devised a specific optimization algorithms founded in *genetic programming* which are able to find optimal choices of discriminating concept committees.

Various optimizations of the measures can be foreseen as concerns its definition. Among the possible sets of features we will prefer those that are able to discriminate the individuals in the ABox.

Since the function is very dependent on the concepts included in the committee of features F, two immediate heuristics can be derived:

- Limit the number of concepts of the committee, including especially those that are endowed with a real discriminating power;

- Find sets of discriminating features, by allowing also their composition employing the specific constructors made available by the representation language of choice.

Both these objectives can be accomplished by means of randomized optimization techniques especially when knowledge bases with large sets of individuals are available. Namely, part of the entire data can be drawn in order to learn optimal F sets, in advance with respect to the successive usage for all other purposes.

Specifically, we experimented the usage of genetic programming for constructing optimal sets of features. Thus we devised the algorithm depicted in Fig. 1. Essentially the algorithm searches the space of all possible feature committees starting from an initial guess (determined by MAKEINITIALFS($\mathcal{K}$)) based on the concepts (both primitive and defined) currently referenced in the knowledge base $\mathcal{K}$.

The outer loop gradually augments the cardinality of the candidate committees. It is repeated until the algorithm realizes that employing larger feature committees would

```
FeatureSet OPTIMIZEFS(𝒦, maxGenerations, minFitness)
input:
    𝒦: current knowledge base
    maxGenerations: maximal number of generations
    minFitness: minimal fitness value
output:
    FeatureSet: FeatureSet
begin
currentBestFitness := 0; formerBestFitness := 0;
currentFSs := MAKEINITIALFS(𝒦); formerFSs := currentFSs;
repeat
    fitnessImproved := false;
    generationNumber := 0;
    currentBestFitness := BESTFITNESS(currentFSs);
    while (currentBestFitness < minFitness) or (generationNumber < maxGenerations)
        begin
        offsprings := GENERATEOFFSPRINGS(currentFSs);
        currentFSs := SELECTFROMPOPULATION(offsprings);
        currentBestFitness := BESTFITNESS(currentFSs);
        ++generationNumber;
        end
    if  (currentBestFitness > formerBestFitness) and (currentBestFitness < minFitness) then
        begin
        formerFSs := currentFSs;
        formerBestFitness := currentBestFitness;
        currentFSs := ENLARGEFS(currentFSs);
        end
    else fitnessImproved := true;
    end
until not fitnessImproved;
return BEST(formerFSs);
end
```

Fig. 1. Feature set optimization algorithm based on Genetic Programming

not yield a better fitness value with respect to the best fitness recorded in the previous iteration (with fewer features).

The inner loop is repeated for a number of generations until a stop criterion is met, based on the maximal value of generations maxGenerations or, alternatively, when an minimal threshold for the fitness value minFitness is reached by some feature set in the population, which can be returned.

As regards the BESTFITNESS() routine, it computes the best feature committee in a vector in terms of their *discernibility* [22, 12]. For instance, given the whole set of individuals $IS = \mathsf{Ind}(\mathcal{A})$ (or just a sample to be used to induce an optimal measure) the fitness function may be:

$$\text{DISCERNIBILITY}(\mathsf{F}) := \frac{1}{|IS|^2} \sum_{(a,b)\in IS^2} \sum_{i=1}^{|\mathsf{F}|} \frac{\mid \pi_i(a) - \pi_i(b) \mid}{2 \cdot |\mathsf{F}|}$$

As concerns finding candidate sets of concepts to replace the current committee (GENERATEOFFSPRINGS() routine), the function was implemented by recurring to simple transformations of a feature set:

- Choose $\mathsf{F} \in$ currentFSs;
- Randomly select $F_i \in \mathsf{F}$;
  - replace F_i with $F_i' \in$ RANDOMMUTATION(F_i) randomly constructed, or
  - replace F_i with one of its refinements $F_i' \in$ REF(F_i)

Refinement of concept description may be language specific. E.g. for the case of $\mathcal{ALC}$ logic, refinement operators have been proposed in [13].

This is iterated till a suitable number of offsprings is generated. Then these offspring feature sets are evaluated and the best ones are included in the new version of the currentFSs array; the minimal fitness value for these feature sets is also computed. As mentioned, when the while-loop is over the current best fitness is compared with the best one computed for the former feature set length; if an improvement is detected then the outer repeat-loop is continued, otherwise (one of) the former best feature set(s) is selected for being returned as the result of the algorithm.

Further methods for performing feature construction by means of randomized approaches are discussed in [8], where we propose a different approach based on *simulated annealing* in a DL framework, employing similar refinement operators.

3 Evolutionary Clustering Around Medoids

The conceptual clustering procedure consists of two phases: one that detects the clusters in the data and the other that finds an intensional definition for the groups of individuals detected in the former phase.

The first clustering phase implements a genetic programming learning scheme, where the designed representation for the competing genes is made up of strings (lists) of individuals of different lengths, where each individual stands as prototypical for one cluster. Thus, each cluster will be represented by its prototype recurring to the notion of *medoid* [15, 14] on a categorical feature-space w.r.t. the distance measure previously defined. Namely, the medoid of a group of individuals is the individual that has the lowest distance w.r.t. the others. Formally. given a cluster $C = \{a_1, a_2, \ldots, a_n\}$, the medoid is defined:

$$m = \text{medoid}(C) := \underset{a \in C}{\text{argmin}} \sum_{j=1}^{n} d(a, a_j)$$

The algorithm performs a search in the space of possible clusterings of the individuals optimizing a fitness measure maximizing discernibility of the individuals of the different clusters (inter-cluster separation) and the intra-cluster similarity measured in terms of our metric.

The second phase is more language dependent. The various cluster can be considered as training examples for a supervised algorithm aimed at finding an intensional DL definition for one cluster against the counterexamples, represented by individuals in different clusters [16, 9].

3.1   The Clustering Algorithm

The proposed clustering algorithm can be considered as an extension of methods based on genetic programming, where the notion of cluster prototypical instance of centroid, typical of the numeric feature-vector data representations, is replaced by that of medoid [15] as in (*Partition Around Medoids* or *PAM*): each cluster is represented by one of the individuals in the cluster, the medoid, i.e., in our case, the one with the lowest average distance w.r.t. all the others individuals in the cluster. In the algorithm, a genome will be represented by a list of medoids $G = \{m_1, \ldots, m_k\}$. Per each generation those that are considered as best w.r.t. a fitness function are selected for passing to the next generation. Note that the algorithm does not prescribe a fixed length of these lists (as, for instance in K-MEANS and its extensions [14]), hence it should be able to detect an optimal number of clusters for the data at hand.

Fig. 2 reports a sketch of the clustering algorithm. After the call to the initialization procedure INITIALIZE() returning the randomly generated initial population of medoid strings (currentPopulation) in a number of popLength, it essentially consists of the typical generation loop of genetic programming.

At each iteration this computes the new offsprings of current best clusterings represented by currentPopulation. This is performed by suitable genetic operators explained in the following. The fitnessVector recording the quality of the various offsprings (i.e. clusterings) is then updated, which is used to select the best offsprings that survive, passing to the next generation.

The quality of a genome $G = \{m_1, \ldots, m_k\}$ is evaluated by distributing all individuals among the clusters ideally formed around the medoids listed in it. Let C_i be the cluster around medoid m_i, $i = 1, \ldots, k$. Then, the measure is computed as follows:

$$\text{UNFITNESS}(G) := \sqrt{k+1} \sum_{i=1}^{k} \sum_{x \in C_i} d_p(x, m_i)$$

```
medoidVector ECM(maxGenerations, minGap)
input:
    maxGenerations: max number of iterations;
    minGap: minimal gap for stopping the evolution;
output:
    medoidVector: list of medoids
begin
INITIALIZE(currentPopulation,popLength);
while (generation ≤ maxGenerations) and (gap > minGap)
    begin
    offsprings := GENERATEOFFSPRINGS(currentPopulation);
    fitnessVector := COMPUTEFITNESS(offsprings);
    currentPopulation := SELECT(offsprings,fitnessVector);
    gap := (UNFITNESS[popLength]−UNFITNESS[1]);
    generation++;
    end
return currentPopulation[0]; // best genome
end
```

Fig. 2. ECM: the EVOLUTIONARY CLUSTERING AROUND MEDOIDS algorithm

This measure is to be minimized. The factor $\sqrt{k+1}$ is introduced in order to penalize those clusterings made up of too many clusters that could enforce the minimization in this way (e.g. by proliferating singletons). This can be considered a measure of incoherence *within* the various clusters, while the fitness function used in the metric optimization procedure measures discernibility as the spread of the various individuals in the derived space independently of their classification.

The loop condition is controlled by two factors the maximal number of generation (the maxGenerations parameter) and the difference (gap) between the fitness of best and of the worst selected genomes in currentPopulation (which is supposed to be sorted in ascending order, 1 through popLength). Thus another stopping criterion is met when this gap becomes less than the minimal gap minGap passed as a parameter to the algorithm, meaning that the algorithm has reached a (local) minimum.

It remains to specify the nature of the GENERATEOFFSPRINGS procedure function and the number of such offsprings, which may as well be another parameter of the ECM algorithm. Three mutation and one crossover operators are implemented:

DELETION(G) drop a randomly selected medoid:
    $G := G \setminus \{m\}, m \in G$
INSERTION(G) select $m \in \mathrm{Ind}(\mathcal{A}) \setminus G$ that is added to G:
    $G := G \cup \{m\}$
REPLACEMENTWITHNEIGHBOR(G) randomly select $m \in G$ and replace it with $m' \in \mathrm{Ind}(\mathcal{A}) \setminus G$ such that $\forall m'' \in \mathrm{Ind}(\mathcal{A}) \setminus G \ d(m, m') \leq d(m, m'')$:
    $G' := (G \setminus \{m\}) \cup \{m'\}$
CROSSOVER(G_A,G_B) select subsets $S_A \subset G_A$ and $S_B \subset G_B$ and exchange them between the genomes:
    $G_A := (G_A \setminus S_A) \cup S_B$ and $G_B := (G_B \setminus S_B) \cup S_A$

A (10+60) selection strategy has been implemented, indicating, resp., the number of parents selected for survival and the number of their offsprings.

3.2   The Supervised Learning Phase

Each cluster may be labeled with a new DL concept definition which characterizes the individuals in the given cluster while discriminating those in other clusters [9]. The process of labeling clusters with concepts can be regarded as solving a number of supervised learning problems in the specific multi-relational representation targeted in our setting. As such, it deserves specific solutions that are suitable for the DL languages employed.

A straightforward solution, for DLs that allow for the computation of (an approximation of) the *most specific concept* (msc) and *least common subsumer* (lcs) [1] (such as $\mathcal{ALC}$) is depicted in Fig. 3.

However, such a solution is likely to produce overly specific definitions which may lack of predictiveness w.r.t. future individuals. Hence, better generalizing operators would be needed. Alternatively, algorithms for learning concept descriptions expressed in DLs may be employed [13]. Further refinement operators for the $\mathcal{ALC}$ DL have been proposed [18] to be employed in an algorithm performing a heuristic search in the refinement tree guided by a fitness function.

3.3   Discussion

For an analysis of the algorithm, the parameters of the methods based on genetic programming have to be considered, namely maximum number of iterations, number of offsprings, number of genomes that are selected for the next generation. However, it should be also pointed out that computing the fitness function requires some inference service (instance-checking) from a reasoner whose complexity may dominate the overall complexity of the process. This depends on the DL language of choice and also on the structure of the concepts descriptions handled, as investigated in the specific area (see [1], Ch. 3).

The representation of centers by means of medoids has two advantages. First, it presents no limitations on attributes types, and, second, the choice of medoids is dictated

---

input   Clustering $= \{C_j \mid j = 1, \ldots, k\}$: set of clusters
$\mathcal{K} = \langle \mathcal{T}, \mathcal{A} \rangle$: knowledge base;
output Descriptions: set of DL concept descriptions
Descriptions $:= \emptyset$;
for each $C_j \in$ Clustering:
    for each individual $a_i \in C_j$:
        do compute $M_i := \mathsf{msc}(a_i)$ w.r.t. $\mathcal{A}$;
    let MSCs$_j := \{M_i \mid a_i \in C_j\}$;
    Descriptions $:=$ Descriptions $\cup \{\mathsf{lcs}(\mathsf{MSCs}_j)\}$;
return Descriptions;

Fig. 3. A basic concept induction algorithm from clusterings

by the location of a predominant fraction of points inside a cluster and, therefore, it is lesser sensitive to the presence of outliers. Density based methods could be also investigated, yet this may be difficult when handling complex data. In K-MEANS case a cluster is represented by its centroid, which is a mean (usually weighted average) of points within a cluster. This works conveniently only with numerical attributes and can be negatively affected by a single outlier.

Together with the density based clustering methods, also the algorithms based on medoids have several favorable properties w.r.t. other methods based on (dis)similarity. Since it performs clustering with respect to any specified metric, it allows for a flexible definition of the similarity function. This flexibility is particularly important in biological applications where researchers may be interested, for example, in grouping correlated or possibly also anti-correlated elements. Many clustering algorithms do not allow for a flexible definition of similarity: mostly they are rather based on a distances in Euclidean spaces. In addition, the algorithm has the advantage of identifying clusters by the medoids which represent more robust representations of the cluster centers that are less sensitive to outliers than other cluster profiles, such as the cluster centers of K-MEANS. This robustness is particularly important in the common context that many elements do not belong exactly to any cluster, which may be the case of the membership in DL knowledge bases, which may be not ascertained given the OWA.

4   Related Work

The unsupervised learning procedure presented in this paper is mainly based on two factors: the semantic dissimilarity measure and the clustering method. To the best of our knowledge in the literature there are very few examples of similar clustering algorithms working on complex representations that are suitable for knowledge bases of semantically annotated resources. Thus, in this section, we briefly discuss sources of inspiration for our procedure and some related approaches.

As previously mentioned, various attempts to define semantic similarity (or dissimilarity) measures for concept languages have been made, yet they have still a limited applicability to simple languages [4] or they are not completely semantic depending also on the structure of the descriptions [5]. OSS is another recent proposal for an asymmetric similarity function for concepts within an ontology [23] based on its structure. Very few works deal with the comparison of individuals rather than concepts.

In the context of clausal logics, a metric was defined [21] for the Herbrand interpretations of logic clauses as induced from a distance defined on the space of ground atoms. This kind of measures may be employed to assess similarity in *deductive databases*. Although it represents a form of fully semantic measure, different assumptions are made with respect to those which are standard for knowledgeable bases in the SW perspective. Therefore the transposition to the context of interest is not straightforward.

Our measure is mainly based on Minkowski's measures [26] and on a method for distance induction developed by Sebag [24] in the context of *machine learning*, where *metric learning* is developing as an important subfield. In this work it is shown that the induced measure could be accurate when employed for classification tasks even though set of features to be used were not the optimal ones (or they were redundant).

Indeed, differently from our unsupervised learning approach, the original method learns different versions of the same target concept, which are then employed in a voting procedure similar to the Nearest Neighbor approach for determining the classification of instances.

A source of inspiration was also *rough sets* theory [22] which aims at the formal definition of vague sets by means of their approximations determined by an indiscernibility relationship. Hopefully, these methods developed in this context will help solving the open points of our framework (see Sect. 6) and suggest new ways to treat uncertainty.

Our algorithm adapts to the specific representations devised for the SW context a combination of evolutionary clustering and the distance-based approaches (see [14]). Specifically, in the methods derived from K-MEANS and K-MEDOIDS each cluster is represented by one of its points.

Early versions of this approach are represented by further algorithms based on PAM such as CLARA [15], and CLARANS [20]. They implement iterative optimization methods that essentially cyclically relocate points between perspective clusters and re-compute potential medoids. The leading principle for the process is the effect on an objective function. The whole dataset is assigned to resulting medoids, the objective function is computed, and the best system of medoids is retained. In CLARANS a graph is considered whose nodes are sets of k medoids and an edge connects two nodes if they differ by one medoid. While CLARA compares very few neighbors (a fixed small sample), CLARANS uses random search to generate neighbors by starting with an arbitrary node and randomly checking maxneighbor neighbors. If a neighbor represents a better partition, the process continues with this new node. Otherwise a local minimum is found, and the algorithm restarts until a certain number of local minima is found. The best node (i.e. a set of medoids) is returned for the formation of a resulting partition. Ester et al. [6] extended CLARANS to deal with very large spatial databases.

Our algorithm may be considered an extension of evolutionary clustering methods [11] which are also capable to determine a good estimate of the number of clusters [10]. Besides, we adopted the idea of representing clusterings (genomes) as strings of cluster centers [17] transposed to the case of medoids for the categorical search spaces of interest.

Other related recent approaches are represented by the UNC algorithm and its extension to the hierarchical clustering case H-UNC [19]. Essentially, UNC solves a multi-modal function optimization problem seeking dense areas in the feature space. It is also able to determine their number. The algorithm is also demonstrated to be noise-tolerant and robust w.r.t. the presence of outliers. However, the applicability is limited to simpler representations w.r.t. those considered in this paper.

Further comparable clustering methods are those based on an *indiscernibility relationship* [12]. While in our method this idea is embedded in the semi-distance measure (and the choice of the committee of concepts), these algorithms are based on an iterative refinement of an equivalence relationship which eventually induces clusters as equivalence classes.

As mentioned in the introduction, the classic approaches to conceptual clustering [25] in complex (multi-relational) spaces are based on structure and logics. Kietz & Morik proposed a method for efficient construction of knowledge bases for the BACK representation language [16]. This method exploits the assertions concerning the roles available in the knowledge base, in order to assess, in the corresponding relationship, those subgroups of the domain and ranges which may be inductively deemed as disjoint. In the successive phase, supervised learning methods are used on the discovered disjoint subgroups to construct new concepts that account for them. A similar approach is followed in [9], where the supervised phase is performed as an iterative refinement step, exploiting suitable refinement operators for a different DL, namely $\mathcal{ALC}$.

5 Experimental Evaluation

A comparative evaluation of the method is not possible yet, since to the best of our knowledge, there is no similar algorithm which can cope with complex DL languages such as those indicated in the following Table 1. The only comparable (logical) approaches to clustering DL KBs are suitable for limited languages only (e.g. see [16, 9]).

The clustering procedure was validated through some standard internal indices [14, 3]. As pointed out in several surveys on clustering, it is better to use a different criterion for the clustering algorithm (e.g. for choosing the candidate cluster to bisection) and for assessing the quality of its resulting clusters.

To this purpose, we propose a generalization of Dunn's index [3] to deal with medoids. Let $P = \{C_1, \ldots, C_k\}$ be a possible clustering of n individuals in k clusters. The index can be defined:

$$V_{GD}(P) := \min_{1 \leq i \leq k} \left\{ \min_{\substack{1 \leq j \leq k \\ i \neq j}} \left\{ \frac{\delta_p(C_i, C_j)}{\max_{1 \leq h \leq k} \{\Delta_p(C_h)\}} \right\} \right\}$$

where δ_p is the Hausdorff distance for clusters derived from d_p (defined: $\delta_p(C_i, C_j) = \max\{d_p(C_i, C_j), d_p(C_j, C_i)\}$, where $d_p(C_i, C_j) = \max_{a \in C_i}\{\min_{b \in C_j}\{d_p(a, b)\}\}$) while the cluster diameter measure Δ_p is defined:

$$\Delta_p(C_h) := \frac{2}{|C_h|} \sum_{c \in C_h} d_p(c, m_h)$$

The other indices employed are more standard: the mean within-cluster square sum error (WSS), a measure of cohesion, and the silhouette measure [15].

Table 1. Ontologies employed in the experiments

ONTOLOGY	DL	#concepts	#object prop.	#data prop.	#individuals
FSM	$\mathcal{SOF}(D)$	20	10	7	37
S.-W.-M.	$\mathcal{ALCOF}(D)$	19	9	1	115
TRANSPORTATION	$\mathcal{ALC}$	44	7	0	250
NTN	$\mathcal{SHIF}(D)$	47	27	8	676
FINANCIAL	$\mathcal{ALCIF}$	60	16	0	1000

Table 2. Results of the experiments: average value (±std. deviation) and min−max value ranges

ONTOLOGY	SILHOUETTE index	DUNN'S index	WSS index
FSM	.998 (±.005)	.221 (±.003)	30.254 (±11.394)
	.985−1.000	.212−.222	14.344−41.724
S.-W.-M.	1.000 (±.000)	.333 (±.000)	11.971 (±11.394)
	1.000−1.000	.333−.333	7.335−13.554
TRANSPORTATION	.976 (±.000)	.079 (±.000)	46.812 (±5.944)
	.976−.976	.079−.079	39.584−57.225
NTN	.986 (±.007)	.058 (±.003)	96.155 (±24.992)
	.974−.996	.056−.063	64.756−143.895
FINANCIAL	.927 (±.034)	.237 (±.000)	130.863 (±24.117)
	.861−.951	.237−.237	99.305−163.259

For the experiments, a number of different ontologies represented in OWL were selected, namely: FSM, SURFACE-WATER-MODEL, TRANSPORTATION and NEWTESTAMENTNAMES from the Protégé library[1], the FINANCIAL ontology[2] employed as a testbed for the PELLET reasoner. Table 1 summarizes important details concerning the ontologies employed in the experimentation. A variable number of assertions per single individual was available in the ontology. For each ontology, the experiments have been repeated for 10 times. The PELLET 1.4 reasoner was employed to compute the projections required for determining the distance between individuals. An overall experimentation (10 repetitions) on a single ontology took from a few minutes up to less than one hour on a 2.5GhZ (512Mb RAM) Linux Machine.

The outcomes of the experiments are reported in Table 2. It is possible to note that the the Silhouette measure is quite close its optimal value (1), thus providing an absolute indication for the quality of the obtained clusterings. The variability is limited thus the performance appears to be quite stable.

Dunn's and WSS indices may be employed as a suggestion on whether to accept or not the (number of) clusters computed by the algorithm. Namely, among the various repetitions, those final clusterings whose values maximize these indices would have to be preferred. The high variance observed for the WSS index (that it is not limited within a range) has to be considered in proportion with its mean values. Besides, this measure is very sensitive to the number of clusters produced by the method. Although the algorithm converges to a stable number of clusters a difference of 1 may yield a sensible variation of the WSS, also because medoids are considered as centers rather than centroids.

6   Conclusions and Future Work

This work has presented a clustering for (multi-)relational representations which are standard in the Semantic Web field. Namely, it can be used to discover interesting

---

[1] http://protege.stanford.edu/plugins/owl/owl-library
[2] http://www.cs.put.poznan.pl/alawrynowicz/financial.owl

groupings of semantically annotated resources in a wide range of concept languages. The method exploits a novel dissimilarity measure, that is based on the resource semantics w.r.t. a number of dimensions corresponding to a committee of features represented by a group of concept descriptions (discriminating features). The algorithm, is an adaptation of clustering procedures employing medoids since complex representations typical of the ontology in the Semantic Web are to be dealt with.

Better fitness functions may be investigated for both the evolutionary distance optimization procedure and the clustering one. In particular, feature selection for inducing a good distance measure deserves an independent investigation in order to make the choice efficient despite the large extent of the search space. As mentioned, we are investigating other stochastic procedures based on local search [8] and also extensions which can treat less uniformly the cases of uncertainty, e.g. evidence combination methods related to rough sets theory.

We are also devising extensions that are able to produce hierarchical clusterings [7] which would suggest new (non necessarily disjoint) concepts. Instead of repeatedly bisecting the target cluster (as in BISECTING K-MEANS [14]) the algorithm would autonomously find an optimal number for the split at each level.

References

[1] Baader, F., Calvanese, D., McGuinness, D., Nardi, D., Patel-Schneider, P. (eds.): The Description Logic Handbook. Cambridge University Press, Cambridge (2003)

[2] Berners-Lee, T., Hendler, J., Lassila, O.: The Semantic Web. Scientific American (May 2001)

[3] Bezdek, J.C., Pal, N.R.: Some new indexes of cluster validity. IEEE Transactions on Systems, Man, and Cybernetics 28(3), 301–315 (1998)

[4] Borgida, A., Walsh, T.J., Hirsh, H.: Towards measuring similarity in description logics. In: Horrocks, I., Sattler, U., Wolter, F. (eds.) Working Notes of the International Description Logics Workshop, Edinburgh, UK, CEUR Workshop Proceedings, vol. 147 (2005)

[5] d'Amato, C., Fanizzi, N., Esposito, F.: Reasoning by analogy in description logics through instance-based learning. In: Tummarello, G., Bouquet, P., Signore, O. (eds.) Proceedings of Semantic Web Applications and Perspectives, 3rd Italian Semantic Web Workshop, SWAP 2006, Pisa, Italy, CEUR Workshop Proceedings, vol. 201 (2006)

[6] Ester, M., Kriegel, H.-P., Sander, J., Xu, X.: A density-based algorithm for discovering clusters in large spatial databases. In: Proceedings of the 2nd Conference of ACM SIGKDD, pp. 226–231 (1996)

[7] Fanizzi, N., d'Amato, C., Esposito, F.: A hierarchical clustering procedure for semantically annotated resources. In: Basili, R., Pazienza, M.T. (eds.) AI*IA 2007. LNCS (LNAI), vol. 4733, pp. 266–277. Springer, Heidelberg (2007)

[8] Fanizzi, N., d'Amato, C., Esposito, F.: Induction of optimal semi-distances for individuals based on feature sets. In: Working Notes of the International Description Logics Workshop, DL 2007, Bressanone, Italy, CEUR Workshop Proceedings, vol. 250 (2007)

[9] Fanizzi, N., Iannone, L., Palmisano, I., Semeraro, G.: Concept formation in expressive description logics. In: Boulicaut, J.-F., Esposito, F., Giannotti, F., Pedreschi, D. (eds.) ECML 2004. LNCS (LNAI), vol. 3201, pp. 99–113. Springer, Heidelberg (2004)

[10] Ghozeil, A., Fogel, D.B.: Discovering patterns in spatial data using evolutionary programming. In: Koza, J.R., Goldberg, D.E., Fogel, D.B., Riolo, R.L. (eds.) Genetic Programming 1996: Proceedings of the First Annual Conference, Stanford University, CA, USA, pp. 521–527. MIT Press, Cambridge (1996)

[11] Hall, L.O., Özyurt, I.B., Bezdek, J.C.: Clustering with a genetically optimized approach. IEEE Trans. Evolutionary Computation 3(2), 103–112 (1999)

[12] Hirano, S., Tsumoto, S.: An indiscernibility-based clustering method. In: Hu, X., Liu, Q., Skowron, A., Lin, T.Y., Yager, R., Zhang, B. (eds.) 2005 IEEE International Conference on Granular Computing, pp. 468–473. IEEE, Los Alamitos (2005)

[13] Iannone, L., Palmisano, I., Fanizzi, N.: An algorithm based on counterfactuals for concept learning in the semantic web. Applied Intelligence 26(2), 139–159 (2007)

[14] Jain, A.K., Murty, M.N., Flynn, P.J.: Data clustering: A review. ACM Computing Surveys 31(3), 264–323 (1999)

[15] Kaufman, L., Rousseeuw, P.J.: Finding Groups in Data: an Introduction to Cluster Analysis. John Wiley & Sons, Chichester (1990)

[16] Kietz, J.-U., Morik, K.: A polynomial approach to the constructive induction of structural knowledge. Machine Learning 14(2), 193–218 (1994)

[17] Lee, C.-Y., Antonsson, E.K.: Variable length genomes for evolutionary algorithms. In: Whitley, L., Goldberg, D., Cantú-Paz, E., Spector, L., Parmee, I., Beyer, H.-G. (eds.) Proceedings of the Genetic and Evolutionary Computation Conference, GECCO 2000, p. 806. Morgan Kaufmann, San Francisco (2000)

[18] Lehmann, J., Hitzler, P.: A refinement operator based learning algorithm for the ALC description logic. In: ILP 2007. LNCS, vol. 4894, pp. 147–160. Springer, Heidelberg (2007)

[19] Nasraoui, O., Krishnapuram, R.: One step evolutionary mining of context sensitive associations and web navigation patterns. In: Proceedings of the SIAM conference on Data Mining, Arlington, VA, pp. 531–547 (2002)

[20] Ng, R., Han, J.: Efficient and effective clustering method for spatial data mining. In: Proceedings of the 20th Conference on Very Large Databases, VLDB 1994, pp. 144–155 (1994)

[21] Nienhuys-Cheng, S.-H.: Distances and limits on herbrand interpretations. In: Page, D.L. (ed.) ILP 1998. LNCS, vol. 1446, pp. 250–260. Springer, Heidelberg (1998)

[22] Pawlak, Z.: Rough Sets: Theoretical Aspects of Reasoning About Data. Kluwer Academic Publishers, Dordrecht (1991)

[23] Schickel-Zuber, V., Faltings, B.: OSS: A semantic similarity function based on hierarchical ontologies. In: Veloso, M.M. (ed.) Proceedings of the 20th International Joint Conference on Artificial Intelligence, IJCAI 2007, Hyderabad, India, pp. 551–556 (2007)

[24] Sebag, M.: Distance induction in first order logic. In: Džeroski, S., Lavrač, N. (eds.) ILP 1997. LNCS, vol. 1297, pp. 264–272. Springer, Heidelberg (1997)

[25] Stepp, R.E., Michalski, R.S.: Conceptual clustering of structured objects: A goal-oriented approach. Artificial Intelligence 28(1), 43–69 (1986)

[26] Zezula, P., Amato, G., Dohnal, V., Batko, M.: Similarity Search: The Metric Space Approach. Springer, Heidelberg (2007)

Feature Selection: Near Set Approach

James F. Peters[1] and Sheela Ramanna[2,*]

[1] Department of Electrical and Computer Engineering,
University of Manitoba,
Winnipeg, Manitoba R3T 5V6 Canada
jfpeters@ee.umanitoba.ca
[2] Department of Applied Computer Science,
University of Winnipeg,
Winnipeg, Manitoba R3B 2E9 Canada
s.ramanna@uwinnipeg.ca

Abstract. The problem considered in this paper is how to select features that are useful in classifying perceptual objects that are qualitatively but not necessarily spatially near each other. The term *qualitatively near* is used here to mean closeness of descriptions or distinctive characteristics of objects. The solution to this problem is inspired by the work of Zdzisław Pawlak during the early 1980s on the classification of objects. In working toward a solution of the problem of the classification of perceptual objects, this article introduces a near set approach to feature selection. Consideration of the nearness of objects has recently led to the introduction of what are known as near sets, an optimist's view of the approximation of sets of objects that are more or less near each other. Near set theory started with the introduction of collections of partitions (families of neighbourhoods) that provide a basis for a feature selection method based on the information content of the partitions of a set of sample objects. A byproduct of the proposed approach is a feature filtering method that eliminates features that are less useful in the classification of objects. This contribution of this article is the introduction of a near set approach to feature selection.

Keywords: Description, entropy, feature selection, filter, information content, nearness, near set, perception, probe function.

1 Introduction

The problem considered in this paper is how to select the features of objects that are useful in classifying perceptual objects that are qualitatively but not necessarily spatially near each other. The term *qualitatively near* is used here to mean closeness of descriptions or distinctive characteristics of objects. The

* We gratefully acknowledge the very helpful comments, insights and corrections concerning this paper by Andrzej Skowron and the anonymous reviewers. This research has been supported by the Natural Sciences and Engineering Research Council of Canada (NSERC) grants 185986 and 194376.

Z.W. Raś, S. Tsumoto, and D. Zighed (Eds.): MCD 2007, LNAI 4944, pp. 57–71, 2008.
© Springer-Verlag Berlin Heidelberg 2008

solution to this problem is inspired by the work of Zdzisław Pawlak during the early 1980s on the classification of objects [16], elaborated in [17,21], and a view of perception that is on the level of classes instead of individual objects [14]. In working toward a solution of the problem of the classification of perceptual objects, this article introduces a nearness description principle. An *object description* is defined by means of a vector of probe function values associated with an object (see, *e.g.*, [15]). Each probe function ϕ_i represents a feature of an object of interest. Sample objects are near each other if, and only if the objects have similar descriptions.

Ultimately, there is interest in selecting the probe functions [15] that lead to descriptions of objects that are *minimally* near each other. This is an essential idea in the near set approach [10,18,20,23] and differs markedly from the minimum description length (MDL) proposed in 1983 by Jorma Rissanen [27]. MDL depends on minimizing the length of a message expressed as a (negative) log-posterior distribution [9]. By contrast, NDP deals with a set X that is the domain of a description used to identify similar objects. The term *similar* is used here to denote the presence of objects that have descriptions that match each other to some degree.

The near set approach leads to partitions of ensembles of sample objects with measurable information content and an approach to feature selection. The proposed feature selection method considers combinations of n probe functions taken r at a time in searching for those combinations of probe functions that lead to partitions of a set of objects that has the highest information content. It is Shannon's measure of the information content [12,30] of an outcome that provides a basis for the proposed feature selection method. In this work, feature selection results from a filtering method that eliminates those features that have little chance to be useful in the analysis of sample data. The proposed approach does not depend on the joint probability of finding a feature value for an input vectors that belong to the same class as in [8]. In addition, the proposed approach to measuring the information content of families of neighbourhoods differs from the rough set-based form of entropy in [29]. Unlike the dominance-relation rough set approach [7], the near set approach does not depend on preferential ordering of value sets of functions representing object features. The contribution of this article is the introduction of a near set approach to feature selection.

This article has the following organization. A brief introduction to the notation and basic approach to object description is given in Sect. 2. A brief introduction to nearness approximation spaces is given in Sect. 4. A nearness description principle is introduced in Sect. 3. A near set-based feature selection method is introduced in Sect. 5.

2   Object Description

Objects are known by their descriptions. An *object description* is defined by means of a tuple of function values $\phi(x)$ associated with an object $x \in X$

Table 1. Description Symbols

Symbol	Interpretation
$\Re$	Set of real numbers,
$\mathcal{O}$	Set of perceptual objects,
X	$X \subseteq \mathcal{O}$, set of sample objects,
x	$x \in \mathcal{O}$, sample object,
$\mathcal{F}$	A set of functions representing object features,
B	$B \subseteq \mathcal{F}$,
ϕ	$\phi : \mathcal{O} \to \Re^L$, object description,
L	L is a description length,
i	$i \leq L$,
ϕ_i	$\phi_i \in B$, where $\phi_i : \mathcal{O} \longrightarrow \Re$, probe function,
$\phi(x)$	$\phi(x) = (\phi_1(x), \phi_2(x), \phi_3(x), \ldots, \phi_i(x), \ldots, \phi_L(x))$.

(see (1)). The important thing to notice is the choice of functions $\phi_i \in B$ used to describe an object of interest.

Object Description: $\phi(x) = (\phi_1(x), \phi_2(x), \ldots, \phi_i(x), \ldots, \phi_L(x))$.   (1)

The intuition underlying a description $\phi(x)$ is a recording of measurements from sensors, where each sensor is modelled by a function ϕ_i. Assume that $B \subseteq \mathcal{F}$ (see Table 1) is a given set of functions representing features of sample objects $X \subseteq \mathcal{O}$. Let $\phi_i \in B$, where $\phi_i : \mathcal{O} \longrightarrow \Re$. The value of $\phi_i(x)$ is a measurement associated with a feature of an object $x \in X$. The function ϕ_i is called a *probe* [15]. In combination, the functions representing object features provide a basis for an *object description* $\phi : \mathcal{O} \to \Re^L$, a vector containing measurements (returned values) associated with each functional value $\phi_i(x)$ in (1), where the description length $|\phi| = L$.

2.1   Sample Behaviour Description

By way of illustration, consider the description of the behaviour observable in biological organisms. For example, a behaviour can be represented by a tuple

$$(s, a, p(s, a), r)$$

Table 2. Sample ethogram

x_i	s	a	$p(s,a)$	r	d
x_0	0	1	0.1	0.75	1
x_1	0	2	0.1	0.75	0
x_2	1	2	0.05	0.1	0
x_3	1	3	0.056	0.1	1
x_4	0	1	0.03	0.75	1
x_5	0	2	0.02	0.75	0
x_6	1	2	0.01	0.9	1
x_7	1	3	0.025	0.9	0

where $s, a, p(s, a), r$ denote organism functions representing state, action, action preference in a state, and reward for an action, respectively. A reward r is observed in state s and results from an action a performed in the previous state. The preferred action a in state s is calculated using

$$p(s, a) \leftarrow p(s, a) + \beta\delta(s, a),$$

where β is the actor's learning rate and $\delta(r, s)$ is used to evaluate the quality of action a (see [24]). In combination, tuples of behaviour function values form the following description of an object x relative to its observed behaviour:

Organism Behaviour: $\phi(x) = (s(x), a(x), r(x), V(s(x)))$.

Table 2 exhibits a sample observed behaviours of an organism.

3   Nearness of Objects

Approximate, a [L. *approximat-us* to draw near to.]
A. *adj.*
1. Very near, in position or in character;
closely situated; nearly resembling.
–Oxford English Dictionary, 1933.

Table 3. Set, Relation, Probe Function Symbols

Symbol	Interpretation
$\sim_B$	$\{(x, x') \mid f(x) = f(x') \, \forall f \in B\}$, indiscernibility relation,
$[x]_B$	$[x]_B = \{x' \in X \mid x' \sim_B x\}$, elementary granule (class),
$\mathcal{O}/\sim_B$	$\mathcal{O}/\sim_B = \{[x]_B \mid x \in \mathcal{O}\}$, quotient set,
ξ_B	Partition $\xi_B = \mathcal{O}/\sim_B$,
$\Delta\phi_i$	$\Delta\phi_i = \phi_i(x') - \phi_i(x)$, probe function difference,

Sample objects $X \subseteq \mathcal{O}$ are near each other if, and only if the objects have similar descriptions. Recall that each description ϕ^1 defines a description of an object (see Table 1). Then let $\Delta\phi_i$ denote

$$\Delta\phi_i = \phi_i(x') - \phi_i(x),$$

where $x, x' \in \mathcal{O}$ (see Table 3). The difference $\Delta\phi$ leads to a definition of the indiscernibility relation $\sim_B$ introduced by Zdzisław Pawlak [16] (see Def. 1).

---

[1] In a more general setting that includes data mining, ϕ_i would be defined to allow for non-numerical values, *i.e.*, let $\phi_i : X \longrightarrow V$, where V is the *value set* for the range of ϕ_i [26]. The more general definition of $\phi_i \in \mathcal{F}$ is also better in setting forth the algebra and logic of near sets after the manner of the algebra and logic of rough sets [5,26]. Real-valued probe functions are used in object descriptions in this article because we have science and engineering applications of near sets in mind.

Definition 1. Indiscernibilty Relation
Let $x, x' \in \mathcal{O}, B \subseteq \mathcal{F}$.

$$\sim_B = \{(x, x') \in \mathcal{O} \times \mathcal{O} \mid \forall \phi_i \in B \,\text{.}\, \Delta\phi_i = 0\}, where\ i \leq\ description\ length\ |\phi|\,.$$

Definition 2. Nearness Description Principle (NDP)
Let $B \subseteq \mathcal{F}$ be a set of functions representing features of objects $x, x' \in \mathcal{O}$. Objects x, x' are minimally near each other if, and only if there exists $\phi_i \in B$ such that $x \sim_{\{\phi_i\}} x'$, i.e., $\Delta\phi_i = 0$.

In effect, objects x, x' are considered *minimally near* each other whenever there is at least one probe function $\phi_i \in B$ so that $\phi_i(x) = \phi_i(x')$. A *probe function* can be thought of as a model for a sensor (see, *e.g.*, [15,21]). Then ϕ_i constitutes a minimum description of the objects x, x' that makes it possible for us to assert that x, x' are near each other. Ultimately, there is interest in identifying the probe functions that lead to partitions with the highest information content. The nearness description principle (NDP) differs markedly from minimum description length (MDL) proposed by Jorma Rissanen [27]. MDL deals with a set $X = \{x_i \mid i = 1, \dots\}$ of possible message lengths required to transmit outputs data models and a set Θ of possible probability models. By contrast, NDP deals with a set X that is the domain of a description $\phi : X \longrightarrow \Re^L$ and the discovery of at least one probe function $\phi_i(x)$ in a particular description $\phi(x)$ used to identify similar objects in X. The term *similar* is used here to denote the presence of objects $x, x' \in X$ and at least one ϕ_i in object description ϕ, where $x \sim_{\phi_i} x'$. In that case, objects x, x' are said to be similar. This leads to a feature selection method, where one considers combinations of n probe functions r in searching for those combinations of probe functions that lead to partitions with the highest information content.

Observation 1. Near Objects in a Class
Let $\xi_B = \mathcal{O}/\sim_B$ denote a partition of $\mathcal{O}$. Let $[x]_B \in \xi_B$ denote an equivalence class. Assume $x, x' \in [x]_B$. From Table 3 and Def. 1, we know that for each $\phi_i \in B, \Delta\phi_i = 0$. Hence, from Def. 2, x, x' are near objects.

Theorem 1. *The objects in a class $[x]_B \in \xi_B$ are near objects.*

Proof. The nearness of objects in a class in ξ_B follows from Obs. 1. □

The basic idea in the near set approach to object recognition is to compare object descriptions. Sets of objects X, X' are considered near each other if the sets contain objects with at least partial matching descriptions.

Definition 3. Near Sets *[19]*
Let $X, X' \subseteq \mathcal{O}, B \subseteq \mathcal{F}$. Set X is near X' if, and only if there exists $x \in X, x' \in X', \phi_i \in B$ such that $x \sim_{\{\phi_i\}} x'$.

Object recognition problems, especially in images [2,10], and the problem of the nearness of objects have motivated the introduction of near sets (see, *e.g.*, [18,22]).

4   Nearness Approximation Spaces

The original generalized approximation space (GAS) model [31] has recently been extended as a result of recent work on nearness of objects (see, *e.g.*, [10,18,20,22,23,32,33]). A nearness approximation space (NAS) is a tuple

$$NAS = (\mathcal{O}, \mathcal{F}, \sim_{B_r}, N_r, \nu_{N_r}),$$

defined using set of perceived objects $\mathcal{O}$, set of probe functions $\mathcal{F}$ representing object features, indiscernibility relation $\sim_{B_r}$ defined relative to $B_r \subseteq B \subseteq \mathcal{F}$, family of neighbourhoods N_r, and neighbourhood overlap function ν_{N_r}. The relation $\sim_{B_r}$ is the usual indiscernibility relation from rough set theory restricted to a subset $B_r \subseteq B$. The subscript r denotes the cardinality of the restricted subset B_r, where we consider $\binom{|B|}{r}$, *i.e.*, $|B|$ functions $\phi_i \in \mathcal{F}$ taken r at a time to define the relation $\sim_{B_r}$. This relation defines a partition of $\mathcal{O}$ into non-empty, pairwise disjoint subsets that are equivalence classes denoted by $[x]_{B_r}$, where

$$[x]_{B_r} = \{x' \in \mathcal{O} \mid x \sim_{B_r} x'\}.$$

These classes form a new set called the quotient set $\mathcal{O}/\sim_{B_r}$, where

$$\mathcal{O}/\sim_{B_r} = \{[x]_{B_r} \mid x \in \mathcal{O}\}.$$

In effect, each choice of probe functions B_r defines a partition ξ_{B_r} on a set of objects $\mathcal{O}$, namely,

$$\xi_{B_r} = \mathcal{O}/\sim_{B_r}.$$

Every choice of the set B_r leads to a new partition of $\mathcal{O}$. The overlap function ν_{N_r} is defined by

$$\nu_{N_r} : \mathcal{P}(\mathcal{O}) \times \mathcal{P}(\mathcal{O}) \longrightarrow [0,1],$$

Table 4. Nearness Approximation Space Symbols

Symbol	Interpretation
B	$B \subseteq \mathcal{F}$,
B_r	$r \leq |B|$ probe functions in B,
$\sim_{B_r}$	Indiscernibility relation defined using B_r,
$[x]_{B_r}$	$[x]_{B_r} = \{x' \in \mathcal{O} \mid x \sim_{B_r} x'\}$, equivalence class,
$\mathcal{O}/\sim_{B_r}$	$\mathcal{O}/\sim_{B_r} = \{[x]_{B_r} \mid x \in \mathcal{O}\}$, quotient set,
ξ_{B_r}	Partition $\xi_{\mathcal{O},B_r} = \mathcal{O}/\sim_{B_r}$,
ϕ_i	Probe function $\phi_i \in \mathcal{F}$,
r	$\binom{|B|}{r}$, *i.e.*, $|B|$ functions $\phi_i \in \mathcal{F}$ taken r at a time,
$N_r(B)$	$N_r(B) = \{\xi_{B_r} \mid B_r \subseteq B\}$, set of partitions,
ν_{N_r}	$\nu_{N_r} : \mathcal{P}(\mathcal{O}) \times \mathcal{P}(\mathcal{O}) \longrightarrow [0,1]$, overlap function,
$N_r(B)_*X$	$N_r(B)_*X = \bigcup_{x:[x]_{B_r} \subseteq X} [x]_{B_r}$, lower approximation,
$N_r(B)^*X$	$N_r(B)^*X = \bigcup_{x:[x]_{B_r} \cap X} [x]_{B_r} \neq \emptyset$, upper approximation,
$Bnd_{N_r(B)}(X)$	$N_r(B)^*X \backslash N_r(B)_*X = \{x \in N_r(B)^*X \mid x \notin N_r(B)_*X\}$.

where $\mathcal{P}(\mathcal{O})$ is the powerset of $\mathcal{O}$. The overlap function ν_{N_r} maps a pair of sets to a number in $[0, 1]$ representing the degree of overlap between sets of objects with features defined by probe functions $B_r \subseteq B$. For each subset $B_r \subseteq B$ of probe functions, define the binary relation $\sim_{B_r} = \{(x, x') \in \mathcal{O} \times \mathcal{O} : \forall \phi_i \in B_r, \phi_i(x) = \phi_i(x')\}$. Since each $\sim_{B_r}$ is, in fact, the usual indiscernibility relation [16], let $[x]_{B_r}$ denote the equivalence class containing x, i.e.,

$$[x]_{B_r} = \{x' \in \mathcal{O} \mid \forall f \in B_r, f(x') = f(x)\}.$$

If $(x, x') \in \sim_{B_r}$ (also written $x \sim_{B_r} x'$), then x and x' are said to be B-indiscernible with respect to all feature probe functions in B_r. Then define a collection of partitions $N_r(B)$ (families of neighbourhoods), where

$$N_r(B) = \{\xi_{B_r} \mid B_r \subseteq B\}.$$

Families of neighborhoods are constructed for each combination of probe functions in B using $\binom{|B|}{r}$, i.e., $|B|$ probe functions taken r at a time. The family of neighbourhoods $N_r(B)$ contains a set of percepts. A *percept* is a byproduct of perception, i.e., something that has been observed [13]. For example, a class in $N_r(B)$ represents *what has been perceived about objects belonging to a neighbourhood*, i.e., observed objects with matching probe function values.

Definition 4. Near Sets. *Let* $X, X' \subseteq \mathcal{O}, B \subseteq \mathcal{F}$. *Set* X *is near* X' *if, and only if there exists* $x \in X, x' \in X', \phi_i \in B$ *such that* $x \sim_{\{\phi_i\}} x'$.

If X is near X', then X is a near set relative to X' and X' is a near set relative to X. Notice that if we replace X' by X in Def. 4, then a set X containing near objects is a near set.

Theorem 2. Families of Neighbourhoods Theorem. *A collection of partitions (families of neighbourhoods)* $N_r(B)$ *is a near set.*

4.1   Sample Families of Neighbourhoods

Let $X \subseteq \mathcal{O}, B \subseteq \mathcal{F}$ denote a set of sample objects $\{x_0, x_1, \ldots, x_7\}$ and set of functions $\{s, a, p, r\}$, respectively. Sample values of the state function $s : X \longrightarrow \{0, 1\}$ and action function $a : X \longrightarrow \{1, 2, 3\}$ are shown in Table 2. Assume reward function $r : A \longrightarrow [0, 1]$ and a preference function $p : S \times A \longrightarrow [0, 1]$. After discretizing the function values in Table 2, we can, for example, extract the collection of partitions $N_1(B)$ for $r = 1$.

$$X = \{x_0, x_1, \ldots, x_7\},$$
$$N_1(B) = \{\xi_{\{s\}}, \xi_{\{a\}}, \xi_{\{p\}}, \xi_{\{r\}}\}, \text{where}$$
$$\xi_{\{s\}} = [x_0]_{\{s\}}, [x_2]_{\{s\}},$$
$$\xi_{\{a\}} = [x_0]_{\{a\}}, [x_1]_{\{a\}}, [x_3]_{\{a\}},$$
$$\xi_{\{p\}} = [x_0]_{\{p\}}, [x_2]_{\{p\}}, [x_3]_{\{p\}}, [x_5]_{\{p\}}, [x_6]_{\{p\}},$$
$$\xi_{\{r\}} = [x_0]_{\{r\}}, [x_2]_{\{r\}}.$$

4.2   Information Content of a Partition

The Shannon information content of an outcome x is defined by

$$h(x) = log_2 \frac{1}{Pr(x)},$$

which is measured in bits, where *bit* denotes a variable with value 0 or 1, and $h(v)$ provides a measure of the information content of the event $x = v$ [12], which differs from the rough set-based form of entropy in [29]. The assumption made here is that the event $x = x_i$ recorded in Table 2 is random and that a sample X has a uniform distribution, *i.e.*, each event in the sample has the same probability. In effect, $Pr(x = x_i) = \frac{1}{|X|}$. The occurrence of a class $[x]_{B_r} = [x_i]_{B_r} \in \xi_{B_r}$ is treated as a random event, and all classes in a partition ξ_{B_r} are assumed to be equally likely. Then, for example, there are 2 classes in the partition $\xi_{\{s\}} \in N_1(B)$ and $Pr([x_0] \in \xi_{\{s\}}) = log_2 \frac{1}{\frac{1}{2}} = log_2 2 = 1$. The information content $H(X)$ of an ensemble X is defined to be the average Shannon information content of the events represented by X.

$$H(X) = \sum_{x \in X} Pr(x) \cdot log_2 \frac{1}{Pr(x)}.$$

Then, for example, the information content of sample partitions in $N_1(B) = \{\xi_{\{s\}}, \xi_{\{a\}}, \xi_{\{p\}}, \xi_{\{r\}}\}$ have the following information content.

$$H(\xi_{\{s\}}) = H(\xi_{\{r\}}) = \frac{1}{2} \cdot log_2 \frac{1}{\frac{1}{2}} + \frac{1}{2} \cdot log_2 \frac{1}{\frac{1}{2}} = log_2 2 = 1,$$

$$H(\xi_{\{a\}}) = \frac{1}{3} \cdot (3 \cdot log_2 3) = \frac{1}{3} \cdot (3 \cdot 1.59) = 1.59,$$

$$H(\xi_{\{p\}}) = \frac{1}{5} \cdot (5 \cdot log_2 5) = 2.3$$

This suggests an approach to feature selection based on information content, which is computationally rather simple.

5   Feature Selection

A practical outcome of the near set approach is a feature selection method. Recall that each partition $\xi_{B_r} \in N_r(B)$ contains classes defined by the relation $\sim_{B_r}$. We are interested in partitions in $\xi_{B_r} \in N_r(B)$ with information content greater than or equal to some threshold th. The basic idea here is to identify probe functions that lead to partitions with the highest information content, which occurs in partitions with high numbers of classes. In effect, as the number of classes in a partition increases, there is a corresponding increase in the information content of the partition.

By sorting Φ based on information content using Alg. 1, we have a means of means of selecting tuples containing probe functions that define partitions having the highest information content.

Algorithm 1. Feature Selection

Input : List Φ of partitions, where $\Phi[i]$ = number of classes in $\xi_{B_r} \in N_r(B)$,
threshold th.
Output: Ordered list Γ, where $\Gamma[i]$ is a winning partition.
Initialize $i = |\Phi|$;
Sort Φ in descending order based on the information content of $\xi_{B_r} \in N_r(B)$;
while ($\Phi[i] \geq th$ *and* $i \geq 1$) **do**
$\quad | \quad \Gamma[i] = \Phi[i]$;
$\quad | \quad i = i - 1$;
end

5.1   Sample Feature Selections

$N_2(B) = \left\{ \xi_{\{s,a\}}, \xi_{\{s,p\}}, \xi_{\{a,p\}}, \xi_{\{s,r\}}, \xi_{\{a,r\}}, \xi_{\{p,r\}} \right\}$, where

$\xi_{\{s,a\}} = \left\{ [x_0]_{\{s,a\}}, [x_1]_{\{s,a\}}, [x_2]_{\{s,a\}}, [x_3]_{\{s,a\}} \right\}$,

$\xi_{\{s,p\}} = \left\{ [x_0]_{\{s,p\}}, [x_2]_{\{s,p\}}, [x_3]_{\{s,p\}}, [x_4]_{\{s,p\}}, [x_5]_{\{s,p\}}, [x_6]_{\{s,p\}}, [x_7]_{\{s,p\}} \right\}$,

$\xi_{\{a,p\}} = \left\{ [x_0]_{\{a,p\}}, [x_1]_{\{a,p\}}, [x_2]_{\{a,p\}}, [x_3]_{\{a,p\}}, [x_4]_{\{a,p\}}, [x_5]_{\{a,p\}}, [x_6]_{\{a,p\}}, [x_7]_{\{a,p\}} \right\}$,

$\xi_{\{s,r\}} = \left\{ [x_0]_{\{s,r\}}, [x_2]_{\{s,r\}}, [x_6]_{\{s,r\}} \right\}$,

$\xi_{\{a,r\}} = \left\{ [x_0]_{\{a,r\}}, [x_1]_{\{a,r\}}, [x_2]_{\{a,r\}}, [x_3]_{\{a,r\}}, [x_7]_{\{a,r\}} \right\}$,

$\xi_{\{p,r\}} = \left\{ [x_0]_{\{p,r\}}, [x_2]_{\{p,r\}}, [x_3]_{\{p,r\}}, [x_4]_{\{p,r\}}, [x_5]_{\{p,r\}}, [x_6]_{\{p,r\}} \right\}$.

This section continues the exploration of the partitions in families of neighbourhoods for each choice of $r \leq |B|$. The observations in Table 2 were produced by an ecosystem simulation reported in [25]. The function values in Table 2 were discretized using ROSE [28]. ROSE was used in this study because it allows the user to use as input a discretized table. This was important for us because Table 2 was discretized using a local Object Recognition System (ORS) toolset. The ORS discretized table was used to define the partition of the sample objects in Table 2. Collections of partitions $N_2(B), N_3(B)$ can be extracted from Table 2 (in this section, $N_2(B)$ is given and the display of $N_3(B)$ is omitted to save space).

Table 5. Partition Information Content Summary

ξ	H	ξ	H	ξ	H
$\xi_{\{s\}}$	1.0	$\xi_{\{s,a\}}$	2.0	$\xi_{\{s,a,p\}}$	3.0
$\xi_{\{a\}}$	1.59	$\xi_{\{s,p\}}$	2.8	$\xi_{\{a,p,r\}}$	3.0
$\xi_{\{p\}}$	2.3	$\xi_{\{a,p\}}$	3.0	$\xi_{\{s,a,r\}}$	1.63
$\xi_{\{r\}}$	1.0	$\xi_{\{s,r\}}$	1.58	$\xi_{\{s,p,r\}}$	2.8
		$\xi_{\{a,r\}}$	1.46		
		$\xi_{\{p,r\}}$	1.63		

In the single feature case (r = 1), functions a and p define partitions with the highest information content for sample X represented by Table 2. Notice that preference p is more important than a in the ecosystem experiments, because p indicates the preferred action in a give state. In the two feature case (r = 2), feature combination a, p define a partition with the highest information content, namely, 3 (see Table 5). If we set the threshold $r \geq 3$, then features a, p (action, preference) are selected because the information content of partition $H(\xi_{\{a,p\}}) = 3$ in $N_2(B)$, which is also one of the reducts found by ROSE [28] using Table 2 as input. Notice that a, p defines all of the high scoring partitions in Table 5. Similarly, for $N_3(B)$, the features s, a, p are selected, where the information content of $H(\xi_{\{s,a,p\}}) = 3$ (highest) and $H(\xi_{\{s,a,r\}}) = 1.58$ (lowest) which is at variance with the findings by ROSE, which selects $\{s, a, r\}$ as a reduct. Finally, notice that ROSE selects a as the core feature and the highest scoring partitions defined by combinations of functions containing a.

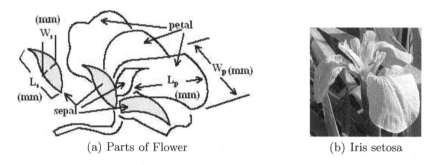

(a) Parts of Flower   (b) Iris setosa

Fig. 1. Fisher Iris Taxonomy

5.2 Feature Selection for Fisher's Iris Taxonomy Measurements

A second illustration of the near set approach to feature selection is based on Fisher's taxonomic measurements [6] of length L and width W (in mm) of the sepals and petals of three species of Iris flowers(Iris setosa canadensis, Iris versicolor, and Iris virginica). Sepals form a protective layer (called calyx) that encloses the petals of a flower (see Fig. 1(a)). Petals form a whorl (corolla) within the sepals and enclose the reproductive organs. A sample Iris setosa is shown in Fig. 1(b). These measurements consist of 50 samples of each type of Iris and are available from many sources[2]. Let L_s, W_s, L_p, W_p denote sepal length, sepal width, petal length and petal width, respectively. Let X, B denote Fisher's set of sample Iris flowers and set of functions representing features of the sepals and petals, respectively. An overview of the partitions for the each of the families of neighbourhoods for the Iris measurements is given, next.

---

[2] UCI Machine Learning archives, `http://mlearn.ics.uci.edu/MLRepository.html`

$$X = \{x_0, x_1, \dots, x_{150}\},$$
$$B = \{L_s, W_s, L_p, W_p\},$$
$$N_1(B) = \left\{\xi_{\{L_s\}}, \xi_{\{W_s\}}, \xi_{\{L_p\}}, \xi_{\{W_p\}}\right\},$$
$$N_2(B) = \left\{\xi_{\{L_s, W_s\}}, \xi_{\{L_s, L_p\}}, \xi_{\{W_s, L_p\}}, \xi_{\{W_s, W_p\}}, \xi_{\{W_p, W_p\}}, \xi_{\{L_p, W_p\}}\right\},$$
$$N_3(B) = \left\{\xi_{\{L_s, W_s, L_p\}}, \xi_{\{L_s, W_s, L_p\}}, \xi_{\{L_s, W_s, L_p\}}, \xi_{\{L_s, W_s, L_p\}}\right\},$$
$$N_4(B) = \left\{\xi_{\{L_s, W_s, L_p, W_p\}}\right\}.$$

The distribution of the feature values for the high scoring feature-triplets $\{W_s, L_p, W_p\}$ is shown in Fig. 2. This visualization of the Iris measurements has been extracted from [3]. Another view of the distribution of Iris measures is given in a textile plot in Fig. 3 from [11]. Each quadrant in the 3D view in Fig. 2 shows the distribution of the Iris measurements (this is made clearer by the separate distributions of the measurement pairs shown in the darkened 2D areas in Fig. 2, *e.g.*, the lower lefthand corner of the 2D views represents the distribution of the measurements for $\{L_s, W_p\}$). Except for the mistaken use of centimeter instead millimeter unit of measurement in the textile plot, the separation of the feature values for Iris setosa from versicolor and virginica measurements corroborates in the textile plot in Fig. 3 confirms the results from [4]. It is this separation that motivated the use of Iris setosa measurements as a basis for feature selection. A summary of the information content of the partitions of the set of Iris setosa measurements is given in Table 6.

For the sake of comparison with the evaluation of the results in Table 5, ROSE was used to find the reducts for the Iris measurements. It was found that the reduced feature sets are $\{L_s, W_s, L_p\}$, $\{L_s, W_s, W_p\}$, $\{W_s, L_p, W_p\}$, and $\{L_s, W_s, L_p\}$ (all combinations of 3 features). Notice that the information content values for the partitions in $N_3(B)$ are approximately the same

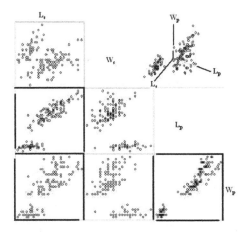

Fig. 2. Visualization of Iris Measurements

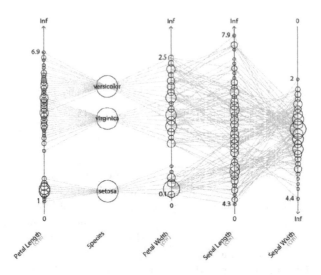

Fig. 3. Textile Visualization of Iris Measurements

Table 6. Iris Setosa Partition Information Content Summary

ξ	H	ξ	H	ξ	H	ξ	H
$\xi_{\{L_s\}}$	4.17	$\xi_{\{L_s,W_s\}}$	5.25	$\xi_{\{L_s,W_s,L_p\}}$	5.52	$\xi_{\{L_s,W_s,L_p,W_p\}}$	5.61
$\xi_{\{W_s\}}$	3.46	$\xi_{\{L_s,L_p\}}$	5.23	$\xi_{\{L_s,W_s,W_p\}}$	5.55		
$\xi_{\{L_p\}}$	3.91	$\xi_{\{W_s,L_p\}}$	5.29	$\xi_{\{W_s,L_p,W_p\}}$	5.61		
$\xi_{\{W_p\}}$	3.32	$\xi_{\{W_s,W_p\}}$	5.21	$\xi_{\{L_s,L_p,W_p\}}$	5.61		
		$\xi_{\{W_p,W_p\}}$	5.00				
		$\xi_{\{L_p,W_p\}}$	5.13				

as the information content of the partition in $N_4(B)$, which corroborates the feature set reduction results from ROSE. Further, notice that the information content of $\xi_{\{L_s\}}$ for the length-of-petal L_p partition (*i.e.*, $H(\xi_{\{L_p\}})$ = 3.91). This feature is a member of each of the highest scoring reducts, namely, $H(\xi_{\{W_s,L_p,W_p\}}) = H(\xi_{\{L_s,L_p,W_p\}}) = 5.61$. Also, notice that the highest scoring feature pair $\{W_s, L_p\}$ ($H(\xi_{\{W_s,L_p\}}) = 5.29$) is also a member of one of the highest scoring reducts, namely, $\{W_s, L_p, W_p\}$, where $H(\xi_{\{W_s,L_p,W_p\}}) = 5.61$.

5.3   Complexity Issue

Notice that the proposed feature selection method is based on collections of partitions (families of neighborhoods) constructed for each combination of probe functions in B using $\binom{|B|}{r}$, *i.e.*, $|B|$ probe functions taken r at a time. Hence, the proposed method has very high complexity for large B. To achieve feature selection with polynomial time complexity, features are selected by considering only the partitions in $N_1(B)$ or in $N_2(B)$. This approach can be used to identify all of those partitions with information content $H(\xi_f)$ for partition $X/\sim_{\{f\}}$ defined relative to a single function f representing a feature of the sample objects or for combinations of pairs of features, *i.e.*, r = 1 or r = 2. For recognition of sample objects with thousands of features, this approach is useful in an unsupervised learning environment[3]. More work needs to be done on the proposed feature selection method to make it useful in discovering reducts suitable for supervised learning for sample objects with many features.

6 Conclusion

The proposed near set approach to feature selection is tractable, since it is always possible to find the information content of partitions of sample objects at the single feature level within a reasonable time. The results reported in this paper corroborate Ewa Orłowska's observation that perception is more feasible at the class level than they are at the object level [14]. This is apparent in the contrast between the diffusion of the 2D clusters or 2D textile plots of Iris measurements and the information content measurements for partitions of Iris measurements. The kernel of the near set approach to feature selection is a direct result of the original approach to the classification of objects introduced by Zdzisław Pawlak during the early 1980s [16]. Feature selection is based on measurement of the information content of the partitions defined by selected combinations of features. Future work will include consideration of the relation between the proposed feature selection method for large numbers of features for objects in various sample spaces.

References

1. Banerjee, M., Mitra, S., Banka, H.: Evolutionary rough feature selection in gene expression data. IEEE Transactions on Systems, Man, and Cybernetics–Part C: Applications and Reviews 37(4), 1–12 (2007)
2. Borkowski, M., Peters, J.F.: Matching 2D image segments with genetic algorithms and approximation spaces. In: Peters, J.F., Skowron, A. (eds.) Transactions on Rough Sets V. LNCS, vol. 4100, pp. 63–101. Springer, Heidelberg (2006)
3. Bradley, J.W., Webster, R.: Interactive Java tools for exploring high dimensional data, http://www.stat.sc.edu/~west/bradley/
4. Bugrien, J.B., Kent, J.T.: Independent component analysis: A approach to clustering. In: Proc. LASR Statistics Workshop, pp. 1–4 (2005)
5. Düntsch, I.: A logic for rough sets. Theoretical Computer Science 179, 427–436 (1997)

---

[3] See, *e.g.* lymphocyte samples with genes representing 4026 features [1]. Assuming that a maximum of m operations are required in each partition in $N_1(B)$, then $4026 \cdot m$ operations are necessary for feature selection for single feature partitions. By choosing a threshold th, it is then possible to select the most discriminating features useful in classifying the lymphocyte samples.

6. Fisher, R.A.: The use of multiple measurements in taxonomic problems. Annals of Eugenics 7 Pt. II, 179–188 (1936)
7. Greco, S., Matarazzo, B., Slowinski, R.: Dominance-based rough set approach to knowledge discovery. In: Zhong, N., Liu, J. (eds.) Intelligent Technologies for Information Analysis, pp. 513–552. Springer, Berlin (2004)
8. Guyon, I., Gunn, S., Nikravesh, M., Zadeh, L.A. (eds.): Feature Extraction. Foundations and Applications. Springer, Berlin (2006)
9. Hastie, T., Tibshirani, R., Friedman, J.: The Elements of Statistical Learning. Data Mining, Inference, and Prediction. Springer, Berlin (2001)
10. Henry, C., Peters, J.F.: Image Pattern Recognition Using Approximation Spaces and Near Sets. In: An, A., Stefanowski, J., Ramanna, S., Butz, C.J., Pedrycz, W., Wang, G. (eds.) RSFDGrC 2007. LNCS (LNAI), vol. 4482, pp. 475–482. Springer, Heidelberg (2007)
11. Kumusaka, N., Shibata, R.: High dimensional data visualization: The textile plot, Research Report KSTS/RR-2006/001, Department of Mathematics, Keio University, Computational Statistics & Data Analysis (February 13, 2006) (submitted)
12. MacKay, D.J.C.: Information Theory, Inference, and Learning Algorithms. Cambridge University Press, Cambridge (2003)
13. Murray, J.A., Bradley, H., Craigie, W., Onions, C.: The Oxford English Dictionary. Oxford University Press, London (1933)
14. Orłowska, E.: Semantics of Vague Concepts. Applications of Rough Sets, Institute for Computer Science, Polish Academy of Sciences, Report 469 (1982); Orłowska, E.: Semantics of Vague Concepts. In: Dorn, G., Weingartner, P. (eds.) Foundations of Logic and Linguistics. Problems and Solutions, pp. 465–482. Plenum Press, London/NY (1985)
15. Pavel, M.: Fundamentals of Pattern Recognition, 2nd edn. Marcel Dekker Inc., New York (1993)
16. Pawlak, Z.: Classification of Objects by Means of Attributes, Institute for Computer Science, Polish Academy of Sciences, Report 429 (1981)
17. Pawlak, Z., Skowron, A.: Rudiments of rough sets. Information Sciences 177, 3–27 (2007)
18. Peters, J.F.: Near sets. Special theory about nearness of objects. Fundamenta Informaticae 76, 1–28 (2007)
19. Peters, J.F.: Near sets. General theory about nearness of objects. Applied Mathematical Sciences 1(53), 2029–2609 (2007)
20. Peters, J.F.: Near Sets. Toward Approximation Space-Based Object Recognition. In: Yao, J., Lingras, P., Wu, W.-Z., Szczuka, M., Cercone, N.J., Ślęzak, D. (eds.) RSKT 2007. LNCS (LNAI), vol. 4481, pp. 22–33. Springer, Heidelberg (2007)
21. Peters, J.F.: Classification of objects by means of features. In: Proc. IEEE Symposium Series on Foundations of Computational Intelligence (IEEE SCCI 2007), Honolulu, Hawaii, pp. 1–8 (2007)
22. Peters, J.F., Skowron, A., Stepaniuk, J.: Nearness in approximation spaces. In: Lindemann, G., Schlilngloff, H., et al. (eds.) Proc. Concurrency, Specification & Programming (CS&P 2006), Informatik-Berichte Nr. 206, Humboldt-Universität zu Berlin, pp. 434–445 (2006)
23. Peters, J.F., Skowron, A., Stepaniuk, J.: Nearness of Objects: Extension of Approximation Space Model. Fundamenta Informaticae 79, 1–24 (2007)
24. Peters, J.F., Henry, C., Gunderson, D.S.: A Biologically-inspired approximate adaptive learning control strategies: A rough set approach. International Journal of Hybrid Intelligent Systems 4(4), 203–216 (2007)

25. Peters, J.F., Henry, C., Ramanna, S.: Rough Ethograms: Study of Intelligent System behaviour. In: Kłopotek, M.A., Wierzchoń, S., Trojanowski, K. (eds.) New Trends in Intelligent Information Processing and Web Mining (IIS 2005), Gdańsk, Poland, pp. 117–126 (2005)
26. Polkowski, L.: Rough Sets. Mathematical Foundations. Springer, Heidelberg (2002)
27. Rissanen, J.J.: A universal prior for integers and estimation by Minimum Description Length. Annals of Statistics 11(2), 416–431 (1983)
28. Rough Sets Data Explorer, Version 2.2, ©1999-2002 IDSS, `http://idss.cs.put.poznan.pl/site/software.html`
29. Shankar, B.U.: Novel classification and segmentation techniques with application to remotely sensed images. In: Peters, J.F., Skowron, A., Marek, V.W., Orłowska, E., Słowiński, R., Ziarko, W. (eds.) Transactions on Rough Sets VII. LNCS, vol. 4400, pp. 295–380. Springer, Heidelberg (2007)
30. Shannon, C.E.: A mathematical theory of communication. Bell Systems Technical Journal 27, 279–423, 623–656 (1948)
31. Skowron, A., Stepaniuk, J.: Generalized approximation spaces. In: Lin, T.Y., Wildberger, A.M. (eds.) Soft Computing, Simulation Councils, San Diego, pp. 18–21 (1995)
32. Skowron, A., Swiniarski, R., Synak, P.: Approximation spaces and information granulation. In: Peters, J.F., Skowron, A. (eds.) Transactions on Rough Sets III. LNCS, vol. 3400, pp. 175–189. Springer, Heidelberg (2005)
33. Skowron, A., Stepaniuk, J., Peters, J.F., Swiniarski, R.: Calculi of approximation spaces. Fundamenta Informaticae 72(1-3), 363–378 (2006)

Evaluating Accuracies of a Trading Rule Mining Method Based on Temporal Pattern Extraction

Hidenao Abe[1], Satoru Hirabayashi[2], Miho Ohsaki[3], and Takahira Yamaguchi[4]

[1] Department of Medical Informatics, Shimane University, School of Medicine
abe@med.shimane-u.ac.jp
[2] Graduate School of Science and Technology, Keio University
and_joy@ae.keio.ac.jp
[3] Faculty of Engineering, Doshisha University
mohsaki@mail.doshisha.ac.jp
[4] Faculty of Science and Technology, Keio University
yamaguti@ae.keio.ac.jp

Abstract. In this paper, we present an evaluation of accuracies of temporal rules obtained from the integrated temporal data mining environment using trading dataset from the Japanese stock market. Temporal data mining is one of key issues to get useful knowledge from databases. However, users often face on difficulties during such temporal data mining process for data pre-processing method selection/construction, mining algorithm selection, and post-processing to refine the data mining process. To get rules that are more valuable for domain experts from a temporal data mining process, we have designed an environment, which integrates temporal pattern extraction methods, rule induction methods and rule evaluation methods with visual human-system interface. Then, we have done a case study to mine temporal rules from a Japanese stock market database for trading. The result shows the availability to find out useful trading rules based on temporal pattern extraction.

1 Introduction

In recent years, KDD (Knowledge Discovery in Databases) [3] has been widely known as a process to extract useful knowledge from databases. In the research field of KDD, 'Temporal (Time-Series) Data Mining' is one of important issues to mine useful knowledge such as patterns, rules, and structured descriptions for a domain expert. However, huge numerical temporal data such as stock market data, medical test data, and sensor data have been only stored to databases.

Besides, many temporal mining schemes such as temporal pattern extraction methods and frequent itemset mining methods have been proposed to find out useful knowledge from numerical temporal databases. Although each method can find out partly knowledge of each suggested domains, there is no systematic framework to utilize each given numerical temporal data through whole of the KDD process.

To above problems, we have developed an integrated temporal data mining environment, which can apply numerical temporal data to find out valuable knowledge

Z.W. Raś, S. Tsumoto, and D. Zighed (Eds.): MCD 2007, LNAI 4944, pp. 72–81, 2008.

systematically. The environment consists of temporal pattern extraction, mining, mining result evaluation support system to attempt numerical temporal data from various domains.

In this paper, we present an evaluation of the integrated temporal data mining environment with Japanese stock market data. Then, we discuss about the availability of the temporal rule mining process based on temporal pattern extraction.

2 Related Work

Many efforts have been done to analyze temporal data at the field of pattern recognitions. Statistical methods such as autoregressive model and ARIMA (AutoRegressive Integrated Moving Average) have been developed to analyze temporal data, which have linearity, periodicity, and equalized sampling rate. As signal processing methods, Fourier transform, Wavelet, and fractal analysis method have been also developed to analyze such well formed temporal data. These methods based on mathematic models restrict input data, which are well sampled. However, temporal data include ill-formed data such as clinical test data of chronic disease patients, purchase data of identified customers, and financial data based on social events. To analyze these ill-formed temporal data, we take another temporal data analysis method such as DTW (Dynamic Time Wrapping)[1], temporal clustering with multiscale matching [5], and finding Motif based on PAA (Piecewise Approximation Aggregation) [6].

To find out useful knowledge to decide orders for stock market trading, many studies have done. For example, temporal rule induction methods such as Das's framework [2] have been developed. Frequent itemset mining methods are also often attempt to the domain [15]. Although they analyze the trend of price movement, many trend analysis indices such as moving average values, Bollinger band signals, MACD signals, RSI and signals based on balance table are often never considered. In addition, these studies aim not to find out decision support knowledge, which directly indicates orders for stock market trading, but useful patterns to think better decision by a domain expert. Therefore, the decision support of trading order is still costly task even if a domain expert uses some temporal data analysis methods. The reason of this problem is that decision criteria of trading called anomaly are obtained from very complex combination of many kinds of indices related to the market by domain experts.

3 An Integrated Temporal Data Mining Environment

Our temporal data mining environment needs temporal data as input. Output rules are if-then rules, which have temporal patterns or/and ordinal clauses, represented in $A=x$, $A<=y$, and $A>z$. Combinations of extracted patterns and/or ordinal clauses can be obtained as if-then rules by a rule induction algorithm. Fig. 1 illustrates a typical output it-then rule visualized with our temporal data mining environment.

To implement the environment, we have analyzed temporal data mining frameworks [2, 10]. Then, we have identified procedures for pattern extraction as data pre-processing, rule induction as mining, and evaluation of rules with visualized rule as

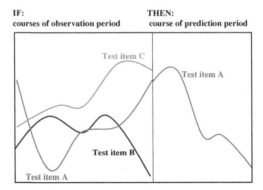

Fig. 1. Typical output if-then rule, which consists of patterns both its antecedent and its consequent

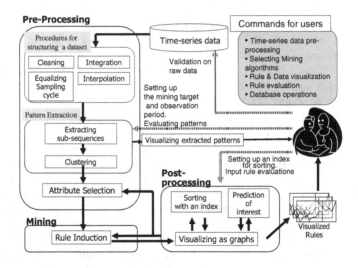

Fig. 2. A system flow view of the integrated temporal data mining environment

post-processing of mined result. The system provides these procedures as commands for users. At the same time, we have designed graphical interfaces, which include data processing, validation for patterns on elemental sequences, and rule visualization as graphs. Fig. 2 shows us a typical system flow of this temporal data mining environment.

3.1   Details of Procedures to Mine Temporal Rules

We have identified procedures for temporal data mining as follows:

Data pre-processing
- pre-processing for data construction
- temporal pattern extraction
- attribute selection

Mining
- rule induction

Post-processing of mined results
- visualizing mined rule
- rule selection
- supporting rule evaluation

Other database procedures
- selection with conditions
- join

As data pre-processing procedures, pre-processing for data construction procedures include data cleaning, equalizing sampling rate, interpolation, and filtering irrelevant data. Since these procedures are almost manual procedures, they strongly depend on given temporal data and a purpose of the mining process. Temporal pattern extraction procedures include determining the period of sub-sequences and finding representative sequences with a clustering algorithm such as K-Means, EM clustering and the temporal pattern extraction method developed by Ohsaki et al. [12]. Attribute selection procedures are done by selecting relevant attributes manually or using attribute selection algorithms [7].

At mining phase, we should choose a proper rule induction algorithm with some criterion. There are so many rule induction algorithms such as Version Space [9], AQ15 [8], C4.5 rule [13], and any other algorithm. To support this choice, we have developed a tool to construct a proper mining application based on constructive meta-learning called CAMLET. However, we have taken PART [4] implemented in Weka [16] in the case study to evaluate improvement of our pattern extraction algorithm.

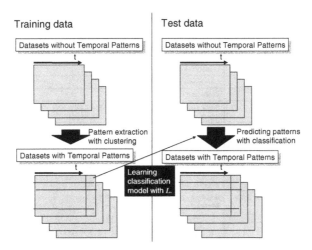

Fig. 3. The process to obtain a test dataset based on temporal patterns of a training dataset using classification learning algorithm

To predict class of a test dataset with learned a classification model, the system should formally predict pattern symbols of the test dataset using some classification learning method L based on the training dataset as shown in Figure 3.

To validate mined rules correctly, users need readability and ease for understand about mined results. We have taken 39 objective rule evaluation indexes to select mined rules [11], visualizing and sorting them depended on users' interest. Although these two procedures are passive support from a viewpoint of the system, we have also identified active system reaction with prediction of user evaluation based on objective rule evaluation indexes and human evaluations.

Other database procedures are used to make target data for a data mining process.

Since the environment has been designed based on open architecture, these procedures have been able to develop separately. To connect each procedure, we have only defined input/output data format.

4 Evaluating Temporal Rule Mining Performances with the Integrated Temporal Data Mining Environment

After implementing the integrated temporal data mining environment described in Section 3, we have done a case study on Japanese stock market database.

In this case study, we firstly gathered temporal price data and its trend index values through Kaburobo SDK [17]. Then, using the environment, we evaluated the performance of if-then rules based on temporal patterns. Finally, with regarding to the results, we discuss about the availability of our temporal rule mining based on temporal pattern extraction.

Table 1. The descripiton about attributes from Kaburobo SDK.

Attribute name		Description
R A W	opening	opening price of the day (O_t)
	high	Highest price of the day (H_t)
	low	Lowest price of the day (L_t)
	closing	Closing price of the day (C_t)
	Volume	Volume of the day (V_t)
T R E N D I N D I C E S	Moving Average	Buy: if $SMA_t - LMA_t < 0 \cap SMA_{t-1} - LMA_{t-1} > 0$, Sell: if $SMA_t - LMA_t > 0 \cap SMA_{t-1} - LMA_{t-1} < 0$ Where $SMA_t = (C_t + C_{t-1} + h + C_{t-12})/13$, and $LMA_t = (C_t + C_{t-1} + h + C_{t-25})/26$
	Bolinger Band	Buy: if $C_t \geq (MA_t + 2\sigma) \times 0.05$, Sell: if $C_t \leq (MA_t - 2\sigma) \times 0.05$ where $MA_t = (C_t + C_{t-1} + h + C_{t-24})/25$
	Envelope	Buy: if $C_t \geq MA_t + (MA_t \times 0.05)$, Sell: if $C_t \leq MA_t - (MA_t \times 0.05)$
	HLband	Buy: if $C_t < LowLine_{t-10days}$, Sell: if $C_t > HighLine_{t-10days}$
	MACD	Buy: if $MACD_t - AvgMACD_{t-9days} > 0 \cap MACD_{t-1} - AvgMACD_{(t-1)-9days} < 0$ Sell: if $MACD_t - AvgMACD_{t-9days} < 0 \cap MACD_{t-1} - AvgMACD_{(t-1)-9days} > 0$ Where $MACD_t = EMA_{t-12days} - EMA_{t-26days}$, $EMA_t = EMA_{t-1} + (2/range + 1)(C_{t-1} - EMA_{t-1})$
	DMI	Buy: if $PDI_t - MDI_t > 0 \cap PDI_{t-1} - MDI_{t-1} < 0$, Sell: if $PDI_t - MDI_t < 0 \cap PDI_{t-1} - MDI_{t-1} > 0$ Where $PDI = \sum_{i=t-n}^{t}(H_i - H_{i-1}) + \sum_{i=t-n}^{t} TR_i \times 100$, $MDI = \sum_{i=t-n}^{t}(L_i - L_{i-1}) + \sum_{i=t-n}^{t} TR_i \times 100$ $\cdot TR_i = \max\{(H_i - C_{i-1}), (C_{i-1} - L_i), (H_i - L_i)\}$
	volumeRatio	$VR_t = \{(\sum_{i=t-25, H_i > L_i}^{t} V_i + \sum_{i=t-25, H_i = L_i}^{t} V_i)/(\sum_{i=t-25, H_i < L_i}^{t} V_i + \sum_{i=t-25, H_i = L_i}^{t} V_i)\} \times 100$
	RSI	$RSI_t = 100 - 100 /\{\sum_{i=t-13, C_{i-1} < C_i}^{t}(C_{i-1} - C_i) / \sum_{i=t-13, C_{i-1} > C_i}^{t}(C_{i-1} - C_i) + 1\}$
	Momentum	$M_t = C_t - C_{t-10}$
	Ichimoku1	Buy: if $C_{t-1} < RL_{t-9days} \cap C_t > RL_{t-9days}$, Sell: if $C_{t-1} > RL_{t-9days} \cap C_t < RL_{t-9days}$ Where $RL_{t-9days} = average(\max(H_t) + \min(L_t))$ $(i = t-8, t-7, h, t)$
	Ichimoku2	Buy: if $C_{t-1} < RL_{t-26days} \cap C_t > RL_{t-26days}$, Sell: if $C_{t-1} > RL_{t-26days} \cap C_t < RL_{t-26days}$ Where $RL_{t-26days} = average(\max(H_t) + \min(L_t))$ $(i = t-25, t-24, h, t)$
	Ichimoku3	Buy: if $RL_{(t-2)-26days} < RL_{(t-2)-9days} \cap RL_{(t-1)-26days} > RL_{(t-1)-9days} \cap RL_{(t-1)-26days} < RL_{t-26days}$ Sell: if $RL_{(t-2)-26days} > RL_{(t-2)-9days} \cap RL_{(t-1)-9days} < RL_{(t-1)-26days} \cap RL_{(t-1)-26days} > RL_{t-26days}$
	Ichimoku4	Buy: if $C_t > AS1_{t-26} \cap C_t > AS2_{t-26}$, Sell: if $C_t < AS1_{t-26} \cap C_t < AS2_{t-26}$ Where $AS1_t = median(RL_{t-9days} - RL_{t-26days})$, $AS2_t = (\max(H_t) - \min(L_t))/2$ $(i = t-51, t-50, h, t)$

4.1 Description About Temporal Datasets

Using Kaburobo SDK, we got four price values, trading volume, and 13 trend indices as attributes of each target dataset as shown in Table 1. Excepting DMI, volume ratio, and momentum, the trend indices are defined as trading signals: buy and sell. The attribute values of these indices are converted from 1.0 to -1.0. Thus, 0 means nothing to do (or hold on the stock) for these attributes.

We obtained temporal data consists of the above mentioned attributes about five financial companies and four telecommunication companies as follows: Credit Saison (Saison), Orix, Mitsubishi Tokyo UFJ Financial Group (MUFJFG), Mitsui Sumitomo Financial Group (MSFG), Mizuho Financial Group (MizuhoFG), NTT, KDDI, NTT Docomo, and Softbank. The period, which we have collected from the temporal stock data, is from 5th January 2006 to 31st May 2006. For each day, we have made decisions as the following: the decision is if the closing value rises 5% within 20 days then 'buy', otherwise if the closing value falls 5% within 20 days then 'sell', otherwise 'hold'. We set these decisions as the class attribute to each target instance. Table 2 shows the class distributions about the nine stocks for the period.

Table 2. The distributions of decisions of the nine stocks during the five months

Finance	Buy	sell	Telecom.	buy	sell
Saison	37	53	NTT	27	32
Orix	43	40	KDDI	42	39
MUFJFG	0	50	NTTdocomo	19	29
MSFG	6	27	Softbank	23	69
MizuhoFG	38	31			

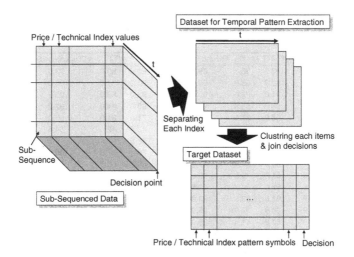

Fig. 4. An illustration of the process to obtain target datasets from temporal data

For each gathered temporal data of the nine stocks, the system extracted temporal patterns for each attribute. Then, the symbols of each pattern and the decision of each day joined as each instance of the target dataset as illustrated in Figure 4.

4.2 Evaluating Temporal Pattern Prediction by Boosted C4.5

To extract temporal patterns, we have used K-Means and EM algorithm, which are implemented in Weka. Then, to predict temporal pattern of each test dataset, we have used Boosted C4.5 [14], which is also implemented in Weka.

Table 4 shows accuracies of temporal pattern prediction using Boosted C4.5 on patterns obtained by each clustering algorithm. These accuracies are averages of 100 times repeated 10-fold cross validation on the 18 datasets of attributes of each target dataset.

Table 3. Accuracies (%) of temporal pattern prediction by Boosted C4.5 on the two clustering algorithm

K-Means	Saison	MUFJFG	MSFG	MizuhoFG	Orix	KDDI	NTT	NTTdocomo	Softbank	AVERAGE
opening	88.0	83.0	86.0	89.0	83.0	93.0	92.0	91.0	93.0	88.7
high	84.0	88.0	94.0	87.0	83.0	93.0	91.0	90.0	95.0	89.4
low	85.0	92.0	90.0	92.0	81.0	93.0	91.0	92.0	91.0	89.7
closing	86.0	86.0	93.0	91.0	74.0	93.0	92.0	89.0	95.0	88.8
volume	70.0	79.0	86.0	72.0	71.0	79.0	80.0	69.0	85.0	76.8
MovingAverage	96.0	94.9	84.8	88.9	81.8	91.9	62.6	94.9	62.6	84.3
BollingerBand	94.9	90.9	79.8	93.9	94.9	80.8	100.0	86.9	100.0	91.4
Envelope	89.9	89.9	93.9	89.9	89.9	85.9	82.8	100.0	80.8	89.2
HLband	91.9	83.8	90.9	89.9	83.8	87.9	76.8	72.7	91.9	85.5
MACD	84.8	91.9	77.8	81.8	91.9	76.8	90.9	71.7	61.6	81.0
DMI	76.8	84.8	88.9	82.8	90.9	85.9	85.9	90.9	77.8	85.0
volumeRatio	87.9	87.9	91.9	88.9	90.9	91.9	91.9	92.9	84.8	89.9
RSI	85.9	88.9	85.9	88.9	83.8	87.9	83.8	89.9	86.9	86.9
Momentum	82.8	85.9	76.8	81.8	86.9	85.9	82.8	85.9	89.9	84.3
Ichimoku1	67.7	92.9	90.9	86.9	74.7	48.5	87.9	57.6	79.8	76.3
Ichimoku2	58.6	87.9	77.8	82.8	83.8	58.6	73.7	86.9	68.7	75.4
Ichimoku3	97.0	97.0	94.9	74.7	100.0	100.0	100.0	75.8	90.9	92.3
Ichimoku4	78.8	84.8	93.9	89.9	91.9	73.7	93.9	81.8	93.9	87.0

EM	Saison	MUFJFG	MSFG	MizuhoFG	Orix	KDDI	NTT	NTTdocomo	Softbank	AVERAGE
opening	88.0	93.0	86.0	89.0	99.0	90.0	90.0	93.0	94.0	91.3
high	90.0	88.0	85.0	96.0	93.0	90.0	91.0	93.0	91.0	90.8
low	94.0	91.0	92.0	90.0	94.0	92.0	80.0	95.0	90.0	90.9
closing	87.0	95.0	83.0	92.0	97.0	93.0	84.0	93.0	89.0	90.3
volume	72.0	58.0	77.0	64.0	64.0	71.0	81.0	60.0	86.0	70.3
MovingAverage	49.5	63.6	65.7	54.5	63.6	43.4	42.4	54.5	46.5	53.8
BollingerBand	74.7	82.8	80.8	59.6	86.9	48.5	100.0	74.7	100.0	78.7
Envelope	85.9	90.9	85.9	60.6	74.7	78.8	57.6	100.0	87.9	80.2
HLband	71.7	89.9	84.8	87.9	79.8	68.7	58.6	57.6	77.8	75.2
MACD	63.6	51.5	49.5	49.5	58.6	64.6	44.4	58.6	40.4	53.4
DMI	55.6	62.6	69.7	45.5	80.8	57.6	38.4	57.6	59.6	58.6
volumeRatio	89.9	81.8	92.9	93.9	81.8	84.8	85.9	92.9	92.9	88.6
RSI	85.9	88.9	92.9	86.9	80.8	89.9	81.8	80.8	84.8	85.9
Momentum	84.8	87.9	81.8	88.9	89.9	85.9	78.8	84.8	88.9	85.7
Ichimoku1	47.5	39.4	45.5	47.5	54.5	56.6	60.6	54.5	47.5	50.4
Ichimoku2	50.5	46.5	51.5	63.6	58.6	54.5	53.5	41.4	65.7	54.0
Ichimoku3	72.7	80.8	75.8	97.0	100.0	100.0	100.0	85.9	97.0	89.9
Ichimoku4	82.8	82.8	87.9	94.9	62.6	76.8	78.8	85.9	97.0	83.3

4.3 Mining Results of the Nine Temporal Stock Data

In this section, we show accuracies of temporal rule mining with PART on each data-sets themselves and cross-stocks.

As shown in Table 4, predicting temporal patterns for test dataset are succeeded, because the accuracies of the nine dataset are satisfactory high scores as a classification task.

Table 4. Accuracies (%) of re-substitution on the two temporal pattern extraction with K-Means and EM algorithm

Finance	K-Means	EM	Telecom.	K-Means	EM
Saison	90.1	88.9	NTT	84.8	90.9
Orix	88.9	84.8	KDDI	86.9	78.8
MUFJFG	90.9	93.9	NTTdocomo	80.8	85.9
MSFG	96.0	90.9	Softbank	93.9	89.9
MizuhoFG	92.9	83.8			

Table 5 shows accuracies (%) of cross stock evaluation on the two temporal pattern extraction algorithms. The cross stock evaluation uses different stocks as training dataset and test dataset. Stocks in rows mean training datasets, and columns mean test datasets. As shown in this table, bolded accuracies go beyond 50%, which means that the mined rules work better than just predicting sell or buy. The result shows the performance of our temporal rules depends on the similarity of trend values rather than the field of each stock.

Table 5. Accuracies (%) of cross stock evaluation with tempral patterns using K-Means and EM algorithm

K-Means	Saison	MUFJFG	MSFG	MizuhoFG	Orix	NTT	KDDI	NTTdocomo	Softbank
Saison		44.4	28.3	31.3	40.4	29.3	35.4	22.2	49.5
MUFJFG	46.5		44.4	30.3	42.4	32.3	39.4	29.3	**55.6**
MSFG	44.4	24.2		38.4	31.3	28.3	27.3	29.3	22.2
MizuhoFG	46.5	31.3	33.3		29.3	22.2	20.2	22.2	**58.6**
Orix	38.4	**50.5**	27.3	31.3		32.3	39.4	19.2	30.3
NTT	14.1	**50.5**	27.3	31.3	14.1		39.4	37.4	6.1
KDDI	12.1	44.4	**56.6**	27.3	31.3	41.4		**55.6**	16.2
NTTdocomo	26.3	40.4	**52.5**	33.3	23.2	30.3	20.2		8.1
Softbank	44.4	28.3	18.2	45.5	34.3	40.4	30.3	26.3	

EM	Saison	MUFJFG	MSFG	MizuhoFG	Orix	NTT	KDDI	NTTdocomo	Softbank
Saison		46.5	28.3	31.3	38.4	**51.5**	**65.7**	21.2	32.3
MUFJFG	31.3		**51.5**	31.3	38.4	29.3	41.4	22.2	46.5
MSFG	23.2	**58.6**		34.3	31.3	43.4	32.3	30.3	29.3
MizuhoFG	35.4	31.3	34.3		31.3	42.4	38.4	43.4	20.2
Orix	41.4	29.3	39.4	34.3		37.4	21.2	28.3	25.3
NTT	41.4	21.2	20.2	42.4	44.4		33.3	23.2	39.4
KDDI	**61.6**	**59.6**	**50.5**	28.3	27.3	42.4		28.3	37.4
NTTdocomo	27.3	42.4	29.3	**52.5**	25.3	30.3	19.2		28.3
Softbank	**52.5**	45.5	27.3	31.3	41.4	33.3	43.4	19.2	

Figure 5 shows an example of temporal rules. These rules are obtained from the training dataset with EM algorithm temporal pattern extraction for Saison. As shown in Table 3, the rule set of Saison works the best to KDDI test dataset.

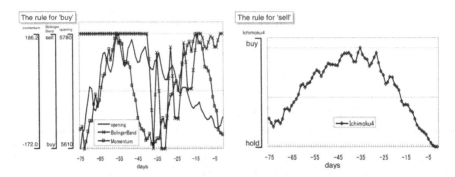

Fig. 5. An example of rule for 'buy' and rule for 'sell'

4.4 Discussion About the Temporal Rule Mining

As for temporal pattern prediction by Boosted C4.5, the algorithm achieved high accuracies to raw valued attributed such as opening, high, low and closing on each clustering algorithm. However, to predict temporal patterns obtained by EM algorithm, Boosted C4.5 did not work well to the other attributes. To predict temporal patterns more exactly, we need to introduce some learning algorithm selection mechanism.

The prediction of decisions for each dataset works satisfactorily with regarding to the result of Table 4, predicting temporal patterns of test dataset with Boosted C4.5. However, mined rules based on temporal patterns are rather over fitting to each training dataset as shown in Table 5. One of the solutions to avoid over fitting will be to mine a temporal rule set from a training dataset, which consists of multiple stocks.

With regarding to Figure 5, our temporal rule mining system can find out adequate combinations of trend index patterns for each stock. To learn adequate trend index pattern combinations is very costly work for trading beginners. Thus, our temporal rule mining can support traders who want to know the adequate combinations of trend indices for each stock.

5 Conclusion

We have designed and implemented a temporal data mining environment, which integrates temporal pattern extraction, rule induction, and rule evaluation.

As the result of the case study on the nine Japanese stock datasets, this environment mines valuable temporal rules to predict different stock decisions based on temporal pattern extraction. The result also indicated the availability to support stock traders to learn adequate combinations of trend index patterns.

In future, we will evaluate trading result with the predictions of decisions by each mined temporal rule set on the stock trading simulator included in Kaburobo SDK.

Although we have not tried to select proper algorithms for the temporal pattern extraction procedure, the attribute selection procedure and the mining procedure, it is also able to connect subsystems for selecting each proper algorithm to this environment.

References

1. Berndt, D.J., Clifford, J.: Using dynamic time wrapping to find patterns in time series. In: Proc. of AAAI Workshop on Knowledge Discovery in Databases, pp. 359–370 (1994)
2. Das, G., King-Ip, L., Heikki, M., Renganathan, G., Smyth, P.: Rule Discovery from Time Series. In: Proc. of International Conference on Knowledge Discovery and Data Mining, pp. 16–22 (1998)
3. Fayyad, U.M., Piatetsky-Shapiro, G., Smyth, P.: From Data Mining to Knowledge Discovery: An Overview. In: Advances in Knowledge Discovery and Data Mining, pp. 1–34. AAAI Press / The MIT Press, CA (1996)
4. Frank, E., Witten, I.H.: Generating accurate rule sets without global optimization. In: Proc. of the Fifteenth International Conference on Machine Learning, pp. 144–151 (1998)
5. Hirano, S., Tsumoto, S.: Mining Similar Temporal Patterns in Long Time-Series Data and Its Application to Medicine. In: Proc. of the 2002 IEEE International Conference on Data Mining, pp. 219–226 (2002)
6. Lin, J., Keogh, E., Lonardi, S., Patel, P.: Finding Motifs in Time Series. In: Proc. of Workshop on Temporal Data Mining, pp. 53–68 (2002)
7. Liu, H., Motoda, H.: Feature selection for knowledge discovery and data mining. Kluwer Academic Publishers, Dordrecht (1998)
8. Michalski, R., Mozetic, I., Hong, J., Lavrac, N.: The AQ15 Inductive Leaning System: An Overview and Experiments, Reports of Machine Leaning and Inference Laboratory, MLI-86-6, George Maseon University (1986)
9. Mitchell, T.M.: Generalization as Search. Artificial Intelligence 18(2), 203–226 (1982)
10. Ohsaki, M., Sato, Y., Yokoi, H., Yamaguchi, T.: A Rule Discovery Support System for Sequential Medical Data - In the Case Study of a Chronic Hepatitis Dataset -. In: ECML/PKDD-2003 Workshop on Discovery Challenge, pp. 154–165 (2003)
11. Ohsaki, M., Kitaguchi, S., Okamoto, K., Yokoi, H., Yamaguchi, T.: Evaluation of Rule Interestingness Measures with a Clinical Dataset on Hepatitis. In: Boulicaut, J.-F., Esposito, F., Giannotti, F., Pedreschi, D. (eds.) PKDD 2004. LNCS (LNAI), vol. 3202, pp. 362–373. Springer, Heidelberg (2004)
12. Ohsaki, M., Abe, H., Kitaguchi, S., Kume, S., Yokoi, H., Yamaguchi, T.: Development and Evaluation of an Integrated Time-Series KDD Environment - A Case Study of Medical KDD on Hepatitis-. In: Joint Workshop of Vietnamese Society of Artificial Intelligence, SIGKBS-JSAI, ICS-IPSJ and IEICE-SIGAI on Active Mining, vol. 23 (2004)
13. Quinlan, J.R.: Programs for Machine Learning. Morgan Kaufmann, San Francisco (1992)
14. Quinlan, J.R.: Bagging, Boosting and C4.5. AAAI/IAAI 1, 725–730 (1996)
15. Raymond, W., Ada, F.: Mining top-K frequent itemsets from data streams. Data Mining and Knowledge Discovery 13(2), 193–217 (2006)
16. Witten, I.H., Frank, E.: Data Mining: Practical Machine Learning Tools and Techniques with Java Implementations. Morgan Kaufmann, San Francisco (2000)
17. KabuRobo: http://www.kaburobo.jp (in Japanese)

Discovering Word Meanings Based on Frequent Termsets*

Henryk Rybinski[1], Marzena Kryszkiewicz[1], Grzegorz Protaziuk[1],
Aleksandra Kontkiewicz[1], Katarzyna Marcinkowska[1], and Alexandre Delteil[2]

[1] ICS, Warsaw University of Technology
{hrb,mkr,gprotazi}@ii.pw.edu.pl, {akontkie,kmarcink}@elka.pw.edu.pl
[2] France Telecome R & D
alexandre.delteil@orange-ft.com

Abstract. Word meaning ambiguity has always been an important problem in information retrieval and extraction, as well as, text mining (documents clustering and classification). Knowledge discovery tasks such as automatic ontology building and maintenance would also profit from simple and efficient methods for discovering word meanings. The paper presents a novel text mining approach to discovering word meanings. The offered measures of their context are expressed by means of frequent termsets. The presented methods have been implemented with efficient data mining techniques. The approach is domain- and language-independent, although it requires applying part of speech tagger. The paper includes sample results obtained with the presented methods.

Keywords: Association rules, frequent termsets, homonyms, polysemy.

1 Introduction

Discovery of word sense is what lexicographers do by profession. Automating this process has a much shorter history. First attempts were made in the 1960s by Sparck Jones [16]. In modern text mining and information retrieval, knowing the exact sense (or meaning) of a word in a document or a query becomes an important issue. It is considered that knowledge of an actual meaning of a polysemous word can considerably improve the quality of the information retrieval process by means of retrieving more relevant documents or extracting relevant information from the documents.

As sufficiently large corpora and efficient computers have become available, several attempts to automate the process have been undertaken. An overview of methods that have used artificial intelligence, machine readable dictionaries (knowledge-based methods) or corpora (knowledge-poor methods) can be found in [8,13].

Many other text processing and knowledge management tasks, such as automatic translation, information extraction or ontology comparison, also require the ability to detect an actual meaning of a word.

---

* The work has been performed within the project granted by France Telecom.

Z.W. Raś, S. Tsumoto, and D. Zighed (Eds.): MCD 2007, LNAI 4944, pp. 82–92, 2008.
© Springer-Verlag Berlin Heidelberg 2008

The importance of polysemy detection becomes even clearer when one looks at the statistics for the English language. As stated in [11]:

"It has been estimated that more than 73% of words used in common English texts are polysemous, i.e., more than 73% of words have more than one sense."

One should also note that many of those 73% words are highly polysemous. As stated in [10], an estimate of the average number of senses per word for all words found in common English texts is about 6.55. Therefore, there is a great need for methods able to distinguish polysemous words, as well as, their meanings.

Unfortunately, manually created dictionaries, which tend to ignore domain specific word senses, are not sufficient for the above mentioned applications. For this reason an amount of research concerning algorithms for automatic discovery of word senses from text corpora was carried out. An excellent survey of the history of ideas used in word sense disambiguation is provided by Ide and Veronis [8].

As for today, no satisfactory method has been described in the literature for discovering word meanings. However, a few methods that address this problem merit attention. Two main strategies for finding homonyms described in the literature are: Word Sense Discrimination (WSDc), i.e. the task of grouping the contexts that represent the senses of a polysemous word, and Word Sense Disambiguation (WSDa), i.e. the task of automatic labeling of polysemous words with a sense tag taken from a predefined set, [13]. There are a couple of different approaches to Word Sense Discrimination, as vector or cluster based, for example. The first one treats words as numeric vectors, whereas the second groups words with similar meaning into clusters. Most methods, however, use word distributional data and statistical analysis.

In the presented paper a novel text mining approach to discovery of homonyms is presented. The method applies the Word Sense Discrimination strategy, carrying out only shallow text analysis, which is restricted to the recognition of parts of speech, and is domain-independent. The method consists in determining atomic contexts of terms of interest by means of maximal frequent termsets, which then are used for determining discriminant contexts. The rest of the paper is organized as follows: some relevant examples of WSDc systems are described in Section 2. Section 3 presents a deeper analysis of the problem as well as the basic ideas and concepts of the proposed method. The experiments and their results are presented in Section 4. Finally, Section 5 concludes the paper.

2   Related Work

As mentioned above, during the past years, a number of various approaches for the discovery of word senses and meanings have been proposed [8]. In [9] and [12] the authors describe a set of problems that are faced, when applying them to real data. The main ones are: data sparseness, infrequent meanings, overlapping meanings and the use of part-of-speech tags. Some of them are described in more detail below.

The method Clustering by Committee, introduced in [12], is based on grammatical relations between words. It requires that all documents are POS tagged and, what is also important, grammatically correct. Each word in this method is represented as a feature vector, where each feature corresponds to a context that the word in question appears in. After computing all vectors, words considered as similar are grouped into clusters. Each cluster represents a single meaning. Finally, for each word the most similar clusters, with respect to a given minimal similarity measure value, are determined. These closest clusters constitute all discovered meanings of a word. The authors claim, that this method discovers also less frequent senses of a word and avoids discovering duplicate senses.

Another group of methods used for discovering word senses are Bayesian networks [17]. These methods are based on the statistical analysis and can be used in the situations, when only small or insufficient amount of data is available, e.g. when larger amount of data would prevent the system from being able to handle it. A Bayesian network is built using local dependencies between words. In such a network, each node represents a word, and an edge represents a conditional probability of the connected nodes. Bayesian networks can be used to determine a probability of co-occurrence of a set of words.

Markov clustering method [2] concentrates on analyzing texts written in a natural language. This method allows discovering the meanings of words used by examined group within a particular range. It follows an assumption, that nouns which frequently appear together in a list, are also semantically related. A result of the Markov clustering is a graph, where nouns co-occurring with appropriate frequency are connected with an edge. The ambiguous words are those that are connected with disjunctive parts of such a graph.

The main concern indicated in many papers is the size of the repositories, on which the experiments are carried out. Too little data leads to data sparseness problems; too much data causes the memory lack problems. In addition, current part-of-speech taggers still occasionally fail to produce the correct results, thus leading to errors.

In most cases, the methods for discovering meanings require the use of a predefined set of meanings for a word, which is their main disadvantage. Usually, the methods require checking found meanings against those contained by a semantic dictionary, thesaurus, ontology etc., which are not easily available. It is therefore desired to develop other methodologies that could also give satisfactory results for the cases of a limited lexical support. Such methodologies are known as "knowledge-poor".

The new method proposed in the paper satisfies this constraint, compromising both simplicity and sufficient efficiency. It also manages to handle the problem of overlapping meanings, as well as data sparseness and infrequent meanings.

3   Homonyms and Maximal Frequent Termset Contexts

Distinct meanings of homonyms are indicated by various distinct contexts in which they appear frequently. This assumption is based on the distributional hypothesis [6], where the underlying idea is that *"a word is characterized by the company it keeps"*. The rule is very intuitive and therefore is applied to the proposed approaches. The problem is, however, how the notion of a context is defined. For example, it can be understood as a set of words surrounding a target word frequently enough in documents, paragraphs, or sentences.

In our approach, a context is evaluated with respect to paragraphs as below.

Let dictionary $D = \{t_1, t_2, \ldots, t_m\}$ be a set of distinct words, called *terms*. In general, any set of terms is called a termset. The set $\mathcal{P}$ is a set of *paragraphs*, where each paragraph P is a set of terms such that $P \subseteq \mathcal{P}$.

Statistical significance of a termset X is called *support* and is denoted by $sup(X)$. $sup(X)$ is defined as the number (or percentage) of paragraphs in $\mathcal{P}$ that contain X. Clearly, the supports of termsets that are supersets of termset X are not greater than $sup(X)$.

A termset is called *frequent* if it occurs in more than *varepsilon* paragraphs in $\mathcal{P}$, where ε is a user-defined support threshold.

In the sequel, we will be interested in maximal frequent termsets, which we will denote by MF and define as the set of all maximal (in the sense of inclusion) termsets that are frequent.

Let x be a term. By $MF(x)$ we denote all maximal frequent termsets containing x. $MF(x)$ will be used for determining *atomic contexts* for x.

A termset X, $x \notin X$, is defined as an *atomic context* of term x if $\{x\} \cup X$ is an element of $MF(x)$. The set of all atomic contexts of x will be denoted by $AC(x)$:

$$AC(x) = \{X \setminus \{x\} \mid X \in MF(x)\}.$$

Clearly, for each two termsets Y, Z in $AC(x)$, Y differs from Z by at least one term and vice versa. In spite of this, Y and Z may indicate the same meaning of x in reality. Let y be a term in $Y \setminus Z$ and z be a term in $Z \setminus Y$ and $\{xyz\}$ be a termset the support of which is significantly less than the supports of Y and Z. This may suggest that Y and Z probably represent different meanings of x. Otherwise, Y and Z are likely to represent the same meaning of x. Please, note that $\{xyz\}$ plays a role of a *potential discriminant* for pairs of atomic contexts. The set of all potential discriminants for Y and Z in $AC(x)$ will be denoted by $\mathcal{D}(x, Y, Z)$:

$$\mathcal{D}(x, Y, Z) = \{\{xyz\} \mid y \in Y \setminus Z \wedge z \in Z \setminus Y\}.$$

Among the potential discriminants, those which are relatively infrequent are called *proper discriminants*. Formally, the set of *all proper discriminants* for Y and Z in $AC(x)$ will be denoted by $\mathcal{PD}(x, Y, Z)$, and defined as follows:

$$\mathcal{PD}(x, Y, Z) = \{X \in \mathcal{D}(x, Y, Z) \mid relSup(x, X, Y, Z) \leq \delta\}, \text{ where}$$

$$relSup(x, X, Y, Z) = sup(X)/min(sup(xY), sup(xZ)), \text{ and}$$

$$\delta \text{ is a user} - \text{defined threshold.}$$

In the sequel, $relSup(x, X, Y, Z)$ is called a *relative support of discriminant* X for term x with respect to atomic contexts Y and Z.

Our proposal of determining the groups of contexts representing separate meanings of x is based on the introduced notion of proper discriminants for pairs of atomic contexts.

Atomic contexts Y and Z in $AC(x)$ are called *discriminable* if there is at least one proper discriminant in $\mathcal{PD}(x, Y, Z)$. Otherwise, Y and Z are called *indiscriminable*.

A *sense-discriminant* context $\mathcal{SDC}(x, X)$ of x for termset X in $AC(x)$ is defined as the family of those termsets in $AC(x)$ that are indiscriminable with X; that is,

$$\mathcal{SDC}(x, X) = \{Y \in AC(x) \mid \mathcal{PD}(x, X, Y) = \emptyset\}.$$

Clearly, $X \in \mathcal{SDC}(x, X)$. Please, note that sense-discriminant contexts of x for Y and Z, where $Y \neq Z$, may overlap, and in particular, may be equal.

The family of all distinct sense-discriminant contexts will be denoted by $\mathcal{FSDC}(x)$:

$$\mathcal{FSDC}(x) = \{\mathcal{SDC}(x, X) \mid X \in AC(x)\}.$$

Please, note that $\mid \mathcal{FSDC}(x) \mid \leq \mid AC(x) \mid$.

A given term x is defined as a *homonym candidate* if the cardinality of $\mathcal{FSDC}(x)$ is greater than 1. Final decision on homonymy is given to the user. Let us also note that the more overlapping are distinct sense-discriminant contexts, the more difficult is reusing the contexts for the meaning recognition in the mining procedures.

In order to illustrate the introduced concepts below we consider an example, which is based on an experimentally prepared set of documents.

Example 1. A special repository has been built based on Google search engine and the AMI-SME software [4]. For the term apple, the following maximal frequent termsets have been found in the repository (see Table 1):

Table 1. Function Maximal frequent termsets for term *apple (MF(apple))*

Maximal frequent termset	Support
apple species breeding	1100
apple eat	14000
apple genetics	1980
apple gmo	1480
apple,botanics	2510
apple cake	3600
apple genome	1800
apple motherboard	2500
apple mouse pad	2500
apple iphone	6000

From this table we can evaluate the set of atomic contents, which is:

$$AC(apple) = \{\{species, breeding\}, \{eat\}, \{genetics\}, \{gmo\}, \{botanics\}, \{cake\},$$
$$\{genome\}, \{motherboard\}, \{mouse, pad\}, \{iphone\}\}.$$

Given the threshold $\delta = 0.2$, we can evaluate the set of proper discriminants from the set of the potential discriminants $\{apple, z, y\}$. For instance, for the atomic contexts Y and Z such that $Y = \{species, breeding\}$ and $Z = \{motherboard\}$, we have the following potential discriminants $\mathcal{D}(apple, Y, Z) = \{\{apple, species, motherboard\}, \{apple, breeding, motherboard\}\}$. Given the supports of the potential discriminants (see Table 2): $sup(\{apple, species, motherboard\}) = 209$ and $sup(\{apple, breeding, motherboard\}) = 78$, we can calculate their relative supports with respect to the supports of Y and Z as follows:

- $relSup(apple, \{apple, species, motherboard\}, Y, Z) =$
  $sup(\{apple, species, motherboard\})/$
  $min(sup(\{apple, species, breeding\}), sup(apple, motherboard)) =$
  $209/min(1100, 2500) = 0.19;$
- $relSup(apple, \{apple, breeding, motherboard\}, Y, Z) =$
  $sup(\{apple, breeding, motherboard\})/$
  $min(sup(\{apple, species, breeding\}), sup(apple, motherboard)) =$
  $78/min(1100, 2500) = 0.07.$

Both discriminants $\{apple, species, motherboard\}$ and $\{apple, breeding, motherboard\}$ have been found as proper, since their relative supports are below the threshold δ . The set of all proper discriminants is provided in Table 2 .

Table 2. Proper discriminants

$\mathcal{PD}$	Support	Relative support
Apple, species, motherboard	209	209/1100=0.19
Apple, breeding, motherboard	78	78/1100=0.07
Apple, eat motherboard	307	307/2500=0.12
Apple, botanics motherboard	68	68/2500=0.03
Apple, botanics, mouse	482	482/2500=0.19
Apple genome motherboard	74	74/1800=0.04
Apple genome pad	192	192/1800=0.10
Apple gmo motherboard	25	25/1800=0.10
Apple, breeding, iphone	0	0/1100=0.00
Apple, botanics, iphone	209	209/2500=0.08

For this set of proper discriminants we have found two sense-discriminant contexts for the term *apple*:

$\mathcal{FSDC}(apple) = \{\{\{species, breeding\}, \{eat\}, \{genetics\}, \{gmo\}, \{botanics\}, \{cake\}\}, \{apple, motherboard, mouse, pad, iphone\}\}.$ □

4   Homonyms Discovering Procedure

The procedure for discovering homonyms and homograms have been implemented within a specialized text mining platform TOM, which has been built at Warsaw University of Technology with the aim to support ontology maintenance and building with text mining techniques [14,15]. TOM provides a variety of tools for text preprocessing. In addition, in various text mining experiments there may be different needs for defining a text unit. We have therefore introduced an option for defining granularity of the text mining process. In particular, TOM allows viewing the whole corpus as a set of documents, paragraphs or sentences. For example, for the experiments aiming at discovering compound terms [14], or synonyms [15] the granularity was set to the sentence level. For discovering homonyms and homograms we have performed experiments with the granularity level set at the paragraph level.

Text preprocessing phase

The text preprocessing phase has been performed with TOM. As for the experiments of discovering homonyms, the granularity was set at the paragraph level, the first step was to generate a set of paragraphs from all the documents in the repository. It means that the context of particular terms is restricted to the paragraphs.

Then we have used the Hepple tagger [7] for part-of-speech tagging of the words in the sentences. In TOM, the tagger is a wrapped code of the Gate part of speech processing resource [5].

Conversion into "transactional database"

The next step is to convert the text corpora into "transactional database" (in terms of [1]). So, every unit of text (i.e. every paragraph) is converted into a transaction containing a set of terms identifiers. The usage of terms identifiers instead of terms themselves leads to speeding up all the data mining operations. Further on, the identifiers of all terms that do not have required minimum support are deleted from all the transactions.

Finding maximal termsets

Having reduced transaction representation of the text, we find the maximal frequent termsets $MF(x)$ for all terms x from the list of terms of interest by means of any efficient data mining algorithm discovering maximal frequent itemsets [18].

Identification of the sense-discriminant contexts

Having $MF(x)$ for each term x, we calculate the atomic contexts $AC(x)$ by simple removal of x from each termset in $MF(x)$. For later use, for each atomic context X in $AC(x)$, we store the support of the maximal frequent termset xX, from which X was derived. Now, we create potential discriminants for all the pairs of atomic contexts. Then, from the set of potential discriminants, we search for proper discriminants. This requires the calculation of the relative supports of the potential discriminants based on the supports of termsets xX, where $X \in AC(x)$, and the supports of the potential discriminants themselves. While the supports of termsets xX, where $X \in AC(x)$, are already known, the supports of the potential discriminants must be calculated. This is achieved with one scan over the transaction dataset [1]. Eventually, we calculate all the distinct sense-discriminant contexts for x. If the number of the distinct sense-discriminant contexts for the term x is higher than 1, we classify x as a polysemous term with the meanings determined by the contexts.

5   Experiments

In order to evaluate the efficiency of the proposed algorithms we have performed a number of experiments. In general, the found homonym candidates can be classified into three different groups:

a) Same meaning: berlin
   [ontology, european, intelligence, th, proceedings, workshop, eds, conference, learning, germany, artificial, august, ecai, hastings, wiemer-]
   [springer-, verlag]
b) Different meaning: background
   [domain, theory]
   [web, pages]
c) Different use: results
   [table, taxonomy, phase, similarity, measures]
   [information, users, queries]

In our case, we are interested in discovering words which belong to the second and third group. The tests were done using different minimal support values and part-of-speech tags. So far, better results were obtained with using smaller values for the support threshold. As the aim was to find homonymous nouns, only those were considered. As for the contexts we took into account nouns, adjectives and verbs.

First, the tests have been performed on a repository composed of scientific papers dealing with ontologies, semantic web and text mining. The results have been rather modest, which can be justified by the fact that the scientific texts in a limited domain use homogeneous and well defined terminology. Additionally, the size of the repository was not sufficient for the experiments. Nevertheless, for the term background two sense discriminant contexts have been found $\{domain, theory\}$ and $\{web, pages\}$.

Another group of tests have been performed on the Reuters repository. In this case the polysemous words are provided in Table 3.

Table 3. Results for the Reuters repository (the threshold set to 11 occurrences)

President	chinese jiang zemin hussein iraqi saddam
Ltd	pictures seagram unit universal corp murdoch news percent shares worth
York	consolidated exchange stock trading city gate prices
France	budget currency deficits brazil cup world

One can see an erroneous candidate *president*, which results from a shallow semantic analysis of the text. With a deeper processing of the text (in this case replacing the proper names by tokens) the candidate would be rejected.

The term *york* has two meanings – one with the stock, and another one with the city. And with the third example, the term *france* has assigned the context referring to the country (economic situation) and the national football team.

The third group of tests referred to a limited part of the repository from the FAOLEX database [3], composed of the national legislations of various countries in the area of agriculture and food technologies. Two spectacular discovered examples are presented in Table 4.

Table 4. Results for the FAOLEX repository (the threshold set to 20 occurrences)

term	Contexts	discriminants
turkey	{marketing production poults} {Merriam,hunting,licence}	{turkey Merriam production}
plant	{protection} {soil, pest, packaging} {plant product organism} {meat processing} {meat industries}	{Plant,meat,soil}

In the table one can see that in the FAOLEX repository the term *turkey* has two meanings: one referring to the wild turkey (Merriam's turkey subspecies, named in 1900 in honor of C. Hart Merriam, the first chief of the US Biological Survey), the other one referring to the domesticated bird, produced for meat. It is worth to note that because of the specificity of the repository no discriminant was found for the meaning Turkey as The Republic of Turkey. For the homogram *plant* two meanings have been discovered: one related to the biological concept (in FAOLEX found in the context of food and agriculture related activities), and another one related to the "industrial factory" for "meat processing".

Still there is a problem with evaluating recall of the method. To this end we have decided to use the AMI-SME system [4] for building specific subject oriented repositories, where a material for predefined homonyms and homograms should be present. From a set of queries we have composed in the AMI-SME system a repository aiming at checking the relevance of the idea, which was used to find out meanings for the term *apple* (Example 1), and the term *turkey*. For both terms the method has proven its efficiency. Now, we plan performing more tests in order to determine the relevance and precision of the method.

6   Conclusions and Future Work

As said above, the tests were carried out on three repositories. As the results have shown, the repository of scientific papers was not well suited for such experiments. The experiments with the Reuters repository have shown strong dependency on the size of the repository. With the too small repository, the results are rather poor. With the extended repository, we were able to find more polysemous words, on the other hand though, with too large repository we had problems with the data mining algorithms. To this end we have decided to define a priori a list of terms for which the experiment is run.

The results obtained so far show that the method presented above is able to distinguish homonymous words correctly. A human expert can easily interpret the contexts generated by the algorithms. Obviously there are among candidate erroneous terms, one can also expect that there are some undiscovered. It results from the following problems: (a) wrong part-of-speech tags assigned to the words by the POSTagger; (b) the values set for the minimal support, which are too high, but on the other hand, if lowered, caused memory problems; (c) the repository is too small and does not cover proper discriminants.

References

1. Agrawal, R., Srikant, R.: Fast algorithms for mining association rules. In: Proc. of the 20th Int'l Conf. on Very Large Databases, pp. 487–499. Morgan Kaufmann, Santiago (1994)
2. Dorow, B., Widdows, D.: Discovering corpus-specific word senses. In: EACL 2003, Budapest, Hungary, pp. 79–82 (2003)
3. FAOLEX Legal Database, FAO, http://faolex.fao.org/faolex
4. Gawrysiak, P., Rybinski, H., Skonieczny, L, Wiech, P.: AMI-SME: An exploratory approach to knowledge retrieval for SME's. In: 3rd Int'l Conf. on Autonomic and Autonomous Systems, ICAS 2007 (2007)
5. General Architecture for Text Engineering, http://gate.ac.uk/projects.html
6. Harris, Z.: Distributional structure. Word 10(23), 146–162 (1954)
7. Hepple, M.: Independence and commitment: Assumptions for rapid training and execution of rule-based POS taggers. In: Proc. of the 38th Annual Meeting of the Association for Computational Linguistics, ACL 2000 (2000)
8. Ide, N., Veronis, J.: Introduction to the special issue on word sense disambiguation: The state of the art. Computational Linguistics 24(1), 1–40 (Special Issue on Word Sense Disambiguation)
9. Lin, D.: Automatic Retrieval and Clustering of Similar Words. In: Proc. of the 17th Int'l Conf. on Computational linguistics, Canada, vol. 2 (1998)
10. Mihalcea, R., Moldovan, D.: Automatic generation of a coarse grained WordNet. In: Proc. of NAACL Workshop on WordNet and Other Lexical Resources, Pittsburgh, PA (2001)
11. Miller, G., Chadorow, M., Landes, S., Leacock, C., Thomas, R.G.: Using a semantic concordance for sense identification. In: Proc. of the ARPA Human Language Technology Workshop, pp. 240–243 (1994)
12. Pantel, P., Lin, D.: Discovering word senses from text. In: Proc. of the Eighth ACM SIGKDD International Conference on Knowledge Discovery and Data Mining, KDD 2002, Edmonton, Alberta, Canada, July 23-26, 2002, pp. 613–619. ACM Press, New York (2002)
13. Portnoy, D.: Unsupervised Discovery of the Multiple Senses of Words and Their Parts of Speech, The School of Engineering and Applied Science of The George Washington University, September 30 (2006)

14. Protaziuk, G., Kryszkiewicz, M., Rybinski, H., Delteil, A.: Discovering compound and proper nouns. In: Kryszkiewicz, M., Peters, J.F., Rybinski, H., Skowron, A. (eds.) RSEISP 2007. LNCS (LNAI), vol. 4585, pp. 505–515. Springer, Heidelberg (2007)
15. Rybinski, H., Kryszkiewicz, M., Protaziuk, G., Jakubowski, A., Delteil, A.: Discovering synonyms based on frequent termsets. In: Kryszkiewicz, M., Peters, J.F., Rybinski, H., Skowron, A. (eds.) RSEISP 2007. LNCS (LNAI), vol. 4585, pp. 516–525. Springer, Heidelberg (2007)
16. Sparck Jones, K.: Synonymy and Semantic Classification. Edinburgh University Press (1986) (originally published in 1964), ISBN 0-85224-517-3
17. Park, Y.C., Han, Y.S., Choi, K.-S.: Automatic thesaurus construction using bayesian networks. In: The Proc. of the fourth international conference on Information and knowledge management, United States (1995)
18. Zaki Mohammed, J., Karam, G.: Efficiently mining maximal frequent itemsets. In: 1st IEEE Int'l Conf. on Data Mining, San Jose (2001)

Quality of Musical Instrument Sound Identification for Various Levels of Accompanying Sounds

Alicja Wieczorkowska[1] and Elżbieta Kolczyńska[2]

[1] Polish-Japanese Institute of Information Technology,
Koszykowa 86, 02-008 Warsaw, Poland
alicja@pjwstk.edu.pl
[2] Agricultural University in Lublin
Akademicka 13, 20-950 Lublin, Poland
elzbieta.kolczynska@ar.lublin.pl

Abstract. Research on automatic identification of musical instrument sounds has already been performed through last years, but mainly for monophonic singular sounds. In this paper we work on identification of musical instrument in polyphonic environment, with added accompanying orchestral sounds for the training purposes, and using mixes of two instrument sounds for testing. Four instruments of definite pitch has been used. For training purposes, these sounds were mixed with orchestral recordings of various levels, diminished with respect to the original recording level. The level of sounds added for testing purposes was also diminished with respect to the original recording level, in order to assure that the investigated instrument actually produced the sound dominating in the recording. The experiments have been performed using WEKA classification software.

1 Introduction

Recognition of musical instrument sound from audio files is not a new topic and research in this area has already been performed worldwide by various groups of scientists, see for example [2], [3], [4], [6], [7], [9], [11], [14], [21]. This research was mainly performed on singular monophonic sounds, and in this case the recognition is quite successful, with the accuracy level at about 70% for a dozen or more instruments, and exceeding 90% for a few instruments (up to 100% correctness). However, the recognition of instrument, or instruments, in polyphonic recording is much more difficult, especially when no spatial clues are used to locate the sound source and thus facilitate the task [20]. The research has already been performed to separate instruments from polyphonic, poly-tymbral recordings, and to recognize instruments in a noisy environment [6], [13]. The noises added to the recording included noises recorded in the museum (footsteps, talking and clatter), wind gusts, old air-conditioner, and steam factory engine. Therefore, some of the noises were rather unnatural to meet in real recordings. Our idea

Z.W. Raś, S. Tsumoto, and D. Zighed (Eds.): MCD 2007, LNAI 4944, pp. 93–103, 2008.

was to imitate the sounds found in real recordings, so we decided to use sounds of other musical instruments, or of the orchestra.

In our paper we aim at training classifiers for the purpose of recognition of predominant musical instrument sound, using various sets of training data. We believe that using for training not only clean singular monophonic musical instrument sound samples, but also the sounds with added other accompanying sounds, may improve classification quality. We are interested in checking how various levels of accompanying sounds (distorting the original sound waves) influence correctness of classification of predominant (louder) instrument in mixes containing two instrumental sounds.

2 Sound Parameterization

Audio data, for example files of .wav or .snd type, represent a sequence of samples for each recorded channel, where each sample is a digital representation of amplitude of digitized sound. Such sound data are usually parameterized for sound classification purposes, using various features describing temporal, spectral, and spectral-temporal properties of sounds. Features implemented in the worldwide research on musical instrument sound recognition so far include parameters based on DFT, wavelet analysis, MFCC (Mel-Frequency Cepstral Coefficients), MSA (Multidimensional Analysis Scaling) trajectories, and so on [3], [4], [9], [11], [14], [21]. Also, MPEG-7 sound descriptors can be applied [10], although these parameters are not dedicated to recognition of particular instruments in recordings.

We are aware that the choice of the feature vector is important for the success of classification process, and that the results may vary if a different feature vector is used for the training of any classifier. Therefore, we decided to use the feature vector already used in a similar research, which yielded good results for musical instrument identification for monophonic sounds [23]. We applied the following 219 parameters, based mainly on MPEG-7 audio descriptors, and also other parameters used for musical instrument sound identification purposes [23]:

– MPEG-7 audio descriptors [10], [16]:
  - *AudioSpectrumSpread* - a RMS value of the deviation of the Log frequency power spectrum with respect to the gravity center in a frame; averaged over all analyzed frames for a given sound;
  - *AudioSpectrumFlatness*, $flat_1, \ldots, flat_{25}$ - describes the flatness property of the power spectrum within a frequency bin; 25 out of 32 frequency bands were used to calculate these parameters for each frame; averaged over all frames for a given sound;
  - *AudioSpectrumCentroid* - computed as power weighted average of the frequency bins in the power spectrum of all the frames in a sound with a Welch method;
  - *AudioSpectrumBasis*: $basis_1, \ldots, basis_{165}$ - spectrum basis function is used to reduce the dimensionality by projecting the spectrum of the analyzed frame from high dimensional space to low dimensional space with

compact salient statistical information; results averaged over all frames of the sound. Spectral basis parameters are calculated for the spectrum basis functions, where total number of sub-spaces in basis function is 33, and for each sub-space, minimum/maximum/mean/distance/ standard deviation are extracted to flat the vector data. Distance is calculated as the summation of dissimilarity (absolute difference of values) of every pair of coordinates in the vector;

- *HarmonicSpectralCentroid* - the average over the sound duration of the instantaneous Harmonic Centroid within a frame. The instantaneous Harmonic Spectral Centroid is computed as the amplitude (in linear scale) weighted mean of the harmonic peak of the spectrum;
- *HarmonicSpectralSpread* - the average over the sound duration of the instantaneous harmonic spectral spread of a frame, i.e. the amplitude weighted standard deviation of the harmonic peaks of the spectrum with respect to the instantaneous harmonic spectral centroid;
- *HarmonicSpectralVariation* - mean value over the sound duration of the instantaneous harmonic spectral variation, i.e. the normalized correlation between the amplitude of the harmonic peaks of two adjacent frames;
- *HarmonicSpectralDeviation* - the average over the sound duration of the instantaneous harmonic spectral deviation in each frame, i.e. the spectral deviation of the log amplitude components from a global spectral envelope;
- *LogAttackTime* - the decimal logarithm of the duration from the beginning of the signal to the time when it reaches its maximum or its sustained part, whichever comes first;
- *TemporalCentroid* - energy weighted mean of the duration of the sound - represents where in time the energy of the sound is focused;

- other audio descriptors:
  - *Energy* - average energy of spectrum in the entire sound;
  - *MFCC* - min, max, mean, distance, and standard deviation of the MFCC vector; averaged over all frames of the sound;
  - *ZeroCrossingDensity*, averaged through all frames of the sound;
  - *RollOff* - measure of spectral shape, used in the speech recognition, where it is used to distinguish between voiced and unvoiced speech. The roll-off is defined as the frequency below which an experimentally chosen percentage of the accumulated magnitudes of the spectrum is concentrated; averaged over all frames of the sound;
  - *Flux* - the difference between the magnitude of the FFT points in a given frame and its successive frame (value multiplied by 10^7 to comply with WEKA requirements); value averaged over all frames of the sound;
  - *AverageFundamentalFrequency*;
  - *TristimulusParameters*: $tris_1, \ldots, tris_{11}$ - describe the ratio of the amplitude of a harmonic partial to the total harmonic partials (average for the entire sound); attributes based on [17].

Frame-based parameters are represented as average value of the attribute calculated using sliding analysis window, moved through the entire sound. The calculations were performed using 120 ms analyzing frame with Hamming window and hop size 40 ms. Such a long analyzing frame allows analysis even of the lowest sounds. In the described research, data from the left channel of stereo sounds were taken for parameterization.

The parameters presented above describe basic spectral, timbral spectral and temporal audio properties, incorporated into the MPEG-7 standard. Also, spectral basis descriptor from MPEG-7 was used. This attribute is actually a non-scalar one - spectral basis is a series of basis functions derived from the singular value decomposition of a normalized power spectrum. Therefore, a few other features were derived from the spectral basis attribute, to avoid too high dimensionality of the feature vector. Other attributes include time-domain and spectrum-domain properties of sound, commonly used in audio research, especially for music data.

3   Experiments

The goal of our research was to check how modification (i.e. sound mixing) of the initial audio data, representing musical instrument sounds, influences the quality of classifiers trained to recognize these instruments. The initial data were taken from McGill University CDs, used worldwide in research on music instrument sounds [15]. The sounds where recorded stereo with 44.1 kHz sampling rate, and 16 bit resolution. We have chosen 188 sounds of the following instruments (i.e. representing 4 classes):

1. B-flat clarinet - 37 sound objects,
2. C-trumpet (also trumpet muted, mute Harmon with stem out) - 65 objects,
3. violin vibrato - 42 objects
4. cello vibrato - 43 objects.

The sounds were parameterized as described in the previous section, thus yielding the clean data for further work. Next, the clean data were distorted in such a way that an excerpt from orchestral recording was added. We initially planned to use the recordings of the chords constant in time (for a few seconds, i.e. as long as the singular sounds from the MUMS recordings), but it is not so easy and fast to find such chords. Finally, we decided to use Adagio from Symphony No. 6 in B minor, Op. 74, Pathetique by P. Tchaikovsky for this purpose. Four short excerpts from this symphony were diminished to 10%, 20%, 30%, 40% and 50% of original amplitude, and added to the initial sound data, thus yielding 5 versions of distorted data, used for training of classifiers. Those disturbing data were changing in time, but since the parameterization was performed applying short analysis window, we did not decide to search through numerous recordings for excerpts with stable spectra (i.e. long lasting chords), especially that the main harmonic contents was relatively stable in the chosen excerpts.

For testing purposes, all clean singular sound objects were mixed with the following 4 sounds:

1. C4 sound of c-trumpet,
2. A4 sound of clarinet,
3. D5 sound of violin vibrato
4. G3 sound of cello vibrato,

where A4 = 440 Hz (i.e. MIDI notation is used for pitch). The added sounds represent various pitches and octaves, and various groups of musical instruments of definite pitch: brass, woodwinds, and stringed instruments producing both low and high pitched sounds. The amplitude of these added 4 sounds was diminished to 10% of the original level, to make sure that the recognized instrument is actually the main, dominating sound in the mixed recording.

As one can see, we decided to use different data for training and for recognition purposes. Therefore, we could check how classifiers perform on unseen data.

The experiments performed worldwide on musical instrument sounds, so far, mainly focused on monophonic sounds, and numerous classifiers were used for this purpose. The applied classifiers include Bayes decision rules, K-Nearest Neighbor (k-NN) algorithm, statistical pattern-recognition techniques, neural networks, decision trees, rough set based algorithms, Hidden Markov Models (HMM) and Support Vector Machines (SVM) [1], [4], [5], [7], [8], [9], [11], [12], [14], [19], [22]. The research on musical instrument recognition in polyphonic environment (without spacial clues) is more recent, and so far just a few classifiers were used for the identification of instruments (or separation) from poly-timbral recordings, including Bayesian, decision trees, artificial neural networks and some others [6], [13]. In our experiments, we decided to use WEKA (Waikato Environment for Knowledge Analysis) software for classification purposes, with the following classifiers: Bayesian Network, decision trees (Tree J48), Logistic Regression Model (LRM), and Locally Weighted Learning (LWL) [18]. Standard settings of the classifiers were used. The training of each classifier was performed three-fold, separately for each level of the accompanying orchestral sound (i.e. for 10%, 20%, 30%, 40%, and 50%):

– on clean singular sound data only (singular instrument sounds)
– on both singular and accompanied sound data (i.e. mixed with orchestral recording)
– on accompanied sound data only

In each case, the testing was performed on the data obtained via mixing of the initial clean data with other instrument sound (diminished to 10% of original amplitude), as described above.

Summary of results for all these experiments is presented in Tables 1–4.

The improvement of correctness for each classifier, trained on both clean singular sound and accompanied sound data, in comparison with the training on clean singular sound data only, is presented in Figure 1. Negative values indicate decrease of correctness, when the mixes with accompanied sounds were added to the training set.

Table 1. Results of experiments for Bayesian network

Classifier	Added sound level	Training on data:	Correctness %
BayesNet	10%	Singular sounds only	73,14%
		Both singular and accompanied sounds	81,91%
		Accompanied sounds only	77,53%
	20%	Singular sounds only	73,14%
		Both singular and accompanied sounds	76,20%
		Accompanied sounds only	69,41%
	30%	Singular sounds only	73,14%
		Both singular and accompanied sounds	77,39%
		Accompanied sounds only	63,56%
	40%	Singular sounds only	73,14%
		Both singular and accompanied sounds	75,40%
		Accompanied sounds only	60,77%
	50%	Singular sounds only	73,14%
		Both singular and accompanied sounds	75,93%
		Accompanied sounds only	55,98%

Table 2. Results of experiments for decision trees (Tree J48)

Classifier	Added sound level	Training on data:	Correctness %
TreeJ48	10%	Singular sounds only	81,65%
		Both singular and accompanied sounds	80,19%
		Accompanied sounds only	74,47%
	20%	Singular sounds only	81,65%
		Both singular and accompanied sounds	82,05%
		Accompanied sounds only	58,78%
	30%	Singular sounds only	81,65%
		Both singular and accompanied sounds	83,64%
		Accompanied sounds only	64,63%
	40%	Singular sounds only	81,65%
		Both singular and accompanied sounds	66,49%
		Accompanied sounds only	62,23%
	50%	Singular sounds only	81,65%
		Both singular and accompanied sounds	76,86%
		Accompanied sounds only	50,53%

As we can see, for LWL classifier adding mixes with the accompanying sounds to the training data always caused decrease of the correctness of the instrument recognition. However, in most other cases (apart from decision trees) we observe improvement of classification correctness, when mixed sound data are added to the training set.

Table 3. Results of experiments for Logistic Regression Model

Classifier	Added sound level	Training on data:	Correctness %
Logistic	10%	Singular sounds only	78,99%
		Both singular and accompanied sounds	84,31%
		Accompanied sounds only	67,95%
	20%	Singular sounds only	78,99%
		Both singular and accompanied sounds	88,56%
		Accompanied sounds only	64,23%
	30%	Singular sounds only	78,99%
		Both singular and accompanied sounds	84,84%
		Accompanied sounds only	63,16%
	40%	Singular sounds only	78,99%
		Both singular and accompanied sounds	85,77%
		Accompanied sounds only	53,06%
	50%	Singular sounds only	78,99%
		Both singular and accompanied sounds	87,37%
		Accompanied sounds only	49,20%

Table 4. Results of experiments for Locally Weighted Learning

Classifier	Added sound level	Training on data:	Correctness %
LWL	10%	Singular sounds only	68,35%
		Both singular and accompanied sounds	67,02%
		Accompanied sounds only	66,62%
	20%	Singular sounds only	68,35%
		Both singular and accompanied sounds	67,15%
		Accompanied sounds only	67,55%
	30%	Singular sounds only	68,35%
		Both singular and accompanied sounds	62,10%
		Accompanied sounds only	62,37%
	40%	Singular sounds only	68,35%
		Both singular and accompanied sounds	62,37%
		Accompanied sounds only	61,70%
	50%	Singular sounds only	68,35%
		Both singular and accompanied sounds	53,86%
		Accompanied sounds only	53,86%

The improvement of correctness for our classifiers, but trained on mixed sound data only, in comparison with the training on clean singular sound data only, is presented in Figure 2.

As we can see, in this case the accuracy of classification almost always decreases, and we only have improvement of correctness for low levels of

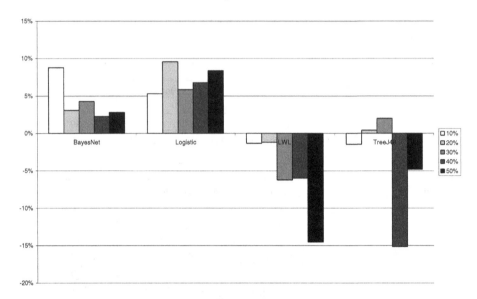

Fig. 1. Change of correctness of musical instrument sound recognition for classifiers built on both clean singular musical instrument sound and accompanied sound data, i.e. with added (mixed) orchestral excerpt of various levels (10%, 20%, 30%, 40%, 50% of original amplitude), and tested on the data distorted through adding other instrument sound to the initial clean sound data. Comparison is made with respect to the results obtained for classifiers trained on clean singular sound data only.

accompanying sounds for the Bayesian network. Therefore we can conclude that clean singular sound data are rather necessary to train classifiers for instrument recognition purposes.

We are aware that the results may depend on the instruments used, and a different choice of instruments may produce different results. Additionally, the loudness of each sounds (both the sounds of interest and accompanying sounds) changes in time, so it may also obfuscate the results of experiments. Also, decision trees are not immune to noise in the data, and since the addition of other sounds can be considered as adding noise to the data, the results are not as good as in case of clean monophonic sounds.

When starting these experiments, we hoped to observe some dependencies between the added disturbances (i.e. accompanying sounds) to the training sound data, the level of the disturbance, and change of the classification correctness. As we can see, there are no such clear linear dependencies. On the other hand, the type of the disturbance/accompaniment (for example, its harmonic contents, and how it overlaps with the initial sound) may also influence the results. Also, when sound mixes were produced, both sounds in any mix were changing in time, what is natural and unavoidable in case of music. Therefore, in some frames the disturbing, accompanying sounds could be louder than the sound of interest, so mistakes regarding identification of the dominant instrument in the mix also may happen as well.

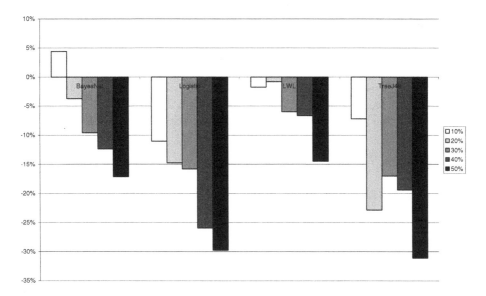

Fig. 2. Change of correctness of musical instrument sound recognition for classifiers built on the mixed sounds, i.e. of singular instruments with added orchestral excerpt of various levels (10%, 20%, 30%, 40%, 50% of original amplitude), and tested on the data with added other instrument sound to the main instrument sound. Comparison is made with respect to the results obtained for classifiers trained on clean singular sound data only.

4   Summary and Conclusions

The experiments described in this paper aimed at observing if (and how) adding disturbance (i.e. accompanying sound added) to the clean musical instrument sound data influences correctness of classification of the instrument, dominating in the polyphonic recording. The clean data represented singular musical instrument sounds of definite pitch and harmonic spectrum. The disturbances added represented various levels of orchestral recordings, added to singular monophonic musical instrument sounds. Tests performed on pairs of instruments sounds have shown that in most cases the use of disturbed (mixed) data, together with initial clean singular sound data, increases the correctness of the classifier, thus increasing its quality. However, no clear linear relationships can be observed. The results for using only distorted data for training showed that clean data are necessary for training purposes.

We plan to continue our experiments, with using various levels of added orchestral sounds for training and for testing the classifiers. Also, since the set of sound features is very important for the correct classification, we plan to check how changes in the feature set influence the quality of classification for distorted in such a way data set.

Acknowledgments

This work was supported by the National Science Foundation under grant IIS-0414815, and also by the Research Center of PJIIT, supported by the Polish National Committee for Scientific Research (KBN).

The authors would like to express thanks to Xin Zhang from the University of North Carolina at Charlotte for help with preparing the initial data.

References

1. Agostini, G., Longari, M., Pollastri, E.: Musical Instrument Timbres Classification with Spectral Features. EURASIP Journal on Applied Signal Processing 1, 1–11 (2003)
2. Ando, S., Yamaguchi, K.: Statistical Study of Spectral Parameters in Musical Instrument Tones. Journal of the Acoustical Society of America 94(1), 37–45 (1993)
3. Aniola, P., Lukasik, E.: JAVA Library for Automatic Musical Instruments Recognition. In: AES 122 Convention, Vienna, Austria (May 2007)
4. Brown, J.C.: Computer identification of musical instruments using pattern recognition with cepstral coefficients as features. Journal of the Acoustical Society of America 105, 1933–1941 (1999)
5. Cosi, P., De Poli, G., Lauzzana, G.: Auditory Modelling and Self-Organizing Neural Networks for Timbre Classification. Journal of New Music Research 23, 71–98 (1994)
6. Dziubinski, M., Dalka, P., Kostek, B.: Estimation of Musical Sound Separation Algorithm Effectiveness Employing Neural Networks. Journal of Intelligent Information Systems 24(2/3), 133–157 (2005)
7. Eronen, A.: Comparison of features for musical instrument recognition. In: Proceedings of the IEEE Workshop on Applications of Signal Processing to Audio and Acoustics WASPAA 2001 (2001)
8. Fujinaga, I., McMillan, K.: Realtime recognition of orchestral instruments. In: Proceedings of the International Computer Music Conference, Berlin, Germany, pp. 141–143 (August 2000)
9. Herrera, P., Amatriain, X., Batlle, E., Serra, X.: Towards instrument segmentation for music content description: a critical review of instrument classification techniques. In: Proc. of International Symposium on Music Information Retrieval ISMIR 2000, Plymouth, MA (2000)
10. ISO/IEC JTC1/SC29/WG11: MPEG-7 Overview (2004), Available at: http://www.chiariglione.org/mpeg/standards/mpeg-7/mpeg-7.htm
11. Kaminskyj, I.: Multi-feature Musical Instrument Sound Classifier w/user determined generalisation performance. In: Proceedings of the Australasian Computer Music Association Conference ACMC 2002, pp. 53–62 (2002)
12. Kostek, B., Czyzewski, A.: Representing Musical Instrument Sounds for Their Automatic Classification. Journal of the Audio Engineering Society 49(9), 768–785 (2001)
13. Lewis, R.A., Zhang, X., Raś, Z.W.: Blind Signal Separation of Similar Pitches and Instruments in a Noisy Polyphonic Domain. In: Esposito, F., Raś, Z.W., Malerba, D., Semeraro, G. (eds.) ISMIS 2006. LNCS (LNAI), vol. 4203, pp. 228–237. Springer, Heidelberg (2006)

14. Martin, K.D., Kim, Y.E.: Musical instrument identification: A pattern-recognition approach. In: 136-th meeting of the Acoustical Society of America, Norfolk, VA (1998)
15. Opolko, F., Wapnick, J.: MUMS - McGill University Master Samples. CD's (1987)
16. Peeters, G., McAdams, S., Herrera, P.: Instrument Sound Description in the Context of MPEG-7. In: Proceedings of the International Computer Music Conference ICMC 2000, Berlin, Germany (August 2000)
17. Pollard, H.F., Jansson, E.V.: A Tristimulus Method for the Specification of Musical Timbre. Acustica 51, 162–171 (1982)
18. The University of Waikato: Weka Machine Learning Project. Internet (2007), Available at: http://www.cs.waikato.ac.nz/~ml/
19. Toiviainen, P.: Optimizing Self-Organizing Timbre Maps: Two Approaches. In: Joint International Conference, II International Conference on Cognitive Musicology, College of Europe at Brugge, Belgium, pp. 264–271 (1996)
20. Viste, H., Evangelista, G.: Separation of Harmonic Instruments with Overlapping Partials in Multi-Channel Mixtures. In: IEEE Workshop on Applications of Signal Processing to Audio and Acoustics WASPAA 2003, New Paltz, NY, USA (October 2003)
21. Wieczorkowska, A.: Towards Musical Data Classification via Wavelet Analysis. In: Ohsuga, S., Raś, Z.W. (eds.) ISMIS 2000. LNCS (LNAI), vol. 1932, pp. 292–300. Springer, Heidelberg (2000)
22. Wieczorkowska, A., Wróblewski, J., Synak, P., Slezak, D.: Application of Temporal Descriptors to Musical Instrument Sound Recognition. Journal of Intelligent Information Systems 21(1), 71–93 (2003)
23. Zhang, X., Ras, Z.W.: Analysis of Sound Features for Music Timbre Recognition. In: Kim, S., Park, J.H., Pissinou, N., Kim, T., Fang, W.C., Slezak, D., Arabnia, H., Howard, D. (eds.) International Conference on Multimedia and Ubiquitous Engineering MUE 2007, Seoul, Korea, April 26-28, 2007, pp. 3–8. IEEE Computer Society, Los Alamitos (2007)

Discriminant Feature Analysis for Music Timbre Recognition and Automatic Indexing

Xin Zhang[1], Zbigniew W. Raś[1,3], and Agnieszka Dardzińska[2]

[1] Univ. of North Carolina, Dept. of Comp. Science, Charlotte, N.C. 28223, USA
[2] Bialystok Technical Univ., Dept. of Comp. Science,
ul. Wiejska 45a 15-351 Bialystok, Poland
[3] Polish-Japanese Institute of Information Technology, ul. Koszykowa 86,
02-008 Warsaw, Poland
{xinzhang,ras,adardzin}@uncc.edu

Abstract. The high volume of digital music recordings in the internet repositories has brought a tremendous need for a cooperative recommendation system to help users to find their favorite music pieces. Music instrument identification is one of the important subtasks of a content-based automatic indexing, for which authors developed novel new temporal features and built a multi-hierarchical decision system S with all the low-level MPEG7 descriptors as well as other popular descriptors for describing music sound objects. The decision attributes in S are hierarchical and they include Hornbostel-Sachs classification and generalization by articulation. The information richness hidden in these descriptors has strong implication on the confidence of classifiers built from S. Rule-based classifiers give us approximate definitions of values of decision attributes and they are used as a tool by content-based Automatic Indexing Systems (AIS). Hierarchical decision attributes allow us to have the indexing done on different granularity levels of classes of music instruments. We can identify not only the instruments playing in a given music piece but also classes of instruments if the instrument level identification fails. The quality of AIS can be verified using precision and recall based on two interpretations: user and system-based [16]. AIS engine follows system-based interpretation.

1 Introduction

The state of art technologies in semantic web and computer storage boost the fast growing of music repositories throughout the internet, which in turn brought the need for intelligent search methods and efficient recommendation systems to help users to find their favorite music pieces.

Mining for knowledge in different representations of musical files (e.g., music recordings, MIDI files, and music notes) involves very different techniques. Research in MIDI files and music notes tackles problems in text mining. Digital recordings contain only sound signals unless manually labelled with semantic descriptions (e.g., author, title, and company). Knowledge mining in digital recordings requires prior retrieval of a large number of sound features from these

Z.W. Raś, S. Tsumoto, and D. Zighed (Eds.): MCD 2007, LNAI 4944, pp. 104–115, 2008.
© Springer-Verlag Berlin Heidelberg 2008

musical sound signals. Timbre identification is one of the important subtasks for mining digital recordings, where timbre is a quality of sound that distinguishes one music instrument from another. Researchers in this area have investigated a number of acoustical features to build computational model for timbre identification. In this paper, authors focus on developing automated indexing solutions for digital recordings based on MIR (Music Information Retrieval) techniques of instruments and their types.

The real use of timbre-based grouping of music is very nicely discussed in [3]. Methods in research on automatic musical instrument sound classification go back to the last few years. We review these methods with respect to monophonic and polyphonic musical sounds.

For monophonic sounds, a number of acoustic features have been explored in [1], [4]. Some of them are quite successful for certain classes of sound data (monophonic, short, limited type of instruments). A digital multimedia file normally contains a huge amount of data, where subtle changes of sound amplitude in time can be critical for human perception system, thus the data-driven timbre identification process demands lots of information to be captured and also demands to describe the patterns among those subtle changes. Since after the dimensional approach to timbre description was proposed in [3], there is no standard parameterization used as a classification basis. Researchers in the area have explored a number of statistical parameters to describe patterns and properties of spectrums of music sounds to distinguish different timbre, such as Tristimulus parameters [14], [6], and irregularity [22], etc.

MPEG-7 standard provides a set of low-level temporal and spectral sound features where some of them are in a form of vector or matrix of a large size. Flattening and summarizing these features for traditional classifiers intuitively increases the number of features but losses some potentially useful information. Therefore, in this paper, authors have proposed a new set of features, sufficient in musical timbre signatures and suitable in format for machine learning classifiers. Authors compare them against popular features in the literature.

For polyphonic sounds, different methods have been investigated by various researchers, such as Independent Component Analysis (ICA) ([8], [21]), Factorial Hidden Markov Models (HMM) ([12], [19]), and Harmonic Sources Separation Algorithms ([2], [25], [9], [26]). ICA requires multiple channels of different sound sources. Most often, HMM works well for sound sources separation where fundamental frequency range is small and the variation is subtle. Harmonic Sources Separation Algorithms can be used to isolate sound sources within a single channel, where efficient solution in one channel can be intuitively applied to other channels and therefore facilitates more types of sound recordings (e.g., mono-channel and stereo with two or more channels).

Our multi-hierarchical decision system is a database of about 1,000,000 musical instrument sounds, each one represented as a vector of approximately 1,100 features. Each instrument sound is labelled by a corresponding instrument. There are many ways to categorize music instruments, such as by playing methods, by instrument type, or by other generalization concepts [23]. Any categorization process is usually represented as a hierarchical schema which can be used by an automatic indexing system and a related cooperative Query Answering System (QAS) [7], [15], [17]. By definition, a cooperative QAS is relaxing a failing query with a goal to find its smallest generalization which does not fail. Two different hierarchical schemas [17] have been used as models of a decision attribute: Hornbostel-Sachs classification of musical instruments and classification of musical instruments by articulation. Each hierarchical classification represents a unique decision attribute, in a database of music instrument sounds, leading to a construction of a new classifier and the same to a different system for automatic indexing of music by instruments and their types [17], [28].

2 Audio Features in Our Research

In their previous work, authors implemented aggregation [28] to the MPEG7 spectral descriptors as well as other popular sound features. This section introduces new temporal features and other popular features used to describe sound objects which we implemented in MIRAI database of music instruments [http://www.mir.uncc.edu]. The spectrum features have two different frequency domains: Hz frequency and Mel frequency. Frame size was carefully designed to be 120ms, so that the 0th octave G (the lowest pitch in our audio database) can be detected. The hop size is 40ms with a overlapping of 80ms. Since the sample frequency of all the music objects is 44,100Hz, the frame size is 5,292. A hamming window is applied to all STFT transforms to avoid jittering in the spectrum.

2.1 Temporal Features Based on Pitch

Pitch trajectories of instruments behave very differently in time. The authors designed parameters to capture the power change in time.

Pitch Trajectory Centroid. PC is used to describe the center of gravity of the power of the fundamental frequency during the quasi-steady state.

$$(1) \quad PC = \frac{\sum_{n=1}^{length(P)} [\frac{n \cdot P(n)}{length(P)}]}{\sum_{n=1}^{length(P)} P(n)}$$

where P is the pitch trajectory in the quasi-steady state, n is the nth frame.

Pitch Trajectory Spread. PS is the RMS deviation of the pitch trajectory with respect to its gravity center.

$$(2) \quad PS = \sqrt{\frac{\sum_{n=1}^{length(P)} [(\frac{n}{length(P)} - PC)^2 \cdot P(n)]}{\sum_{n=1}^{length(P)} P(n)}}$$

Pitch Trajectory Max Angle. PM is an angle of the normalized power maximum vs. its normalized frame position along the trajectory in the quasi-steady state.

(3) $PM = \dfrac{[\dfrac{MAX(P(n))-P(0)}{\frac{1}{length(P)}\cdot\sum_{n=1}^{length(P)}P(n)}]}{[\frac{F(n)-F(0)}{length(P)}]}$

where $F(n)$ is the position of nth frame in the steady state.

Harmonic Peak Relation. HR is a vector describing the relationship among the harmonic partials.

(4) $HR = \frac{1}{m}\sum_{j=1}^{m}\frac{H_j}{H_0}$

where m is the total number of frames in the steady state, H_j is the jth harmonic peak in the ith frame.

2.2 Aggregation Features

MPEG7 descriptors can be categorized into two types: temporal and spectral. The authors applied aggregation among all the frames per music object for all the following instantaneous spectral features.

MPEG7 Spectrum Centroid [29] describes the center-of-gravity of a log-frequency power spectrum. It economically indicates the pre-dominant frequency range. Coefficients under 62.5Hz have been grouped together for fast computation.

MPEG7 Spectrum Spread is the root of mean square value of the deviation of the Log frequency power spectrum with respect to the gravity center in a frame [29]. Like spectrum centroid, it is an economic way to describe the shape of the power spectrum.

MPEG7 Harmonic Centroid is computed as the average over the sound segment duration of the instantaneous harmonic centroid within a frame [29].

The instantaneous harmonic spectral centroid is computed as the amplitude in linear scale weighted mean of the harmonic peak of the spectrum.

MPEG7 Harmonic Spread is computed as the average over the sound segment duration of the instantaneous harmonic spectral spread of frame [29].

The instantaneous harmonic spectral spread is computed as the amplitude weighted standard deviation of the harmonic peaks of the spectrum with respect to the instantaneous harmonic spectral centroid.

MPEG7 Harmonic Variation is defined as the mean value over the sound segment duration of the instantaneous harmonic spectral variation [29].

The instantaneous harmonic spectral variation is defined as the normalized correlation between the amplitude of the harmonic peaks of two adjacent frames.

MPEG7 Harmonic Deviation is computed as the average over the sound segment duration of the instantaneous harmonic spectral deviation in each frame.

The instantaneous harmonic spectral deviation is computed as the spectral deviation of the log amplitude components from a spectral envelope.

MPEG7 Harmonicity Rate is the proportion of harmonics in the power spectrum. It describes the degree of harmonicity of a frame. It is computed by the normalized correlation between the signal and a lagged representation of the signal.

MPEG7 Fundamental Frequency is the frequency that best explains the periodicity of a signal. The ANSI definition of psycho-acoustical terminology says that "pitch is an auditory attribute of a sound according to which sounds can be ordered on a scale from low to high".

MPEG7 Upper Limit of Harmonicity describes the frequency beyond which the spectrum cannot be considered harmonic. It is calculated based on the power spectrum of the original and a comb-filtered signal.

Tristimulus and similar parameters describe the ratio of the amplitude of a harmonic partial to the total harmonic partials [26]. They are first modified tristimulus parameter, power difference of the first and the second tristimulus parameter, grouped tristimulus of other harmonic partials, odd and even tristimulus parameters.

Brightness is calculated as the proportion of the weighted harmonic partials to the harmonic spectrum [10].

$$(4)\ B = \frac{\sum_{n=1}^{N}[n \cdot A_n]}{\sum_{n=1}^{N} A_n}$$

Transient, steady and decay duration. In this research, the transient duration is considered as the time to reach the quasi-steady state of fundamental frequency. At this duration the sound contains more timbre information than pitch information that is highly relevant to the fundamental frequency. Thus differentiated harmonic descriptors values in time are calculated based on the subtle change of the fundamental frequency [27].

Zero crossing counts the number of times that the signal sample data changes signs in a frame [20]

$$(5)\ ZC_j = 0.5 \sum_{n=1}^{N} |\ sign(s_j[n]) - sign(s_j[n-1])\ |$$

$$(6)\ sign(x) = [\text{if } x \geq 0 \text{ then } 1, \text{ else -1}]$$

where s_j is the nth sample in the jth frame, N is the frame size.

Spectrum Centroid describes the gravity center of the spectrum [24]

$$(7)\ C_j = \frac{\sum_{k=1}^{\frac{N}{2}} f(k) \cdot |X_j(k)|}{\sum_{k=1}^{\frac{N}{2}} |X_j(k)|}$$

where N is the total number of the FFT points, $X_j(k)$ is the power of the kth FFT point in the ith frame, $f(k)$ is the corresponding frequency of the FFT point.

Roll-off is a measure of spectral shape, which is used to distinguish between voiced and unvoiced speech [11]. The roll-off is defined as the frequency below which C percentage of the accumulated magnitudes of the spectrum is concentrated, where C is an empirical coefficient.

Flux is used to describe the spectral rate of change [18]. It is computed by the total difference between the magnitude of the FFT points in a frame and its successive frame.

(8) $F_j = \sum_{k=1}^{\frac{N}{2}} (\mid X_j(k) \mid - \mid X_{j-1}(k) \mid)^2$

2.3 Statistical Parameters

In order to flatten the matrix data to suitable format for the classifiers, statistical parameters (e.g., maximum, minimum, average, distance of similarity, standard deviation) are applied to the power of each spectral band.

MPEG7 Spectrum Flatness describes the flatness property of the power spectrum within a frequency bin, which is ranged by edges in the corresponding formula (see [29]). The value of each bin is treated as an attribute value in the database. Since the octave resolution in our research is $1/4$, the total number of bands is 32.

MPEG7 Spectrum Basis Functions are used to reduce the dimensionality by projecting the spectrum from high dimensional space to low dimensional space with compact salient statistical information (see [29]).

Mel Frequency Cepstral Coefficients describe the spectrum according to the human perception system in the Mel scale. They are computed by grouping the STFT points of each frame into a set of 40 coefficients by a set of 40 weighting curves with logarithmic transform and a discrete cosine transform (DCT).

2.4 MPEG7 Temporal Descriptors

The temporal descriptors in MPEG7 [29] have been applied directly into the feature database. **MPEG7 Spectral Centroid** is computed as the power weighted average of the frequency bins in the power spectrum of all frames in a sound segment with Welch method. **MPEG7 Log Attack Time** is defined as the logarithm of the time duration between the time when the signal starts to the time it reaches its stable part, where the signal envelope is estimated by computing the local mean square value of the signal amplitude in each frame. **MPEG7 Temporal Centroid** is calculated as the time average over the energy envelope.

3   Discriminant Analysis for Feature Selection

Logistic regression model is a popular statistical approach of analyzing multinomial response variables. It does not assume normally distributed conditional attributes which can be continuous, discrete, dichotomous or a mix of any of these; it can handle nonlinear relationships between the discrete responses and the explanatory attributes. It has been widely used to investigate the relationship between decision attribute and conditional attributes, using the most economical model. An ordinal response logit model has a form:

(9) $(\frac{Pr(Y=i|x)}{Pr(Y=k+1|x)}) = \alpha_i + \beta_i \cdot x$, $i = 1, 2, ..., k$

where the $k + 1$ possible responses have no natural ordering and $\alpha_1,..., \alpha_k$ are k intercept parameters, $\beta_1,..., \beta_k$ are k vectors of parameters, and Y is the response. For details, see [5]. The system fits a common slopes cumulative model which is a parallel lines regression model based on the cumulative probabilities of the response categories. The significance of an attribute is calculated with the likelihood ratio or chi-square difference test by the Fisher's Score algorithm. A final model is selected, where adding another variable would not improve the model significantly.

4   Experiments

The authors used a subset of their feature database [http://www.mir.uncc.edu] containing 1,569 music recording sound objects of 74 instruments. The authors discriminated instrument types on different levels of a classification tree. The tree consists of three levels: the top level (e.g., aerophone, chordophone, and idiophone), the second level (e.g., lip-vibrated, side, reed, composite, simple, rubbed, shaken, and struck), and the third level (e.g., piano, violin, and flute). All classifiers were 10-fold cross validation with a split of 90% training and 10% testing. We used WEKA for all classifications and SAS LOGISTIC procedure for discriminant analysis. In each experiment, a 99% confidence level was used. Feature extraction was implemented in .NET C++ with connection to MS SQL Server. In LISP notation, we used the following *Music Instrument Classification Tree*:

(Instrument(Aerophone(Lip-vibrated (-,-,-), Side(-,-), Reed(-,-)), Chordophone(Composite, Simple), Idiophone(Rubbed(-), Shaken(-,-), Struck(-,-))))

For classification on the first level in the music instrument family tree, the selected feature set was stored in List I: {PeakRelation8, PeakRelation16, PeakRelation24, MPEGFundFreq, MPEGHarmonicRate, MPEGULHarmonicity, MPEGHarmoVariation, MPEGHarmoDeviation, MPEGFlat3, MPEGFlat8, MPEGFlat18, MPEGFlat30, MPEGFlat36, MPEGFlat46, MPEGFlat55, MPEGFlat56, MPEGFlat66, MPEGFlat67, MPEGFlat76, MPEGFlat77, MPEGFlat83, MPEGFlat85, MPEGFlat94, MPEGFlat96, MPEGSpectrum Centroid, MPEGTC, MPEGBasis59, MPEGBasis200, TristimulusRest, Zero-Crossing, MFCCMaxBand1, MFCCMaxBand3, MFCCMaxBand5, MFCCMax Band6, MFCCMaxBand7, MFCCMaxBand8, MFCCMaxBand10, MFCCMax Band13, MFCCMinBand1, MFCCMinBand13, PitchSpread, MaxAngle}. Experiment was also performed on the rest of features after List I was removed from the whole feature set, which was stored in List II. In the table below, "All" stands for all the attributes used for classifier construction.

Table 1. Results of three groups of features at the top level of the music family tree

Class	Precision			Recall		
	List I	All	List II	List I	All	List II
Idiophone	87.00%	91.10%	95.10%	82.10%	91.40%	94.80%
Chordophone	86.80%	91.30%	88.50%	88.60%	88.90%	84.70%
Aerophone	91.50%	91.30%	87.30%	91.80%	93.50%	90.90%

Table 1 shows the precisions of the classifiers constructed with selected features at the family level. After the less significant features, elected by the logistic model, have been removed, the group of List I slightly improved the precision for aerophone instruments. However, the selected significant feature group (List I) significantly outperformed in precision for aerophone instruments and in recall for both chordophone and aerophone instruments.

For classification at the second level in the music instrument family tree, the selected feature set was stored in List I: {PeakRelation8, PeakRelation16, PeakRelation30, MPEGFundFreq, MPEGHarmonicRate, MPEGULHarmonicity, MPEGHarmoDeviation, MPEGFlat3, MPEGFlat11, MPEGFlat14, MPEGFlat18, MPEGFlat22, MPEGFlat26, MPEGFlat36, MPEGFlat44, MPEGFlat46, MPEGFlat58, MPEGFlat67, MPEGFlat81, MPEGFlat82, MPEGFlat83, MPEGFlat85, MPEGFlat93, MPEGFlat94, MPEGFlat95, MPEGSpectrumCentroid, MPEGTC, MPEGBasis50, MPEGBasis57, MPEGBasis59, MPEGBasis69, MPEGBasis73, MPEGBasis116, MPEGBasis167, MPEGBasis206, Tristimulus1, TristimulusRest, TristimulusBright, ZeroCrossing, SpectrumCentroid2, RollOff, MFCCMaxBand1, MFCCMaxBand3, MFCCMaxBand4, MFCCMaxBand6, MFCCMaxBand7, MFCCMaxBand9, MFCCMinBand2, MFCCMinBand5, MFCCMinBand10, MFCCMinBand13, MFCCAvgBand10, MFCCAvgBand11, PitchSpread, MaxAngle}. Experiment was also performed on List II obtained by removing List I from the whole feature set.

Table 2 shows the precisions of the classifiers constructed with the selected features, all features, and the rest of the features after selection at the second level of the instrument family tree. After the less significant features, elected by

Table 2. Results of three groups of features at the second level of the music family tree

Class	Precision			Recall		
	List I	All	List II	List I	All	List II
Lip − Vibrated	83.80%	84.40%	77.30%	84.70%	88.80%	82.30%
Side	74.30%	73.20%	66.40%	75.70%	64.00%	64.00%
Reed	77.10%	78.30%	70.50%	78.40%	80.10%	70.50%
Composite	84.50%	86.20%	84.90%	86.70%	84.30%	83.90%
Simple	71.20%	74.10%	72.20%	67.20%	80.00%	72.80%
Rubbed	85.30%	82.10%	75.00%	78.40%	86.50%	73.00%
Shaken	79.20%	91.00%	89.50%	64.80%	92.00%	87.50%
Struck	78.20%	86.30%	85.40%	80.40%	79.00%	77.60%

Table 3. Results of three groups of features at the third level of the music family tree

	Precision			Recall		
Class	List I	All	List II	List I	All	List II
Flute	92.90%	67.70%	70.40%	89.70%	72.40%	65.50%
Tubular Bells	86.70%	60.00%	52.40%	72.20%	66.70%	61.10%
Tuba	85.70%	81.80%	85.70%	90.00%	90.00%	90.00%
Electric Bass	83.10%	87.50%	89.10%	80.60%	83.60%	85.10%
Trombone	80.60%	80.60%	76.30%	69.20%	82.10%	74.40%
Marimba	79.20%	89.50%	90.00%	71.40%	89.50%	86.50%
Piano	78.50%	82.40%	83.00%	81.60%	78.40%	74.40%
French Horn	78.00%	83.70%	82.90%	87.70%	88.90%	84.00%
Bass Flute	77.40%	75.50%	71.20%	68.30%	61.70%	61.70%
Alto Flute	76.70%	82.80%	78.60%	79.30%	82.80%	75.90%
Double Bass	75.40%	60.80%	60.00%	75.40%	54.40%	52.60%
Piccolo	74.50%	69.20%	62.00%	71.70%	67.90%	58.50%
CTrumpet	72.00%	68.90%	69.10%	83.10%	78.50%	72.30%
Violin	71.00%	75.00%	77.10%	78.00%	72.70%	76.50%
Oboe	70.30%	71.00%	35.90%	81.30%	68.80%	43.80%
Vibraphone	69.30%	91.40%	85.70%	73.20%	90.10%	93.00%
Bassoon	68.80%	66.70%	45.50%	61.10%	55.60%	27.80%
Cello	67.00%	63.20%	63.50%	61.50%	62.50%	68.80%
Saxophone	66.70%	51.70%	53.60%	46.70%	50.00%	50.00%

the logistic model, have been removed, the group of List I improved the precision for side and rubbed instruments and recall for the side, composite, and struck instruments. Also, the selected significant feature group (List I) significantly outperformed in precision for lip-vibrated, side, reed, and rubbed instruments and in recall for all the types except for simple and shaken instruments.

For classification at the third level in the music instrument family tree, the selected feature set was stored in List I: {MPEGTristimulusOdd, MPEG-FundFreq, MPEGULHarmonicity, MPEGHarmoVariation, MPEGFlatness6, MPEGFlatness14, MPEGFlatness27, MPEGFlatness35, MPEGFlatness43, MPEGFlatness52, MPEGFlatness63, MPEGFlatness65, MPEGFlatness66, MPEGFlatness75, MPEGFlatness76, MPEGFlatness79, MPEGFlatness90, MPEGFlatness91, MPEGSpectrumCentroid, MPEGSpectrumSpread, MPEG-Basis41, MPEGBasis42, MPEGBasis69, MPEGBasis87, MPEGBasis138, MPEGBasis157, MPEGBasis160, MPEGBasis170, MPEGBasis195, Tristim-ulusBright, TristimulusEven, TristimulusMaxFd, ZeroCrossing, Spectrum-Centroid2, Flux, MFCCMaxBand2, MFCCMaxBand3, MFCCMaxBand6, MFCCMaxBand7, MFCCMaxBand9, MFCCMaxBand10, MFCCMinBand1, MFCCMinBand2, MFCCMinBand3, MFCCMinBand6, MFCCMinBand7, MFCCMinBand10, MFCCAvgBand1, MFCCAvgBand12, SteadyEnd, Length}. Experiment was also performed on List II obtained by removing List I from the whole feature set.

Table 3 shows statistics of the precision of the classifiers constructed with the selected features, all features, and the rest of the features for some instruments used in the experiment. The overall accuracy of all the features was slightly better than that of the selected features. The computing time for List I, All, and List II is 7.33, 61.59, and 54.31 seconds respectively.

5   Conclusion and Future Work

A large number of attributes is generated in a table during fattening the features into a single value attributes for classical classifiers by statistical and other feature design methods. Some of the derived attributes may not significantly contribute to the classification models, or sometimes may distract the classification. In the light of the results from the experiments, we conclude that attributes have different degree of influence on the classification performance for different instrument families. The new temporal features related to harmonic peaks significantly improved the classification performance when added into the database with all other features. However, the new features were not suitable to replace the MPEG7 harmonic peak related features and Tristimulus parameters as the logistic studies shows. We also noticed that classifications at a higher level of granularity tended to use more features for correct prediction than those at the lower level. This may especially benefit a cooperative query answering system to choose suitable features for classifiers at different levels.

Acknowledgements

This research was supported by the National Science Foundation under grant IIS-0414815.

References

1. Balzano, G.J.: What are musical pitch and timbre? Music Perception, an interdisciplinary Journal 3, 297–314 (1986)
2. Bay, M., Beauchamp, J.W.: Harmonic source separation using prestored spectra. In: Rosca, J.P., Erdogmus, D., Príncipe, J.C., Haykin, S. (eds.) ICA 2006. LNCS, vol. 3889, pp. 561–568. Springer, Heidelberg (2006)
3. Bregman, A.S.: Auditory scene analysis, the perceptual organization of sound. MIT Press, Cambridge (1990)
4. Cadoz, C.: Timbre et causalite, unpublished paper, Seminar on Timbre, Institute de Recherche et Coordination Acoustique/Musique, Paris, France (April 13-17, 1985)
5. Cessie, S., Houwelingen, J.C.: Ridge Estimators in Logistic Regression. Applied Statistics 41(1), 191–201 (1992)
6. Fujinaga, I., McMillan, K.: Real time recognition of orchestral instruments. In: Proceedings of the International Computer Music Conference, pp. 141–143 (2000)
7. Gaasterland, T.: Cooperative answering through controlled query relaxation. IEEE Expert 12(5), 48–59 (1997)

8. Kinoshita, T., Sakai, S., Tanaka, H.: Musical sound source identification based on frequency component adaptation. In: Proceedings of IJCAI Workshop on Computational Auditory Scene Analysis (IJCAI-CASA 1999), Stockholm, Sweden, pp. 18–24 (July-August 1999)

9. Kitahara, T., Goto, M., Komatani, K., Ogata, T., Okuno, H.G.: Instrument identification in polyphonic music: feature weighting to minimize influence of sound overlaps. EURASIP Journal on Advances in Signal Processing 1, 155–155 (2007)

10. Lewis, R., Zhang, X., Ras, Z.W.: Knowledge discovery based identification of musical pitches and instruments in polyphonic sounds. Journal of Engineering Applications of Artificial Intelligence 20(5), 637–645 (2007)

11. Lindsay, A.T., Herre, J.: MPEG-7 and MPEG-7 Audio-An Overview. J. Audio Eng. Soc. 49, 589–594 (2001)

12. Ozerov, A., Philippe, P., Gribonval, R., Bimbot, F.: One microphone singing voice separation using source adapted models. In: Proc. IEEE Workshop on Applications of Signal Processing to Audio and Acoustics (WASPAA), pp. 90–93 (2005)

13. Pawlak, Z.: Information systems - theoretical foundations. Information Systems Journal 6, 205–218 (1991)

14. Pollard, H.F., Jansson, E.V.: A tristimulus Method for the specification of Musical Timbre. Acustica 51, 162–171 (1982)

15. Ras, Z.W., Dardzińska, A.: Solving Failing Queries through Cooperation and Collaboration. World Wide Web Journal, Special Issue on Web Resources Access 9(2), 173–186 (2006)

16. Ras, Z.W., Dardzińska, A., Zhang, X.: Cooperative Answering of Queries based on Hierarchical Decision Attributes. CAMES Journal, Polish Academy of Sciences, Institute of Fundamental Technological Research 14(4), 729–736 (2007)

17. Ras, Z.W., Zhang, X., Lewis, R.: MIRAI: Multi-hierarchical, FS-tree based Music Information Retrieval System. In: Kryszkiewicz, M., Peters, J.F., Rybinski, H., Skowron, A. (eds.) RSEISP 2007. LNCS (LNAI), vol. 4585, pp. 80–89. Springer, Heidelberg (2007)

18. Scheirer, E., Slaney, M.: Construction and Evaluation of a Robust Multi-feature Speech/Music Discriminator. In: Proc. IEEE Int. Conf. on Acoustics, Speech and Signal Processing (ICASSP) (1997)

19. Smith, J.O., Serra, X.: PARSHL: An Analysis/Synthesis Program for Non Harmonic Sounds Based on a Sinusoidal Representation. In: Proc. Int. Computer Music Conf., Urbana-Champaign, Illinois, pp. 290–297 (1987)

20. Tzanetakis, G., Cook, P.: Musical Genre Classification of Audio Signals. IEEE Trans. Speech and Audio Processing 10, 293–302 (2002)

21. Vincent, E.: Musical source separation using time-frequency source priors. IEEE Transactions on Audio, Speech and Language Processing 14(1), 91–98 (2006)

22. Wieczorkowska, A.: Classification of musical instrument sounds using decision trees. In: Proceedings of the 8th International Symposium on Sound Engineering and Mastering, ISSE 1999, pp. 225–230 (1999)

23. Wieczorkowska, A., Ras, Z., Zhang, X., Lewis, R.: Multi-way Hierarchic Classification of Musical Instrument Sounds. In: Proceedings of the International Conference on Multimedia and Ubiquitous Engineering (MUE 2007), Seoul, South Korea, pp. 897–902. IEEE Computer Society, Los Alamitos (2007)

24. Wold, E., Blum, T., Keislar, D., Wheaton, J.: Content-Based Classification, Search and Retrieval of Audio. IEEE Multimedia, Fall, 27–36 (1996)

25. Zhang, X., Marasek, K., Ras, Z.W.: Maximum likelihood study for sound pattern separation and recognition. In: Proceedings of the IEEE CS International Conference on Multimedia and Ubiquitous Engineering (MUE 2007), Seoul, Korea, April 26-28, 2007, pp. 807–812 (2007)
26. Zhang, X., Ras, Z.W.: Sound isolation by harmonic peak partition for music instrument recognition, in the Special Issue on Knowledge Discovery. Fundamenta Informaticae Journal 78(4), 613–628 (2007)
27. Zhang, X., Ras, Z.W.: Differentiated Harmonic Feature Analysis on Music Information Retrieval For Instrument Recognition. In: Proceeding of IEEE International Conference on Granular Computing, Atlanta, Georgia, May 10-12, 2006, pp. 578–581 (2006)
28. Zhang, X., Ras, Z.W.: Analysis of sound features for music timbre recognition. In: Proceedings of the IEEE CS International Conference on Multimedia and Ubiquitous Engineering (MUE 2007), Seoul, Korea, April 26-28, 2007, pp. 3–8 (2007)
29. ISO/IEC JTC1/SC29/WG11, MPEG-7 Overview (2002),
    http://mpeg.telecomitalialab.com/standards/mpeg-7/mpeg-7.htm

Contextual Adaptive Clustering of Web and Text Documents with Personalization

Krzysztof Ciesielski[1], Mieczysław A. Kłopotek[1,3],
and Sławomir T. Wierzchoń[1,2]

[1] Institute of Computer Science, Polish Academy of Sciences,
ul. Ordona 21, 01-237 Warszawa, Poland
[2] Institute of Informatics, Univ. of Gdansk, Wita Stwosza 57, 80-952 Gdansk
[3] Institute of Informatics, Univ. of Podlasie in Siedlce
{kciesiel,klopotek,stw}@ipipan.waw.pl

Abstract. We present a new method of modeling of cluster structure of a document collection and outline an approach to integrate additional knowledge we have about the document collection like prior categorization of some documents or user defined / deduced preferences in the process of personalized document map creation.

1 Introduction

Web document clustering, especially in large and heterogeneous collections, is a challenging task, both in terms of time and space complexity as well as resulting clustering quality. But the most challenging aspect is the way how the clustering information is conveyed to the end user and how it meets his expectations (so-called personalization).

From the point of view of humans dealing with a given documents collection, each document is rather a complex information structure. Further, a computer system fully understanding the document contents is beyond technological possibilities. Therefore some kinds of "approximation" to the content are done. When processed, documents are treated as "bags of words" or as points in term-document vector space.

To get a deeper insight into the documents content and their mutual relationships, more complex representations are investigated. Particularly, a new form of cluster description – a visual document map representation has been proposed and developed [13,14,16,8]. In a two-dimensional space, divided into quadratic or hexagonal cells, the split into clusters is represented as an assignment of documents to cells in such a way, that documents assigned to cells are as homogenous as possible and cells (clusters) containing similar documents are placed close to one another on the map, and map regions are labeled with best-fitting terms from the documents. An inversion of the clustering (that is clustering of terms instead of documents) is also possible.

It is generally believed that individual information needs of users may differ and there is a general feeling that therefore also the data processing results

Z.W. Raś, S. Tsumoto, and D. Zighed (Eds.): MCD 2007, LNAI 4944, pp. 116–130, 2008.

should accommodate to the profile of a specific user. Countless methods and ways of user profile representation and acquisition have been designed so far.

The problem with map-like representation of document collections, however, relies upon expensive processing / high complexity (in terms of time and space), so that a personalized *ad-hoc* representation is virtually impossible.

In this paper we contest this view and claim that personalization of map representation of large scale document collections is possible by a careful separation of the concept of individual needs from the concept of common knowledge. This leads to the possibility of separation of computationally intense tasks of identification of the structure of clustering space from the relatively less resource consuming pure presentation part.

The paper is organized as follows: In section 2 the concept of cluster space is introduced. Section 3 describes practical ways of cluster space approximation from data. Section 4 outlines possible ways of personalizing the maps presentation. The experimental section 5 presents some evidence justifying utility of the idea of contextual clustering for practical purposes[1]. Last section, 6, summarizes the paper.

2   Clustering Space

It is usually assumed that personalization is needed because of cultural, ethnical etc. differences that influence the "world of values", the "attitudes" and the "views".[2]

So, in the particular case of clustering, the differences (or dissimilarities) between the objects may change from person to person, and – as a consequence – personalization reduces to a total re-clustering of the objects.

Human beings possess to, a large extent, an "objectivised" world perception which can be characterized by a common set of concepts (a vocabulary) that is intended for them to communicate to other human beings. The vast majority of concepts is shared and their meaning not determined by "values", "attitudes" etc. What differs the human beings, is the current needs they are focused on. So, if discussing e.g. an issue in biology, one does not care about concepts important for chemical engineering. Hence, not the personal attitude, but rather the context in which an issue is discussed impacts the feeling of dissimilarity of "opinions" (documents in our case).

---

[1] The paper is primarily concerned with outlining the general concept, while the Reader may also refer to our earlier papers (e.g. [5,3,12]) for experimental results supporting validity of partial clams from which the current exposition is derived.

[2] This idea is particularly believed in the commercial world. See e.g. D. Gibson: New Directions in e-Learning: Personalization, Simulation and Program Assessment. `ali.apple.com/ali_media/Users/1000507/files/others/New_Directions_in_elearning.doc` or InterSight Technologies, Inc. information: iMatterTM Suite - Customer Intelligence: Attitude and Behavior `www.intersighttechnologies.com/behavior/customer-behavior.html`

Therefore, in the sequel we will assume that the proper approach to personalization has to be well founded on the proper representation of document information.

To keep the representation simple and manageable, we assume that documents are treated as bags of words (without bothering about words ordering and sentence structure). They are represented in the space of documents (with dimensions being spread by words and phrases, that is terms) as the points with coordinates being a function of the frequency of terms occurring in these documents. Furthermore, it has been early recognized, that some terms should be weighed more than other [17], because of their varying importance. The similarity between two documents is measured as a cosine of the angle[3] between the vectors drawn from the origin of the coordinate system to these points. It turns out, that this vision of document similarity works quite well in practice, agreeing with human view of text similarity. Furthermore, dimensionality reduction may be of high importance [6] It is generally agreed that dropping the non-important terms will not lead to any loss of information from human point of view, so that one can ignore the respective dimensions, reducing frequently the computational burden significantly, but also removing some "noise" in the data.

The weight $w_{t,d}$ of a term t in the document d may be, among others, calculated as the so-called *tfidf* (term frequency times the inverse document frequency) index:

$$w_{t,d} = f_{t,d} \times \log \frac{|D|}{f_D^{(t)}} \qquad (1)$$

where $f_{t,d}$ is the number of occurrences of term t in document d, $|D|$ is the cardinality of the set of documents, and $f_D^{(t)}$ is the number of documents in collection D containing at least one occurrence of term t (see also [15]).

Each document d is represented by a vector $d = (w_{t_1}, \ldots, w_{t_{|T|}})$, where T is the set of terms. Usually, this vector is normalized, i.e. it is replaced by the vector $d' = (w'_{t_1}, \ldots, w'_{t_{|T|}})$ of unit length. In such a case d' can be viewed as a point of the unit hyper-sphere.

2.1   From Clusters to a Continuous Clustering Space

On the traditional *tfidf*, the weights of terms in a document are influenced by their distribution in the entire document collection. But one would calculate the weights by the very same method, but within any reasonable (homogenous) cluster, the results would be usually different. This fact seems to be ignored by most researchers.

So our first methodological step is to relax the rigid term weighing scheme (called hereafter *global* weighting scheme) by introducing flexible, i.e. *local*

---

[3] Let us draw the attention to the obvious fact that though cosine ranges from -1 to 1, it is never negative for traditional document representation, as term weights are always chosen as non-negative values, hence the angle between documents never exceeds 90°.

scheme [5]. We want to enable local differentiation of term importance within identified clusters (i.e. local context).

But we do not want to replace global term weighing scheme with a local weighing scheme. Rather than this, we consider the impact of the documents that are far away from cluster core, on the term weighing, as compared to the close ones. Another important point is that the terms specific for a given cluster should weight more than terms not specific for any cluster. Last not least, we replace the notion of a cluster to the concept of clustering space. Each document p, i.e. a point in unit hyper-sphere, can be treated as cluster center in a ("continuous") clustering space. We can then define, for each document d, a membership function $m_{d,C(p)}$ in the style of fuzzy set membership function [2], for example as[4]

$$m_{d,C(p)} = \sum_{t \in T} w'_t(d') \cdot w'_t(p) \tag{2}$$

that is the dot product of coordinates of the normalized document d' and the point p on the unit hyper-sphere. In this way we define a continuous function over the unit hyper-sphere HS, which is finite everywhere. Therefore, there exists in particular the integral $M_d = \int_{HS^+} m_{d,C(p)} dp$, where HS^+ is the "positive quarter" (with nonnegative coordinates) of the unit hyper-sphere[5]. By restricting ourselves to the integration over HS^+ we obtain values that are generally lower for documents lying in the "corners" of HS^+ (with a couple of distinguishing significant terms), whereas those in the "middle" (that is containing only many terms with uniformly low significance) will have higher value of the HS^+ integral. Subsequently we will use the integral as a divisor, so that documents with significant words would be promoted.

Given this, we can define the specificity $s_{t,C}$ of a term t in a cluster $C(p)$ as

$$s_{t,C(p)} = |C(p)| \cdot \frac{\sum_{d \in D} \left(f_{t,d} \cdot m_{d,C(p)} \right)}{f_{t,D} \cdot \sum_{d \in D} m_{d,C(p)}} \tag{3}$$

where $f_{t,d}$ is (as earlier) the number of occurrences of term t in document d, $f_{t,D}$ is the number of occurrences of term t in document collection D, and $|C(p)|$ is the "fuzzy cardinality" ("density") of documents at point p, defined as

$$|C(p)| = \sum_{d \in D} \mu_{d,C(p)} \tag{4}$$

where $\mu_{d,C}$ is the normalized membership:

$$\mu_{d,C(p)} = \frac{m_{d,C(p)}}{\int_{HS^+} m_{d,C(p)} dp}$$

---

[4] $m_{d,C(p)}$ tends to decrease with the number of dimensions, but this has no serious impact as it is later subject to normalization.

[5] The reason, why we do not consider HS, but restrict ourselves to HS^+ is that $\int_{HS} m_{d,C(p)} dp$ is independent of the document d: because of the symmetry of the unit hyper-sphere it is equal to zero. As already said, documents (in most term-weighing systems, see [7]) lie in HS^+ and therefore only this part of the unit hyper-sphere is of interest.

In this way we arrive at a new (contextual [5]) term weighing formula for term t in the document d from the point of view of the "local context" $C(p)$

$$w_{t,d,C(p)} = s_{t,C(p)} \times f_{t,d} \times \log \frac{|C(p)|}{f_{C(p)}^{(t)}} \tag{5}$$

where $f_{C(p)}^{(t)}$ is the fuzzy count of documents in collection $C(p)$ containing at least one occurrence of term t,

$$f_{C(p)}^{(t)} = \sum_{\{d:f_{t,d}>0\}} m_{d,C(p)} \tag{6}$$

For consistency, we assume that $w_{t,d,C(p)} = 0$ if $f_{C(p)}^{(t)} = 0$.

The universal weight $tfidf$ given by equation (1) will be replaced by the concept of an "averaged" local weight

$$w_{t,d} = \frac{\int_{HS^+} m_{d,C(p)} \cdot w_{t,d,C(p)} dp}{\int_{HS^+} m_{d,C(p)} dp} \tag{7}$$

Note that the definition of term weights $w_{t,d}$ becomes recursive in this way ($m_{d,C(p)}$ is used here, which is computed in the equation (2) based on $w_{t,d}$ itself) and the fixpoint of this recursion is the intended meaning of term weight. While we do not present here a mathematical proof of the existence of the fixpoint, we show in the experimental section, that the iterative process converges, and it converges to a useful state.

If we look at the process of creation of a document, then we see that the same content could have been expressed in a slightly different way. So, one can view the collection of documents as a sample generated from a complex (mixed) probability distribution. Hence we can think that any point in HS^+ has a term importance just as an instantiation of a random process with an underlying mixture of distributions. And these distributions we want to consider as an underlying feature of the clustering process, so that clusters are sets of points in HS with "similar" term importance distribution.

2.2   Histogram Characterization of a Context

Let us analyze the way how typical hierarchical (or other multistage) algorithms handle lower level clusters. The cluster is viewed as a kind of "averaged" document, eventually annotated with standard deviation of term frequencies and/or term weights. In our opinion, the distribution (approximated in our approach by a discrete histogram [4]) of the term weight (treated as a random variable) reflects much better the linguistic nature of data than hyperspheres around some center. Instead it creates clusters of documents with terms used in a similar way. This was confirmed by our reclassification experiments [4], showing higher stability of histogram-based cluster description versus centroid-based representation[6].

---

[6] Reclassification measure evaluates consistency of the model-derived clustering with the histogram-based clustering space description (cf. [4]).

So for any point p in the clustering space HS^+ and any term t we define a term-weight distribution as one approximated by the histogram in the following manner: Let $\Delta(w',t)$ be a discretization of the normalized weights for the term t assigning a weight for a term the integer identifier of the interval it belongs to (higher interval identifiers denote higher weights). Let $\chi(d,t,q,p)$ be the characteristic function of the term t in the document d and the discretization interval identifier q at point p, equal to $m_{d,C(p)}$ if $q = \Delta(w'_{t,d,C(p)},t)$, and equal zero otherwise. Then the histogram $h(t,p,q)$ is defined as

$$h(t,p,q) = \sum_{d \in D} \chi(d,t,q,p) \tag{8}$$

With h' we denote a histogram normalized in such a way that the sum over all intervals q for a given t and p is equal 1:

$$h'(t,p,q) = \frac{h(t,p,q)}{\sum_q h(t,p,q)} \tag{9}$$

For a more detailed exposition of the concept of histograms see [4].

We can easily come to the conclusion, when looking at typical term histograms that terms significant for a cluster would be ones that do not occur too frequently nor too rarely, have diversified range of values and have many non-zero intervals, especially with high indices.

Hence the significance of term t for the clustering point p may be defined as

$$m_{t,C(p)} = \frac{\sum_q [q \cdot log\,(h'(t,p,q))]}{Q_t} \tag{10}$$

where Q_t is the number of intervals for the term t under discretization.

Let us denote with $H(t,p,q)$ the "right cumulative" histograms, that is $H(t,p,q) = \sum_{k \geq q} h(t,p,k)$. The "right cumulative" histograms are deemed to reflect the idea, that terms with more weight should be "more visible". For technical reasons H' is a histogram normalized in the same way as h'.

Let us measure the dissimilarity between clustering points p_i, p_j with respect to term t as

$$Hell_k(p_i,p_j,t) = \sqrt{\sum_q \left(H'(t,p_i,q)^{(1/k)} - H'(t,p_j,q)^{(1/k)}\right)^k} \tag{11}$$

known as Hellinger divergence, and also as Hellinger-Matsushita-Bhattacharya divergence, [1]. Hellinger divergence, measuring dissimilarity between probability distributions, is well-studied in the literature and frequently used, whenever distributions are to be compared. $Hell_2$ is known to behave like a proper distance.

Finally let us measure the dissimilarity between clustering points p_i, p_j by "averaging over terms" as

$$dst(p_i,p_j) = \frac{\sum_{t \in T} m_{t,C(p_i),C(p_j)} \cdot Hell_k(p_i,p_j,t)}{\sum_{t \in T} m_{t,C(p_i),C(p_j)}} \tag{12}$$

where

$$m_{t,C(p_i),C(p_j)} = \sqrt{(m_{t,C(p_i} + 1) \cdot (m_{t,,C(p_j)} + 1)} - 1$$

With this definition, we can speak of a general notion of a cluster as "islands" in the clustering space such that the divergence within them differs insignificantly, and there exist at least n documents belonging predominantly to such an island. Thus, it can be treated as a dissimilarity measure.

It may be easily deduced that equation (10) gives also interesting possibilities of labeling of cluster space with meaningful sets of terms (concepts).

2.3   User-Related Sources of Information

Let us now turn to the user related information. Some documents may be pre-labeled by the user (with category, liking), there may be past queries available.

Note that the contextual document space, as described in the previous section, may be viewed as a "pure space" with some "material objects" causing a kind of curvature of this space.

The user-related sources can be viewed as consisting of two types of "documents": "material objects" (all the positively perceived, relevant information) and the "anti-material objects" (all the negatively perceived information).

The user-related documents may be also represented in a clustering space, in at least two different ways:

- In separate user-material, user-anti-material and proper document clustering spaces - in this case a "superposition" of these spaces would serve as an additional labeling of the proper document space, beside the original labels derived from document collection content.
- In a joint space – in this case user-related information will transform the document space of the document collection.

While the second approach may be considered as a stronger personalization, it will be more resource consuming and raises the issue of pondering the impact of user related documents against the entire collection, and also that of the relation between positive and negative user information. The first approach will be for sure much less resource consuming, because the processing of the big entire document collection has to be done only once, and the user related information is usually of marginal size and can be processed in a speedy way.

3   Clustering Space Approximation

In the previous section we equipped the clustering space with continuously changing descriptors in terms of term importance distributions. Under this provision the traditional clusters may be viewed as an element of discrete approximations of the clustering space. To represent the continuum more realistically, relations between clusters have to be taken into account.

The approximation makes legitimate ignoring the usually wide areas of clustering space of next to zero proximity to the documents as well as ignoring those terms that within the given subspace are of marginal significance. This is reasonable as each document modifies the space close to it having marginal impact of the rest. So the space may be greedy subdivided into non-empty subspaces that are deemed to be linked if they adhere to one another, and not, if they are of next to zero similarity.

The process, that we apply to approximate the clustering space [12], which we call Adaptive Clustering Algorithm, starts with splitting of the document collection into a set of roughly equally sized sub-collections using the expression (1) as an approximation of term weights for document similarity computation in a traditional clustering algorithm. We work in a hierarchical divisive mode, using the algorithm to split the collection in a small number of subcollections and apply further splitting to sub-collections of too big size. At the end too small clusters are merged with most similar ones. As a next iteration for each sub-collection, being now treated as a context (as it is now feasible), an iterative recomputation of term weights according to equation (7) with respect to cluster center, making the simplifying assumption that documents from other contexts have no impact. Within each context, the dictionaries of terms are reduced removing insignificant terms in a given context (different terms may be "zeroed" in different contexts). Subsequently the inter-document structure is formed. For this purpose one of the known networking clustering algorithms is used, either the growing neural gas (GNG) [9] or idiotypic (artificial immune) network (aiNet) [18,3]. Finally we turn back to the global set of contexts and apply a networking clustering algorithm to representatives of each context. This time, the histograms of contexts are applied to compute a measure of similarity between contexts – see equation (12). While applying the networking clustering, we additionally compute so-called "major topics", that is a split of the (sub)collection into up to 6 sub-clusters, the representatives of which are deemed to be major topics of the collection.

In this way, an approximation of the clustering space is obtained. In case of visualization, the WebSOM algorithm is applied to context representatives, in case one wants to view the global map, and to neural gas cells, or immune network cells in case of detailed view of a context. The computation of the map given the clustering space model is drastically simplified because e.g. with a collection of 12,000,000 documents we need to cluster only 400 representatives. So given such a cluster network, its projection onto a flat rigid document map structure, with treating each whole cluster as a single "document", is a dramatically simpler task than the map creation process for individual documents.

Our implementation of WebSOM differs from the original one in a number of ways, accelerating the processing significantly. One of the features is the topic-sensitive initialization. While WebSOM assigns random initial cluster centers for map cells, we distribute evenly the vectors of major topics over the map and initialize the remaining cells with in-between values (with slight noise). In this way the maps are learned usually quicker and are more stable (no drastic changes from projection to projection).

We have demonstrated in our earlier work [3,4,5,12] that such an approach to document space modeling is stable, scalable and can be run in an incremental manner.

3.1   Exploiting User-Related Sources

With this background we can explain our approach to personalization. We treat the document collection as a piece of knowledge that is esteemed by any user in the same way. So the identified clusters and the identified interrelationships between them are objective, independent of the user. The user at a given moment may be, however, interested to view the collection from a different direction. So the personalization may be reduced to the act of projection of the cluster network onto the flat map, that is, contrary to projection of document collection, a speedy process, to be managed within seconds. In this process, we can proceed in two distinct ways:

- Instead of using the topical vectors of a context / global collection, the user profile topical vector is applied, or
- The user related "documents" are attached to the collection clusters prior to projection (and may or may not influence the projection process) and serve as a source of additional labeling.

3.2   Another View of the Adaptive Clustering Algorithm

Our incremental textual data clustering algorithm relies on merging two known paradigms of clustering: the fuzzy clustering and the subspace clustering. The method differs essentially from Fuzzy C-Means in that it is designed solely for text data and is based on contextual vector representation and histogram-based description of vector subspaces.

Like Fuzzy-C-Means, we start with an initial split into subgroups, represented by a matrix $U(\tau_0)$, rows of which represent documents, and columns representing groups, they are assigned to. Iteratively, we adapt (a) the document representation, (b) the histogram description of contextual groups, (c) membership degree of documents and term significance in the individual groups.

These modifications can be viewed as a recursive relationship leading to a precise description of a contextual subspace in terms of the membership degree of documents and significance of terms in a context and on the other hand improving the understanding of document similarity.

So we can start without any knowledge of document similarity, via a random assignment of documents to a number of groups and global term weighing. But through the iterative process some terms specific for a group would be strengthened, so that class membership of documents would be modified, hence also their vector representation and indirectly similarity definition.

So we can view the algorithm as a kind of reinforcement learning. The usage of histogram approach makes this method incremental.

4   Personalization

The outlined approach to document map oriented clustering enables personalization among others along the following lines:

- Personalized topic -oriented initialization of map like visualization of the selected document space model (also rebuilding of a component model is possible, treating the user profile as a modifier of term weights of all documents)
- Personalized identification of key words, document space / map cell labeling, query expansion
- Document recommendation based on document membership degree in client profile context
- Recommendation of map cells
- Recommendation of other users (measuring the histogram distances between user profiles)
- Clustering of users as well as users and contexts

Present day search engines are characterized by a static information model, that is textual data bases are updated in a heavily discontinuous way which results in abrupt changes of query results (after each cycle of indexing new documents). Also the data organization and search model does not take into account the user profile information for the given document base and the given user query. Hence the reply is frequently identical, independent of the user.

The experimental search engine BEATCA [12] exhibits several capabilities that can become a starting point for a radical change of this situation.

- Reduced processing time, scalability of the adaptive contextual approach, reduced memory requirements of the implemented clustering algorithms (contextual reduction of the vector space) and search (inverted lists compression)
- Possibility of construction and maintenance of multiple models/maps representing diverse views of the same document collection (and fitting the map to the query)
- Possibility of inclusion of system-user interaction history into the algorithm of map initialization (e.g. by strengthening / weakening of terms from documents evaluated by the user as more or less interesting)
- Possibility of inclusion of user preference profiles into the modeling process itself by taking into account the automatically collected information on user walk through the collection or provided externally.

5 Experiments

5.1 Histogram-Based Clustering Quality Measure

Our new approach to document clustering introduces new aspect of evaluation related to comparison of various clusterings constructed over partially different subspaces of the document space. This issue can be partially resolved by applying histogram-based statistics of term weighing function distributions

Both individual context groups and nodes of correctly built graph model (and map cells) have to describe a consistent fragment of the vector space. This fragment is related to a subset of the document collection (assigned to the context or the cell during model training). Document cluster used to be traditionally represented by its central element (centroid, medoid, or a set of elements like in CURE algorithm [10]). SOM and GNG models use so-called reference vector as cell centroid, while immunological models use here individual antibody.

But another possibility to represent a group of documents is to describe it by a set of histograms (one for each term in the document group), as described in section 2. It turns out, that time and space cost of such a representation is low. and we profit from departure from the assumption of spherical shape of vector space fragments related to a single cluster (as centroidal representation assumes) and from the possibility of dynamic modification of such a description when the cluster membership changes.

We gain also a new way of construction of a measure evaluating the quality of obtained clustering (both of the initial identification of contextual groups and of various graph models, like GNG, aiNet).

In the experimental section we shall investigate the stability of groups obtained in the clustering process (contexts and cells). Based on a fixed split we investigate the histogram descriptions of each group and then reclassify all documents choosing for each document the cluster (cell or context) into which the document belongs to the highest degree. Next the agreement of the new clustering with the clustering obtained in the process of model learning is checked.

5.2   Experimental Results

The idea of personalization promoted in this paper heavily assumes that the relationships are intrinsic and do not change under change of perspective. Therefore we evaluated the stability of the split into clusters (contextual groups) in case of the classical cluster representation based on fuzzified centroids (Fuzzy C-Means) and based on histogram descriptions (Fuzzy C-Histograms). Evaluation technique was based on histogram reclassification[7] measure (see section 5.1).

At the same time the quality of the histogram method itself was investigated (its sensitivity to the number of histogram intervals). Influence of contextual vector representation, based on weights $w_{t,d}$, o and the global one, based on $tfidf$ was compared. We checked also the impact of dimensionality reduction (number of different terms in the dictionary) and the number of intervals of histograms describing the distribution of term weights on the results of reclassification. We have also tested the impact of using *contextual (frequent) phrases* (which are derived locally within the contexts in analogy to methods of global phrase identification). Experiments were carried out for document collections *20 Newsgroups*, *12 Newsgroups*, *WebKb* and *Reuters* and the collection *100K Polish Web* containing 96908 pages from Polish Internet (see also [4]).

---

[7] Clustering methods generally attempt to make more or less local decisions when assigning an element to a cluster. Instability of clustering under reclassification would mean that the change of perspective from a local one to a global one rearranges the cluster membership, or stated in a different way, the cluster membership is not the intrinsic property of the document, but rather of the perspective.

The results are presented in table 1. The best result was achieved for the full contextual representation (with contextual phrases) for the set *20 Newsgroups*, for a dictionary reduced to 7363 terms. The reclassification quality was very high (close to 100%) - 19937 documents out of 20000 were correctly classified. The majority of 63 reclassification errors was caused by documents that did not contain any significant term after the reduction of dictionary dimension (they were classified into the default context marked as *"null"*). On the other hand, for the standard method *tfidf* only 90% of documents were correctly reclassified. It is interesting that the reclassification based on less reduced dictionary (leaving A dictionary of 15000 terms, that is twice as high as previously) leads to significantly worse result (85% correctly reclassified documents in case of contextual representation). This confirms the hypothesis that the too many low quality terms (their exponential explosion) may hide the inner structure of natural clusters of the document collection. On the other hand introduction of term weighting and contextual representation supports the reclassification capability.

In case of all other supervised collections significant differences between the stability of clustering, when the clusters were represented by a single representative (centroid, in the table Means) and the representation by histogram

Table 1. Stability of the structure of contextual groups: Reclassification measure

Reclassification	12News	20News	Reuters	WebKb	100K www
Means / $tfidf$	0.665	0.666	0.247	0.704	0.255
Means / $w_{t,d}$ no phrases	0.743	0.938	0.608	0.752	0.503
Means / $w_{t,d}$ + phrases	0.871	0.946	0.612	0.768	0.557
Histograms / $tfidf$	0.878	0.898	0.54	0.697	0.67
Hist. / $w_{t,d}$ no phrases	0.926	0.988	0.849	0.98	0.829
Hist. / $w_{t,d}$ + phrases	0.965	0.997	0.861	0.982	0.969

Table 2. Quality of histogram-based reclassification for the *12 Newsgroups* collection

Histograms / $w_{t,d}$	Precision	Recall	Category size	Reclassification
"null"	–	–	0	55
comp.windows.x	0.992	0.961	408	421
rec.antiques.radio+photo	0.973	0.972	612	613
rec.models.rockets	0.923	0.971	999	950
rec.sport.baseball	0.96	0.985	999	974
rec.sport.hockey	0.98	0.942	365	380
sci.math	0.985	0.933	624	659
sci.med	0.96	0.981	326	319
sci.physics	0.96	0.974	861	849
soc.culture.israel	0.974	0.998	666	650
talk.politics.mideast	0.968	0.988	1000	979
talk.politics.misc	0.966	0.996	872	846
talk.religion.misc	0.977	0.885	357	394

Table 3. Quality of centroid-based reclassification for the *12 Newsgroups* collection

Centroids / $tfidf$	Precision	Recall	Category size	Reclassification
"null"	–	–	0	55
comp.windows.x	0.629	0.883	408	291
rec.antiques.radio+photo	0.584	0.888	612	403
rec.models.rockets	0.762	0.598	999	1274
rec.sport.baseball	0.806	0.609	999	1322
rec.sport.hockey	0.358	0.85	365	154
sci.math	0.692	0.707	624	611
sci.med	0.226	0.961	326	77
sci.physics	0.774	0.545	861	1222
soc.culture.israel	0.597	0.929	666	428
talk.politics.mideast	0.849	0.548	1000	1548
talk.politics.misc	0.692	0.945	872	639
talk.religion.misc	0.131	0.723	357	65

descriptions. The differences were visible both for the global representation $tfidf$ and the contextual representation, based on weights $w_{t,d}$. Particularly important was the impact of departure from the representation by a single gravity center in case of those collections that consist of many fuzzy and thematically similar categories with varying cardinality: *Reuters* and *100K Polish Web*. In the case of representation based on geometrical gravity centers this is related on the one hand with the already mentioned inadequacy of the hyperbolical cluster shape assumption, on the other hand to the overshadowing of low cardinality categories by larger ones. For the *12 Newsgroups* collection with lower number of categories but with varying category cardinalities this is shown by the results in tables 2 and 3. It is visible that for histogram-based representation the cluster stability does not depend on their cardinality. For classical centroidal representation, lower cardinality categories suffer from loss of *precision*, because the documents originally belonging to them are reclassified into bigger contextual groups, represented by more general centroids. For the large groups, their *recall* is impaired, for the clusters contain documents belonging to diverse categories.

The results for the largest collection discussed here, 96908 web pages from Polish Internet (*100K Polish Web*), support the claim of contextual approach. The HTML documents were clustered into over 200 contexts with diverse cardinalities (from several pages to over 2100 pages in a single context) so the reclassification should not be easy. But the contextual weighting resulted in 93848 (97%) correctly reclassified pages, while the $tfidf$ weighting gave only 64859 (67%) pages. The classical representation (centroidal with global weighting via $tfidf$) of that many contexts proved to be totally inadequate, leading to unstable clusters, not reflecting topical similarity. In all cases the dictionary reduced via the histogram method consisted of 42710 terms selected out of over 1,200,000.

In case of all document collections, both when dealing with centroidal clusters and histogram-based clusters, we observed positive influence of *contextual phrases* on the stability of clusters, though various document collections were affected to different degree, and in some cases (especially for *Reuters* collections) it was marginal. This is clearly related to the quality of phrase identification itself, that is poor in case of *Reuters* collection of homogeneous economical news, as phrases really differentiating clusters are hard to find.

Finally, the impact of the number of histogram intervals on the stability was investigated. Obviously, a too low number of them would have a negative impact, as we reduce then the histogram method to centroidal one. Hence we are not surprised to have only 17374 correct reclassifications for contextual method with 3-interval histograms on the *20 Newsgroups*). But any higher meaningful number of intervals lead to high quality reclassification: 19992 for 6-, 19997 for 10-, 19991 for 20-, 19937 for 50- and 19867 correct reclassifications for 100-interval histograms). Similar results were observed for other collections. The lower the number of documents and the lower the document diversity, the lower the number of intervals has to be used to obtain high quality results. However, a too high number of intervals may lead to a poorer approximation. Usage of cumulative histograms may be cure here.

6   Conclusions

We presented a new concept of document cluster characterization via term (importance) distribution histograms. This idea allows the clustering process to have a deeper insight into the role played by each term in formation of a particular cluster. So a full profit can be taken from our earlier idea of "contextual clustering", that is of representing different document clusters in different subspaces of a global vector space. Such an approach to mining high dimensional datasets proved to be an effective solution to the problem of massive data clustering. Contextual approach leads to dynamic adaptation of the document representation, enabling user-oriented, contextual data visualization as a major step on the way to information retrieval personalization in map search engines.

Acknowledgement

This research has been partly funded by the European Commission and by the Swiss Federal Office for Education and Science with the 6th Framework Programme project REWERSE no. 506779 (cf. http://rewerse.net). Also co-financed by Polish state budget funds for scientific research under grants No. N516 00531/0646 and No. N516 01532/1906. Thanks to M. Dramiński and D. Czerski for programming support.

References

1. Basu, A., Harris, I.R., Basu, S.: Minimum distance estimation: The approach using density-based distances. In: Maddala, G.S., Rao, C.R. (eds.) Handbook of Statistics, vol. 15, pp. 21–48. North-Holland, Amsterdam (1997)

2. Bezdek, J.C., Pal, S.K.: Fuzzy Models for Pattern Recognition: Methods that Search for Structures in Data. IEEE, New York (1992)
3. Ciesielski, K., Wierzchoń, S., Kłopotek, M.: An Immune Network for Contextual Text Data Clustering. In: Bersini, H., Carneiro, J. (eds.) ICARIS 2006. LNCS, vol. 4163, pp. 432–445. Springer, Heidelberg (2006)
4. Ciesielski, K., Kłopotek, M.: Towards adaptive Web mining. Histograms and contexts in text data clustering. In: Berthold, M.R., Shawe-Taylor, J., Lavrač, N. (eds.) IDA 2007. LNCS, vol. 4723, pp. 284–295. Springer, Heidelberg (2007)
5. Ciesielski, K., Kłopotek, M.: Text Data Clustering by Contextual Graphs. In: Todorovski, L., Lavrač, N., Jantke, K.P. (eds.) DS 2006. LNCS (LNAI), vol. 4265, pp. 65–76. Springer, Heidelberg (2006)
6. Deerwester, S.C., Dumais, S.T., Landauer, T.K., Furnas, G.W., Harshman, R.A.: Indexing by Latent Semantic Analysis. Journal of the American Society of Information Science 41(6), 391–407 (1990)
7. Dhillon, I.S., Modha, D.S.: Concept Decompositions for Large Sparse Text Data Using Clustering. Machine Learning 42(1-2), 143–175 (2001)
8. Dittenbach, M., Rauber, A., Merkl, D.: Uncovering hierarchical structure in data using the Growing Hierarchical Self-Organizing Map. Neurocomputing 48(1-4), 199–216 (2002)
9. Fritzke, B.: A growing neural gas network learns topologies. In: Tesauro, G., Touretzky, D.S., Leen, T.K. (eds.) Advances in Neural Information Processing Systems, vol. 7, pp. 625–632. MIT Press, Cambridge (1995)
10. Guha, S., Rastogi, R., Shim, K.: CURE: an efficient clustering algorithm for large databases. In: Proceedings of ACM SIGMOD International Conference on Management of Data, pp. 73–84 (1998)
11. Hung, C., Wermter, S.: A constructive and hierarchical self-organising model in a non-stationary environment. In: Int. Joint Conference in Neural Networks (2005)
12. Kłopotek, M., Wierzchoń, S., Ciesielski, K., Draminski, M., Czerski, D.: Techniques and Technologies Behind Maps of Internet and Intranet Document Collections. In: Jie, L., Da Ruan, Guangquan, Z. (eds.) E-Service Intelligence – Methodologies, Technologies and Applications. Studies in Computational Intelligence, vol. 37, pp. 169–190. Springer, Heidelberg (2007)
13. Kohonen, T.: Self-Organizing Maps. Springer Series in Information Sciences, vol. 30. Springer, Heidelberg (2001)
14. Kohonen, T., Kaski, S., Somervuo, P., Lagus, K., Oja, M., Paatero, V.: Self-organization of very large document collections, Helsinki University of Technology technical report (2003)
15. Kolda, T.G.: Limited-Memory Matrix Methods with Applications. PhD thesis, The Applied Mathematics Program, University of Maryland, College Park, MA (1997)
16. Rauber, A.: Cluster Visualization in Unsupervised Neural Networks, Diplomarbeit, Technische Universität Wien, Austria (1996)
17. Salton, G., Buckley, C.: Term-weighting approaches in automatic text retrieval. Information Processing and Management: an International Journal 24(5), 513–523 (1988)
18. Timmis, J.: aiVIS: Artificial Immune Network Visualization. In: Proceedings of EuroGraphics UK 2001 Conference, University College London, pp. 61–69 (2001)

Improving Boosting by Exploiting Former Assumptions

Emna Bahri, Nicolas Nicoloyannis, and Mondher Maddouri

Laboratory Eric, University of Lyon 2,
5 avenue Pierre Mendès France, 69676 Bron Cedex
{e.bahri,nicolas.nicoloyannis}@univ-lyon2.fr,
mondher.maddouri@fst.rnu.tn
http://eric.univ-lyon2.fr

Abstract. The error reduction in generalization is one of the principal motivations of research in machine learning. Thus, a great number of work is carried out on the classifiers aggregation methods in order to improve generally, by voting techniques, the performance of a single classifier. Among these methods of aggregation, we find the Boosting which is most practical thanks to the adaptive update of the distribution of the examples aiming at increasing in an exponential way the weight of the badly classified examples. However, this method is blamed because of overfitting, and the convergence speed especially with noise. In this study, we propose a new approach and modifications carried out on the algorithm of AdaBoost. We will demonstrate that it is possible to improve the performance of the Boosting, by exploiting assumptions generated with the former iterations to correct the weights of the examples. An experimental study shows the interest of this new approach, called hybrid approach.

Keywords: Machine learning, Data mining, Classification, Boosting, Recall, convergence.

1 Introduction

The great emergence of the modern databases and their evolution in an exponential way as well as the evolution of transmission systems result in a huge mass of data which exceeds the human processing and understanding capabilities. Certainly, these data are sources of relevant information and require means of synthesis and interpretation. As a result, researches were based on powerful systems of artificial intelligence allowing the extraction of useful information helping us in decisions making. Responding to this need, data mining was born. It drew its tools from the statistics and databases. The methodology of data mining gives the possibility to build a model of prediction. This model is a phenomenon starting from other phenomena more easily accessible, based on the process of the knowledge discovery from data which is a process of intelligent data classification. However, the built model can sometimes generate errors of classification that even a random classification does not make. To reduce these

Z.W. Raś, S. Tsumoto, and D. Zighed (Eds.): MCD 2007, LNAI 4944, pp. 131–142, 2008.
© Springer-Verlag Berlin Heidelberg 2008

errors, a great amount of research in data mining and specifically in machine learning has been carried out on classifiers aggregation methods having as goal to improve by voting techniques the performance of a single classifier. These aggregation methods are good for compromised Skew-variance, thanks to the three fundamental reasons explained in [6]. These methods of aggregation are divided into two categories. The first category refers to those which merge preset classifiers, such as simple voting [2], the weighted voting [2], and the weighted majority voting [13]. The second category consists of those which merge classifiers according to data during the training, such as adaptive strategies (Boosting) and the basic algorithm AdaBoost [22] or random strategies (Bagging) [3].

We are interested in the method of Boosting, because of the comparative study [7] that shows, in little noise, AdaBoost is seemed to be working against the overfitting. In fact, AdaBoost tries to optimize directly the weighted votes. This observation has been proved not only by the fact that the empirical error on the training set decreases exponentially with iterations, but also by the fact that the error in generalization also decreases, even when the empirical error reached its minimum. However, this method is blamed because of overfitting, and the speed of convergence especially with noise. In the last decade, many studies focused on the weaknesses of AdaBoost and proposed its improvement. The important improvements were carried on the modification of the weight of examples [20], [19], [1], [21], [15], [9], the modification of the margin [10], [21], [18], the modification of the classifiers' weight [16], the choice of weak learning [5], [25] and the speed of convergence [23], [14], [19]. In this paper, we propose a new improvement to the basic Boosting algorithm AdaBoost. This approach aims exploiting assumptions generated with the former iterations of AdaBoost to act both on the modification of the weight of examples and the modification of the classifiers' weight. By exploiting these former assumptions, we think that we will avoid the re-generation of a same classifier within different iterations of AdaBoost. Thus, consequently, we expect a positive effect on the improvement of the speed of convergence. The paper is organized in three sections. In the following section, we describe the studies whose purpose is to improve the Boosting against its weaknesses. In the third section, we describe our improvement of boosting by exploiting former assumptions. In the fourth section, we present an experimental study of the proposed improvement by comparing its error in generalization, its recall and its speed of convergence with AdaBoost, on many real databases. We study also the behavior of the proposed improvement on noisy data. We present also comparative experiments of our proposed method with BrownBoost (a new method known that it improves AdaBoost M1 with noisy data). Lastly, we give our conclusions and perspectives.

2   State of Art

Due to the finding of some weaknesses, such as the overfitting and the speed of convergence, met by the basic algorithm of boosting AdaBoost, several researchers have tried to improve it. Therefore, we make a study of main methods having as purpose to improve boosting relatively to these weaknesses. With this intention, the researchers try to use the strong points of Boosting such as the update of the badly classified examples, the maximization of the margin, the significance of the weights that AdaBoost associates the hypothesis and finally the choice of weak learning.

2.1 Modification of the Examples' Weight

The distributional adaptive update of the examples, aiming at increasing the weight of those badly learned by the preceding classifier, makes it possible to improve the performance of any training algorithm. Indeed, with each iteration, the current distribution supports the examples having been badly classified by the preceding hypothesis, which characterizes the adaptivity of AdaBoost. As a result, several researchers proposed strategies related to a modification of weight update of the examples, to avoid the overfitting.

Indeed, we can quote for example MadaBoost [9] whose aim is to limit the weight of each example by its initial probability. It acts thus on the uncontrolled growth of the weight of certain examples (noise) which is the problem of overfitting.

Another approach which make the algorithm of boosting resistant to the noise is Brownboost [15], an algorithm based on Boost-by-Majority by incorporating a time parameter. Thus for an appropriate value of this parameter, BrownBoost is able to avoid the overfitting. Another approach, which adapts to AdaBoost a logistic regression model, is Logitboost [19].

An approach, which produces less errors of generalization compared with the traditional approach but with the cost of an error of training slightly more raised, is the Modest boost [1]. In fact, its update is based on the reduction in the contribution of classifier, if that functions "too well" on the data correctly classified. This is why the method is called Modest AdaBoost - it forces the classifiers to be "modest" and it works only in the field defined by a distribution.

An approach, which tries to reduce the effect of overfitting by imposing limitations on the distribution produced during the process of boosting is used in SmoothBoost [21]. In particular, a limited weight is assigned to each example individually during each iteration. Thus, the noisy data can be excessively underlined during the iterations since they are assigned to the extremely large weights.

A last approach, Iadaboost [20], is based on the idea of building around each example a local information measurement, making it possible to evaluate the overfitting risks, by using neighboring graph to measure information around each example. Thanks to these measurements, we have a function which translates the need for updating the example. This function makes it possible to manage the outliers and the centers of clusters at the same time.

2.2 Modification of the Margin

Certain studies, analyzing the behavior of Boosting, showed that the error in generalization still decreases even when the errors in training are stable. The explanation is that even if all the examples of training are already well classified, Boosting tends to maximize the margins [21].

Following this, some studies try to modify the margin either by maximizing it or by minimizing it with the objective of improving the performance of Boosting against overfitting.

Several approaches followed such as AdaBoostReg [18] which tries to identify and remove badly labeled examples, or to apply the constraint of the maximum margin to examples supposed to be badly labeled, by using the Soft Margin.

In the algorithm, proposed by [10], the authors use a weighting diagram which exploits a margin function that grows less quickly than the exponential function.

2.3   Modification of the Classifiers' Weight

During the performance evaluation of Boosting, researchers wondered about the significance of the weights $\alpha(t)$ that AdaBoost associates with the produced hypotheses.

However, they noted at the time of experiments on very simple data that the error in generalization decreased further whereas the weak learning had already provided all the possible hypotheses. In other words, when a hypothesis appears several times, it votes finally with a weight, office sum of all $\alpha(t)$, which is perhaps absolute. So several researchers hoped to approach these values by a nonadaptive process, such as locboost [16] an alternative to the construction of the whole representations of experts which allows the coefficients $\alpha(t)$ to depend on the data.

2.4   Choice of Weak Learner

A question that several researchers posed against the problems of boosting is that of weak learner and how to make a good choice of this classifier?

A lot of research moves towards the study of choosing the basic classifier of boosting, such as GloBoost [25]. This approach use a weak learner which produces only correct hypotheses. RankBoost [5] is also an approach which is based on weak learner which accepts as data attributes functions of rank.

2.5   The Speed of Convergence

In addition to the problem of overfitting met by boosting in the modern databases mentioned above, we find another problem : the speed of convergence of Boosting especially AdaBoost.

Indeed, in the presence of noisy data, the optimal error of the training algorithm used is reached after a long time. In other words, AdaBoost "loses" iterations, and thus time, with reweighing examples which do not deserve in theory any attention, since it is a noise.

Thus research was made to detect these examples and improve the performance of Boosting in terms of convergence such as: iBoost [23] which aims at specializing weak hypotheses on the examples supposed to be correctly classified.

The IAdaBoost approach also contributes to improve AdaBoost against its speed of convergence. In fact, the basic idea of the improvement is the modification of the theorem [19]. This modification is carried out in order to integrate the risk of Bayes. The effects of this modification are a faster convergence towards the optimal risk and a reduction of the number of weak hypotheses to build. Finally, RegionBoost [14] is a new weighting strategy of each classifier. This weighting is evaluated at the voting time by a technique based on K Nearest Neighbors of the example to label. This approach makes it possible to specialize each classifier on areas of the training data.

3 Boosting by Exploiting Former Assumptions

To improve the performance of AdaBoost and to avoid forcing it to learn either from the examples that contain noise, or from the examples which would become too difficult to learn during the process of Boosting, we propose a new approach. This approach is based on the fact that for each iteration, Adaboost, builds hypotheses on a defined sample, it makes its updates and it calculates the error of training according to the results given only by these hypotheses. In addition, it does not exploit the results provided by the hypotheses already built on other samples to the former iterations. This approach is called AdaBoostHyb.

Program Code. Input X_0 to classify, $S = (x_1, y_1), ..., (x_n, y_n)$ Sample

- For i=1,n Do
- $p_0(x_i) = 1/n$;
- End FOR
- $t \leftarrow 0$
- While $t \leq T$ Do
- Learning sample S_t from S with probabilities p_t.
- Build a hypotheses h_t on S_t with weak learning A.
- ϵ_t apparent error of h_t on S with $\epsilon_t = \sum weight \ of \ examples$
  such that $argmax(\sum_{i=1}^t \alpha_i h_i(x_i) \neq y_i)$. $\alpha_t = 1/2ln((1 - \epsilon_t)/\epsilon_t)$.
- For i=1, m Do
- $P_{t+1}(x_i) \leftarrow (p_t(x_i)/Z_t)e^{-\alpha_t}$ **if** $argmax(\sum_{i=1}^t \alpha_i h_i(x_i)) = y_i$ (**correctly classified**
  $P_{t+1}(xi) \leftarrow (p_t(x_i)/Z_t)e^{+\alpha_t}$ **if** $argmax(\ \sum_{i=1}^t \alpha_i h_i(x_i)) \neq y_i$ (**badly classified**
  (Z_t normalized to $\sum_{i=1}^n p_t(x_i) = 1$)
- End For
  $t \leftarrow t + 1$
- End While
- Final hypotheses :
  $H(x) = argmax \ y \in Y \sum_{t=1}^T \alpha_t$

The modification within the algorithm is made through two ways:

The first way is during the modification of the weights of the examples: Indeed, this strategy, with each iteration, is based on the opinion of the experts already used (hypotheses of the former iterations) for the update of the weight of the examples.

In fact, we do not compare only the class predicted by the hypothesis of the current iteration with the real class but also the sum of the hypotheses balanced from the first iteration to the current iteration. If this sum votes for a class different from the real class, an exponential update such as in the case of AdaBoost is applied to the badly classified example. Thus, this modification lets the algorithm be interested only in the examples which are either badly classified or not classified yet. So, results related to the improvement the speed of convergence are awaited, similarly for the reduction of the error of generalization, because of the richness of the space of hypotheses to each iteration.

The second way is during the error analysis $\epsilon(t)$ of the hypothesis to the iteration T: Indeed, this other strategy is rather interested in the classifiers' coefficient (hypothesis) to each iteration $\alpha(t)$.

In fact, this coefficient depends on the apparent error analysis $\epsilon(t)$. This method, with each iteration, takes into account hypotheses preceding the current iteration during the calculation of $\epsilon(t)$. So the apparent error with each iteration is the weight of the examples voted badly classified by the hypotheses weighted of the former iterations by comparison to the real class.

Results in improving the error of generalization are expected since the vote of each hypothesis (coefficient $\alpha(t)$) is calculated from the other hypotheses.

4   Experiments

The objective of this part is to compare our new approach and especially its contribution with the original approach of Adaboost and to look further into this comparison by the choice of a version improved of Adaboost (BrownBoost [15]).

Our Choice of BrownBoost was based on its robustness against the problems of noisy data. In fact,BrownBoost is an adaptive algorithm which incorporates a time parameter that corresponds to the proportion of noise in the training data. So by a good estimation of this parameter BrownBoost is capable of avoiding overfitting. The comparison criterions chosen in this article are the error rate, the recall, the speed of convergence and the sensitivity to noise.

Table 1. Databases Description

Databases	Nb. Inst	Attrib	Cl. Pred	Miss.VaL
IRIS	150	4 numeric	3	no
NHL	137	8 numeric and symbolic	2	yes
VOTE	435	16 boolean valued	2	yes
WEATHER	14	4 numeric and symbolic	2	no
CREDIT-A	690	16numeric and symbolic	2	yes
TITANIC	750	3 symbolic	2	no
DIABETES	768	8 numeric	2	no
HYPOTHYROID	3772	30 numeric and symbolic	4	yes
HEPATITIS	155	19 numeric and symbolic	2	yes
CONTACT-LENSES	24	4 nominal	3	no
ZOO	101	18 numeric and boolean	7	no
STRAIGHT	320	2 numeric	2	no
IDS	4950	35 numeric and symbolic	12	no
LYMPH	148	18 numeric	4	no
BREAST-CANCER	286	9 numeric and symbolic	2	yes

To do this experimental comparison, we used the C4.5 algorithm as a weak learner (according to the study of Dietterich [6]). To estimate without skew the theoretical success rate, we used a procedure of cross-validation in 10 folds (according to the study [12]). In order to choose the databases for our experiments, we considered the principle of diversity. We have considered 15 databases of the UCI [8]. Some databases are characterized by theirs missing values (NHL, Vote, Hepatitis, Hypothyroid). Some others concern the problem of multi-class prediction (Iris: 3 classes, Diabetes: 4 classes, Zoo: 7 classes, IDS: 12 classes). We choose the IDS database [24] especially because it has 35 attributes. Table 1 describes the 15 databases used in the experimental comparison.

4.1 Comparison of Generalization Error

Table 2 indicated the error rates in 10-fold cross-validation corresponding to the algorithm AdaBoost M1,BrownBoost and the proposed one. We used the same samples for the tree algorithms in cross-validation for comparison purposes. The results are obtained while having chosen for each algorithm to carry out 20 iterations. The study of the effect of the number of iterations on the error rates of the tree algorithms will be presented in the section 4.3, where we will consider about 100 iterations.

The results in table 2 show already that the proposed modifications improve the error rates of AdaBoost. Indeed, for 14 databases out of 15, the proposed algorithm shows an error rate lower or equal to AdaBoost M1. We remark, also, a significant improvement of the error rates corresponding to the three databases NHL, CONTACT-LENS and BREAST-CANCER. For example, the error rate corresponding to the BREAST-CANCER database goes from 45.81% to 30.41%.

Even, if we compare the proposed algorithm with BrownBoost, we remark that for 11 databases out of 15 the proposed algorithm shows an error rate lower or equal to BrownBoost.

This gain shows that by exploiting hypotheses generated with the former iterations to correct the weights of the examples, it is possible to improve the

Table 2. Rate of error of generalization

Databases	AdaBoost M1	BrownBoost	AdaBoostHyb
IRIS	6.00%	3.89	**3.00%**
NHL	35.00%	30.01	**28.00%**
VOTE	4.36%	4.35	4.13%
WEATHER	21,42%	**21.00**	21.00%
CREDIT-A	15.79%	**13.00**	13.91%
TITANIC	**21.00 %**	24.00	21.00%
DIABETES	27.61%	**25.05**	25.56%
HYPOTHYROID	0.53%	0.6	**0.42%**
HEPATITIS	15,62%	14.10	**14.00%**
CONTACT-LENSES	25.21%	**15.86**	16.00%
ZOO	**7.00%**	7.23	**7.00%**
STRAIGHT	2,40%	**2.00**	**2.00%**
IDS	1,90%	0.67	**0,37%**
LYMPH	19.51%	**18.54**	20.97%
BREAST-CANCER	45.81%	31.06	**30.41%**

performance of the Boosting. This can be explained by the calculation of the precision of the error analysis $\epsilon(t)$ and consequently the calculation of the coefficient of the classifier $\alpha(t)$ as well as the richness of the space of the hypotheses to each iteration since it acts on the whole of the hypotheses generated by the preceding iterations and the current iteration.

4.2   Comparison of Recall

The encouraging results, found previously, enable us to proceed further within the study of this new approach. Indeed, in this part we try to find out the impact of the approach on the recall, since our approach does not really improve Boosting if it acts negatively on the recall.

Table 3 indicates the recall for the algorithms AdaBoost M1, Brownboost and the proposed one. We remark that the proposed algorithm has the best recall overall the 14 for 15 studied databases. This result confirms the preceding ones. We remark also that it increases the recall of the databases having less important error rates.

Considering Brownboost, we remark that it improves the recall of AdaBoostM1, overall the data sets (except the TITANIC one). However, the recall rates given by our proposed algorithm are better than those of Brown-Boost. Except, with the zoo dataset.

It is also noted that our approach improves the recall in the case of the Lymph base where the error was more important. It is noted though that the new approach does not act negatively on the recall but it improves it even when it can not improve the error rates.

Table 3. Rate of recall

Databases	AdaBoost M1	BrownBoost	AdaBoostHyb
IRIS	0,93	0.94	**0,96**
NHL	0,65	0,68	**0,71**
VOTE	0,94	0.94	**0,95**
WEATHER	0,63	0.64	**0,64**
CREDIT-A	0,84	0.85	**0,86**
TITANIC	**0,68**	0.54	**0,68**
DIABETES	0,65	0.66	**0,68**
HYPOTHYROID	0,72	0.73	**0,74**
HEPATITIS	0,69	0.70	**0,73**
CONTACT-LENSES	0,67	0.75	**0,85**
ZOO	0,82	**0.9**	0,82
STRAIGHT	0,95	0.95	**0,97**
IDS	0,97	0.97	**0,98**
LYMPH	0,54	0.62	**0,76**
BREAST-CANCER	0,53	0.55	**0,6**

4.3   Comparison with Noisy Data

In this part, we are based on the study already made by Dietterich [6] by adding random noise to the data. This addition of noise of 20% is carried out, for each one of these databases, by changing randomly the value of the predicted class by another possible value of this class.

Table 3 shows us the behavior of the algorithms with noise. We notice that the hybrid approach is also sensitive to the noise since the error rate in generalization is increased for all the databases.

However this increase remains always inferior with that of the traditional approach except for the databases such as Credit-A, Hepatitis and Hypotyroid.

So, we studied these databases and we observed that all these databases have missing values. In fact, Credited, Hepatitis and Hypothyroid have respectively 5%, 6% and 5,4% of missing values. It seems that our improvement loses its effect with accumulation of two types of noise: missing values and artificial noise, although the algorithm AdaBoostHyb improves the performance of AdaBoost against the noise. Considering Brownboost, we remark that it gives better error rates that AdaboostM1 on all the noisy data sets. However, It gives better error rates than our proposed method, only with 6 data sets. Our proposed method gives better error rates with the other 9 data sets. This encourages us to study in details the behavior of our proposed method on noisy data.

4.4   Comparison of Convergence Speed

In this part, we are interested in the number of iterations that allow the algorithms to converge, i.e. where the error rate is stabilized. Tables 4, 5 and 6 shows us that the hybrid approach allows AdaBoost to converge more quickly. Indeed, the error rate of AdaBoost M1 is not stabilized even after 100 iterations, whereas Adaboost Hyb converges after 20 iterations or even before.

For this reason we choose for the first part 20 iterations to carry out the comparison in terms of error and recall. These results are also valid for the database Hepatitis. In fact, This database has a lot of missing values (Rate 6%). These missing values always present a problem of convergence. Moreover, the same results appear on databases of various types (several attributes, the class to be predicted with K modalities, important sizes).

Table 4. Rate of error on Noisy data

Databases	AdaBoost M1	BrownBoost	AdaBoostHyb
IRIS	33.00%	**26.00**	28.00%
NHL	45.00%	40.00	**32.00%**
VOTE	12.58%	**7.00**	7.76%
WEATHER	25.00%	22	**21%**
CREDIT-A	22.56%	**20.99**	24.00%
TITANIC	34.67%	28.08	**26.98%**
DIABETES	36.43%	32.12	**31.20%**
HYPOTHYROID	0.92%	**0.86**	2.12%
HEPATITIS	31.00%	**27.38**	41.00%
CONTACT-LENSES	33%	30.60	**25%**
ZOO	18.84%	14.56	**11.20%**
STRAIGHT	3.45%	**2.79**	2.81%
IDS	2.40%	1.02	**0.50%**
LYMPH	28.73%	24.57	**24.05%**
BREAST-CANCER	68.00%	50.98	**48.52%**

Table 5. Comparison of speed convergence

-	AdaBoost M1				BrownBoost				AdaBoost hyb			
Nb. iterations	10	20	100	1000	10	20	100	1000	10	20	100	1000
Iris	7,00	6,00	5,90	5,85	3.96	3.89	3,80	3,77	3,50	3,00	3,00	3,00
Nhl	37,00	35,00	34,87	34,55	30,67	30,01	29,89	29,76	31,00	28,00	28,00	28,00
Weather	21,50	21,42	21,40	14,40	21,10	21,00	20,98	21,95	21,03	21,00	21,00	21,00
Credit-A	15,85	15,79	15,75	14,71	13,06	13,00	12,99	12,97	14,00	13,91	13,91	13,91
Titanic	21,00	21,00	21,00	21,00	24,08	24,00	23,89	23,79	21,00	21,00	21,00	21,00
Diabetes	27,70	27,61	27,55	27,54	25,09	25,05	25,03	25,00	25,56	25,56	25,56	25,56
Hypothyroid	0,60	0,51	0,51	0,50	0,62	0,60	0,59	0,55	0,43	0,42	0,42	0,42
Hepatitis	16,12	15,60	14,83	14,19	14,15	14,10	14,08	14,04	14,03	14,00	14,00	14,00
Contact-Lenses	26,30	24,80	24,50	16,33	15,90	15,86	15,83	15,80	16,00	16,00	16,00	16,00
Zoo	7,06	7,00	7,00	7,00	7,25	7,23	7,19	7,15	7,00	6,98	7,00	7,00
Straight	2,50	2,46	2,45	2,42	2,12	2,00	1,98	1,96	0,42	0,42	0,42	0,42
IDS	2,00	1,90	1,88	1,85	0,7	0,67	0,65	0,63	0,7	0,67	0,65	0,63
Lymph	19,53	19,51	19,51	19,50	18,76	18,54	18,50	18,45	18,76	18,54	18,50	18,45
Breast-Cancer	45,89	45,81	45,81	45,79	31,10	31,06	31,04	31,00	31,10	31,06	31,04	31,00

This makes us think that due to the way of calculating the apparent error, the algorithm reaches stability more quickly. Finally, we remark that BrownBoost does'nt converge even after 1000 iterations. This remark prove the fact that the BrownBoost problem is the speed of convergence.

5 Conclusion

In this paper, we proposed an improvement of AdaBoost which is based on the exploitation of the hypotheses already built with the preceding iterations. The experiments carried out and the results show that this approach improves the performance of AdaBoost in error rate, in recall, in speed of convergence and in sensibility to the noise. However, it proved that this same approach remains sensitive to the noise.

We did an experimental comparison of the proposed method with BrownBoost (a new method known that it improves AdaBoost M1 with noisy data). The results show that our proposed method improves the recall rates and the speed of convergence of BrownBoost overall the 15 data sets. The results show also that BrownBoost gives better error rates with some datasets, and our method gives better error rates with other data sets. The same conclusion is reached with noisy data.

To confirm the experimental results obtained, more experimentations are planned. We are working on further databases that were considered by other researchers in theirs studies of the boosting algorithms. We plan to choose weak learning methods other than C4.5, in order to see whether the obtained results are specific to C4.5 or general. We plan to compare the proposed algorithm to new variants of boosting, other than AdaBoost M1. We can consider especially those that improve the speed of convergence like IAdaBoost and RegionBoost. In the case of encouraging comparisons, a theoretical study on convergence will be done to confirm the results of the experiments.

Another objective which seems important to us consists in improving this approach against the noisy data. In fact, the emergence and the evolution of the modern databases force the researchers to study and improve the boosting's capacities of tolerance to the noise. Indeed, these modern databases contain a lot of noise, due to new technologies of data acquisition such as the Web. In parallel, studies such as [5], [17] and [19], show that AdaBoost tends to overfit the data and especially the noise. So, a certain number of recent work tried to limit these risks of overfitting. These improvements are based primarily on the concept that AdaBoost tends to increase the weight of the noise in an exponential way. Thus two solutions were proposed to reduce the sensibility to noise. One is by detecting these data and removing them based on the heuristic and selection of prototypes such as research presented in [4]and [26]. The other solution is by detecting these data through the process of boosting, in which case we speak about a good management of noise. According to the latest approach, we plan to improve the proposed algorithm against the noisy data, by using neighboring graphs or using update parameters.

Finally, a third perspective work aims at studying the Boosting with a weak learner that generates several rules (Rule learning [11]). Indeed, the problem of this type of learners is the production of conflicting rules within the same iteration of boosting. These conflicting rules will have the same weights (attributed by the boosting algorithm). In the voting procedure, we are thinking about a combination of the global weights (those attributed by the boosting algorithm) and the local weights (those attributed by the learning algorithm).

References

1. Vezhnevets, V., Vezhnevets, A.: Modest adaboost: Teaching adaboost to generalize better, Moscow State University (2002)
2. Bauer, E., Kohavi, R.: An empirical comparison of voting classification algorithms: Bagging, boosting, and variants. Machine Learning 24, 173–202 (1999)
3. Breiman, L.: Bagging predictors. Machine Learning 26, 123–140 (1996)
4. Brodley, C.E., Friedl, M.A.: Identifying and eliminating mislabeled training instances. In: AAAI/IAAI, vol. 1, pp. 799–805 (1996)
5. Dharmarajan, R.: An efecient boosting algorithm for combining preferences. Technical report, MIT, Septembet (1999)
6. Dietterich, T.G.: An experimental comparison of three methods for constructing ensembles of decision trees: bagging, boosting, and randomization. Machine Learning, 1–22 (1999)
7. Dietterich, T.G.: Ensemble methodes in machine learning. In: First International Workshop on Multiple ClassifierSystems, pp. 1–15 (2000)
8. Blake, C.L., Newman, D.J., Hettich, S., Merz, C.J.: Uci repository of machine learning databases (1998)
9. Domingo, C., Watanabe, O.: Madaboost: A modification of adaboost. In: Proc. 13th Annu. Conference on Comput. Learning Theory, pp. 180–189. Morgan Kaufmann, San Francisco (2000)
10. Friedman, J., Hastie, T., Tibshirani, R.: Additive logistic regression: a statistical view of boosting. Dept. of Statistics, Stanford University Technical Report (1998)

11. Friedman, J.H., Popescu, B.E.: Predictive learning via rule ensembles (technical report). Stanford University (7) (2005)
12. Kohavi, R.: A study of cross-validation and bootstrap for accuracy estimation and model selection. In: International Joint Conference on Artificial Intelligence (IJCAI) (1995)
13. Littlestone, N., Warmuth, M.K.: The weighted majority algorithm. Information and computation 24, 212–261 (1994)
14. Maclin, R.: Boosting classifiers regionally. In: AAAI/IAAI, pp. 700–705 (1998)
15. McDonald, R., Hand, D., Eckley, I.: An empirical comparison of three boosting algorithms on real data sets with artificial class noise. In: Fourth International Workshop on Multiple Classifier Systems, pp. 35–44 (2003)
16. Meir, R., El-Yaniv, R., Ben-David, S.: Localized boosting. In: Proc. 13th Annu. Conference on Comput. Learning Theory, pp. 190–199. Morgan Kaufmann, San Francisco (2000)
17. Rätsch, G.: Ensemble learning methods for classification. Master's thesis, Dep of computer science, University of Potsdam (April 1998)
18. Rätsch, G., Onoda, T., Müller, K.-R.: Soft margins for adaboost. Mach. Learn. 42(3), 287–320 (2001)
19. Schapire, R.E., Singer, Y.: Improved boosting algorithms using confedence rated predictions. Machine Learning 37(3), 297–336 (1999)
20. Sebban, M., Suchier, H.-M.: Étude sur amélioration du boosting: réduction de l'erreur et accélération de la convergence. Journal électronique d'intelligence artificielle (submitted, 2003)
21. Servedio, R.A.: Smooth boosting and learning with malicious noise. In: Helmbold, D.P., Williamson, B. (eds.) COLT 2001 and EuroCOLT 2001. LNCS (LNAI), vol. 2111, pp. 473–489. Springer, Heidelberg (2001)
22. Shapire, R.: The strength of weak learnability. Machine Learning 5, 197–227 (1990)
23. Kwek, S., Nguyen, C.: iboost: Boosting using an instance-based exponential weighting scheme. In: Elomaa, T., Mannila, H., Toivonen, H. (eds.) ECML 2002. LNCS (LNAI), vol. 2430, pp. 245–257. Springer, Heidelberg (2002)
24. Stolfo, S.J., Fan, W., Lee, W., Prodromidis, A., Chan, P.K.: Cost-based modeling and evaluation for data mining with application to fraud and intrusion detection (1999)
25. Torre, F.: Globoost: Boosting de moindres généralisés. Technical report, GRAppA - Université Charles de Gaulle - Lille 3 (September 2004)
26. Wilson, D.R., Martinez, T.R.: Reduction techniques for instance-based learning algorithms. Machine Learning 38(3), 257–286 (2000)

Discovery of Frequent Graph Patterns That Consist of the Vertices with the Complex Structures

Tsubasa Yamamoto[1], Tomonobu Ozaki[2], and Takenao Ohkawa[1]

[1] Graduate School of Engineering, Kobe University
[2] Organization of Advanced Science and Technology, Kobe University
1-1 Rokkodai, Nada, Kobe 657-8501, Japan
{yamamoto@cs25.scitec.,tozaki@cs.,ohkawa@}kobe-u.ac.jp

Abstract. In some real world applications, the data can be represented naturally in a special kind of graphs in which each vertex consists of a set of (structured) data such as item sets, sequences and so on. One of the typical examples is metabolic pathways in bioinformatics. Metabolic pathway is represented in a graph structured data in which each vertex corresponds to an enzyme described by a set of various kinds of properties such as amino acid sequence, enzyme number and so on. We call this kind of complex graphs *multi-structured graphs*. In this paper, we propose an algorithm named FMG for mining frequent patterns in multi-structured graphs. In FMG, while the external structure will be expanded by the same mechanism of conventional graph miners, the internal structure will be enumerated by the algorithms suitable for its structure. In addition, FMG employs novel pruning techniques to exclude uninteresting patterns. The preliminary experimental results with real datasets show the effectiveness of the proposed algorithm.

1   Introduction

Graphs are widely used to represent complicated structures such as proteins, LSI circuits, hyperlinks in WWW, XML data and so on. Discovering frequent subgraphs from a graph database is one of the most important problems of the graph mining[4]. In recent years, several efficient algorithms of graph mining have been developed[2,6,10,11,16,17]. However, since the structure of data is becoming complex more and more, these algorithms might not be sufficient in some application domains. One typical example of such a complex database is KEGG* , which is a database of metabolic pathways. Metabolism denotes total chemical reactions in the body of organisms. The chemical reaction is risen up by some enzyme and it translates a compound into another one. Pathway is a large network of these reactions, so it can be regarded as a graph structured data. In addition, vertices in the pathway consist of several types of data such as compounds, enzymes, genes and so on. Therefore, a metabolic pathway can

---

* http://www.genome.ad.jp/kegg/

Z.W. Raś, S. Tsumoto, and D. Zighed (Eds.): MCD 2007, LNAI 4944, pp. 143–156, 2008.
© Springer-Verlag Berlin Heidelberg 2008

be naturally represented in a graph in which each vertex consists of a set of (structured) data such as item sets, sequences and so on. Since most of current graph mining algorithms treat the element of vertices as an item, they can not discover frequent subgraphs in which sub-patterns of vertex elements are considered. Because such kind of complex graph is expected to be going to rapidly increase, it is important to establish a flexible technique that can inclusively treat such kind of data.

In this paper, as one of the techniques to deal with such kind of complex graphs, we propose a new frequent graph mining algorithm named **FMG**. While we describe it in detail later, the target of FMG is a special kind of graphs, called *multi-structured graphs*, that consist of vertices holding a set of (structured) data such as item sets, sequences and so on. Given a database of multi-structured graphs, FMG will discover frequent graph patterns that consist of vertices with several complex structures. In order to enumerate frequent patterns completely, on one hand, the external structures will be expanded in the manner of general graph mining algorithms. On the other hand, the internal structures, *i.e.* vertices, will be enumerated by some algorithm suitable for those structures. In addition, FMG employs several novel optimization techniques to exclude uninteresting patterns.

The rest of this paper is organized as follows. In section 2, we introduce some basic notations and define our data mining problem formally. In section 3, our frequent pattern miner FMG is proposed and explained in detail. Preliminary experimental results with pathways in KEGG are reported in section 4. After describing related work in section 5, we conclude this paper in section 6.

2   Preliminaries

A *multi-structured graph* G consists of a set of vertices $V(G)$ and a set of edges $E(G)$. An edge between two vertices v and v' is denoted as $e(v, v')$. Edge labels are not considered in this paper. Each vertex $v \in V(G)$ consists of a length n list of attributes of plural kinds of structured data. We denote the list in v as $list(v) = [elm_1^v, \cdots, elm_n^v] \in [dom(A_1), \cdots, dom(A_n)]$ where $dom(A_i)$ denotes the domain of structure of ith attribute A_i. We show an example of multi-structured graphs in Fig. 1. For example, for $v_{13} \in V(G_1)$, $list(v_{13})$ is $[\{a, b, c\}, \langle AACC \rangle]$ and the domain of the first and second attributes are item sets and sequences, respectively. A spanning tree of a graph is considered with depth first search for numbering the vertices. The first visited vertex is called root, while the last visited vertex is called rightmost vertex. The path from the root to the rightmost vertex in the spanning tree is called rightmost path. In G_1, if we set v_{11} to the root, then the rightmost vertex is v_{14} and the rightmost path becomes v_{11}–v_{13}–v_{14}.

For example, from an attribute whose domain is graph, we can extract several classes of pattern, such as paths, trees and graphs. So, as a bias, we have to give the class of pattern $\mathcal{P}_{A_i}$ to be extracted from each attribute A_i. Given two patterns $p, q \in \mathcal{P}_{A_i}$, $p \preceq q$ denotes that p is more general than or equals to q. Given a vertex v with $list(v) = [elm_1^v, \cdots, elm_n^v]$ and a list of patterns $lp = [p_1, \cdots, p_n] \in [\mathcal{P}_{A_1}, \cdots, \mathcal{P}_{A_n}]$, if all p_i cover its corresponding elm_i^v, i.e. $\forall i\ p_i \preceq elm_i^v$, then we say that lp covers v and denote it as $lp \prec v$. Note that, we assume that each combination of $\mathcal{P}_{A_i}$ and $dom(A_i)$ gives the definition of the *cover relation*.

A subgraph isomorphism of two multi-structured graphs G and G', denoted as $G \subseteq G'$, is an injective function $f : V(G) \Rightarrow V(G')$ such that $(1) \forall v \in V(G)\ v \prec f(v)$, and $(2) \forall e(u,v) \in E(G)\ e(f(u), f(v)) \in E(G')$. If there exists a subgraph isomorphism from G to G', G *is called a subgraph of G' and G' is called a supergraph of G.* Let $D = \{G_1, G_2, \cdots, G_M\}$ be a database of multi-structured graphs. The *support* of a multi-structured subgraph pattern P, hereafter *graph pattern* in short, is defined as follows.

$$sup_D(P) = \frac{\sum_{G \in D} O_P(G)}{M} \qquad where\ \ O_P(G) = \begin{cases} 1\ (P \subseteq G) \\ 0\ (otherwise) \end{cases}$$

Given a user defined threshold σ, a graph pattern P is called *frequent* in D if $sup_D(P) \geq \sigma$ holds. The problem discussed in this paper is stated formally as follows : Given a database D of multi-structured graphs and a positive number $\sigma(0 < \sigma \leq 1)$ called the *minimum support*. Then, the problem of frequent multi-structured subgraph pattern mining is to find all frequent graph patterns P such that $sup_D(P) \geq \sigma$.

3   Mining Frequent Multi-structured Subgraph Patterns

In this section, we propose a frequent multi-structured subgraph pattern miner named FMG. Throughout this section, we use a database shown in Fig. 1 as a running example for explaining the behavior of FMG.

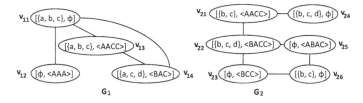

Fig. 1. Target database

FMG employs a kind of the pattern-growth approach [12,17]. Initial patterns of FMG are multi-structured graphs with one vertex. That is to say, they are the sets of all sub-patterns of the vertex element whose size is 1. Note that each attribute defines the size. While the size of item sets is the number of items, the size of sequences is the length. For example, initial patterns taken from G_1 are $\{a\}, \{b\}, \{c\}, \{d\}, \langle A \rangle, \langle B \rangle$ and $\langle C \rangle$. FMG employs two kinds of procedures for expansion. The first one is for expanding internal structures, *i.e.* sub-patterns in a vertex, and the second one is for overall structure or topology of graph patterns. In this paper, the overall structure of a graph pattern is referred to as external structure. By applying these two kinds of expansions to the initial patterns repeatedly, all frequent multi-structured subgraph patterns are to be enumerated.

In the following subsections, internal and external expansions are described in detail. After that, the novel optimization techniques are introduced.

3.1   Expansion of Internal Structures

While we describe it in detail later, FMG employs general graph mining algorithms for expanding external structures. This constrains the enumeration strategy for internal structures, *i.e.* patterns in a vertex. The expansion of the internal structure in FMG is limited only in the rightmost vertex of the graph pattern. If not, many duplicated patterns will be generated.

As described before, an internal structure of a multi-structured graph consists of a list of attributes of plural kinds of structured data such as item sets, sequences and so on. In order to enumerate *single* sub-patterns within each attribute efficiently, FMG employs several existing algorithms. For example, [14] for item sets, and [12] for sequences. In addition to the single patterns in a vertex, the combinations of patterns taken from different attributes have to be considered. To avoid the duplications, the enumerations have to obey the ordering in the attribute list. Consider the case where an attribute A_0 is ahead of another attribute A_1. In this case, given a single pattern $p \in \mathcal{P}_{A_1}$, then we avoid the generation of patterns $q, p \in [\mathcal{P}_{A_0}, \mathcal{P}_{A_1}]$ because it is against the order.

3.2   Expansion of External Structures

Since the external structure of a multi-structured graph is a graph structure, FMG employs the enumeration strategy for the external structure in the manner of general graph mining algorithms[1,17]. To put it concretely, the only vertices on the rightmost path are extended by adding an edge and, if necessary, a vertex.

In order to check the isomorphism and to identify the canonical form of a graph, the concept of code words for simple graphs is introduced [1,17]. The core idea underlying the canonical form is to construct a code word that uniquely identifies a graph up to isomorphism and symmetry. In FMG, the similar code words for the graph patterns are employed. The code word of a multi-structured graph pattern P is in the form as follows.

$$code(P) = l(list(v)) \ (i_d \ [-i_s] \ l(list(v)))^m$$

In this code word, i_s is the index of the source vertex and i_d is the index of the destination vertex, respectively. The index of the source vertex is smaller than that of the destination vertex with respect to an edge. m is the total number of edges. $[-i_s]$ is a negative number. $l(list(v))$ is the string description of $list(v)$. A

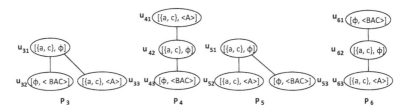

Fig. 2. Four isomorphic graph patterns

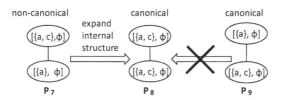

Fig. 3. The internal expansion from a non-canonical pattern to a canonical pattern

symbol ϕ is defined to be the lexicographically smallest string. The code words of graph patterns P_3–P_6 in Fig. 2 are shown below.

$$
\begin{aligned}
code(P_3) &= \ [\{a,c\},\phi] \quad 2-1\ [\phi,\langle BAC\rangle] \quad 3-1\ [\{a,c\},\langle A\rangle] \\
code(P_4) &= [\{a,c\},\langle A\rangle] \quad 2-1\ [\{a,c\},\phi] \quad 3-2\ [\phi,\langle BAC\rangle] \\
code(P_5) &= \ [\{a,c\},\phi] \quad 2-1\ [\{a,c\},\langle A\rangle] \quad 3-1\ [\phi,\langle BAC\rangle] \\
code(P_6) &= [\phi,\langle BAC\rangle] \quad 2-1\ [\{a,c\},\phi] \quad 3-2\ [\{a,c\},\langle A\rangle]
\end{aligned}
$$

The canonical form of a graph pattern, or canonical pattern, is determined to be the lexicographically smallest code word in the set of isomorphic patterns. In Fig. 2, $code(P_6)$ is the lexicographically smallest among the set of isomorphic graph patterns P_3–P_6. Thus, P_6 can be identified as a canonical pattern.

The canonical form of the simple graph pattern satisfies the anti-monotone property, *i.e.* no canonical pattern will be generated by expanding non-canonical patterns. Thus, we need not to expand non-canonical patterns. On the other hand, in case of mining multi-structured graphs, while no canonical pattern will be generated by expanding external structures, some canonical patterns can be generated from non-canonical patterns by expanding those internal structures. For example, in Fig. 3, P_7 is non-canonical pattern, while P_8 and P_9 are canonical patterns respectively. Since the expansion of the internal structures is limited in the rightmost vertex, P_8 is not generated from P_9. On the other hand, P_8 will be generated by the internal expansion of the rightmost vertex of P_7.

Therefore, we cannot prune non-canonical patterns immediately. Non-canonical patterns have to be expanded in the internal structures until they become the canonical. However, external structures of non-canonical patterns need not to be expanded because canonical patterns will be never obtained.

3.3   Pruning Based on the Internal Closedness

It is easy to imagine that the number of frequent graph patterns grows exponentially since all sub-patterns of a frequent graph pattern are also frequent. To alleviate the explosion of frequent patterns, we introduce a pruning technique based on the *internal closedness*.

An occurrence of a graph pattern is represented as a set of edges. The set of all occurrences of a graph pattern P in a graph G is denoted as $emb_G(P)$. Furthermore, we define the occurrence of P in a database D as $Emb^D(P) = \cup_{G \in D} emb_G(P)$.

Suppose that a graph pattern P' is obtained from P by expanding the internal structure. If, for each occurrence $occ_p \in Emb^D(P)$, there exists at least one corresponding occurrence $occ_{p'} \in Emb^D(P')$, i.e. $occ_{p'}$ overlaps occ_p completely, then we denote it as $OM_D(P, P')$. If $OM_D(P, P')$, then $sup_D(P) = sup_D(P')$ holds by definition. Because P is a subgraph of P' and they have the same support value, P can be regarded as redundant. In addition, for any graph pattern Q obtained from P by expanding the external structure, a graph pattern Q' such that $sup_D(Q) = sup_D(Q')$ can be obtained from P' by expanding external

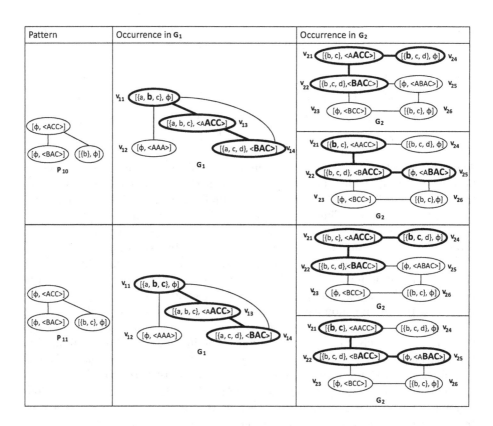

Fig. 4. Two patterns and their occurrences

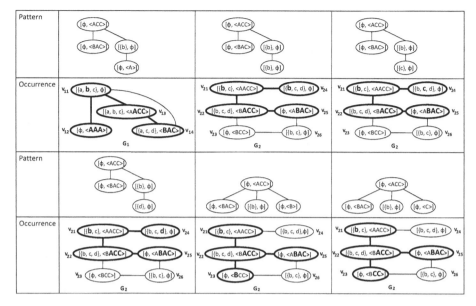

Fig. 5. Patterns obtained by expanding external structure of P_{10} and their occurrences

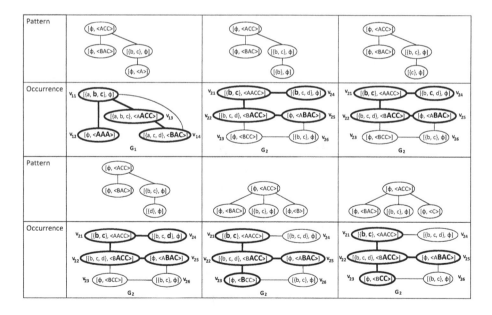

Fig. 6. Patterns obtained by expanding external structure of P_{11} and their occurrences

structure. Thus, the generation of Q is redundant. Therefore, the expansion of the external structure is not to be applied to P. In other words, P will be pruned. We call this pruning *internal closedness pruning*.

Fig. 4 shows frequent patterns and their occurences when σ is 2. P_{10} is a frequent pattern between G_1 and G_2. $emb_{G_1}(P_{10}) = \{\{e(v_{13}, v_{14}), e(v_{13}, v_{11})\}\}$, and $emb_{G_2}(P_{10}) = \{\{e(v_{21}, v_{22}), e(v_{21}, v_{24})\}, \{e(v_{22}, v_{25}), e(v_{22}, v_{21})\}\}$. At the same time, P_{11} is also a frequent pattern, and it is obtained by expanding the internal structure of P_{10}. $emb_{G_1}(P_{11}) = \{\{e(v_{13}, v_{14}), e(v_{13}, v_{11})\}\}$, and $emb_{G_2}(P_{11}) = \{\{e(v_{21}, v_{22}), e(v_{21}, v_{24})\}, \{e(v_{22}, v_{25}), e(v_{22}, v_{21})\}\}$. Because occurrences of P_{11} overlap those of P_{10} completely, it is clear that $OM(P_{10}, P_{11})$ holds. Showing in Fig. 5 and Fig. 6, since internal expansions do not influence the external structure, the search spaces after expanding external structures of P_{10} and P_{11} are the same. So more general pattern P_{10} is redundant, and this pattern is not allowed to expand the external structure. Introducing internal closedeness pruning, patterns shown in Fig. 5 are not enumerated actually.

3.4   Pruning Based on Monotone Constraints

The ability to handle monotone constraints will be incorporated into FMG. In this paper, we divide the monotone constraints into two kinds. The first one is called *external* monotone constraint which gives the restrictions on external structures. The requirement of the minimum number of vertices is an example of this kind of constraint. The second kind of constraint is called *internal* monotone constraint. It constrains patterns in a vertex, *e.g.* the minimum length of a sequence in a vertex.

As similar to the traditional top-down graph mining algorithms, it does not influence the generation of candidate patterns whether a certain pattern satisfies the given external monotone constraints in FMG. On the other hand, internal monotone constraints can be utilized for the effective pruning. In FMG, the

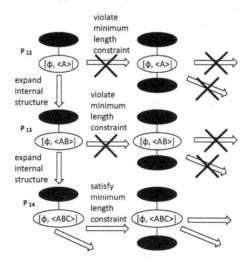

Fig. 7. Expanding with internal monotone constraint

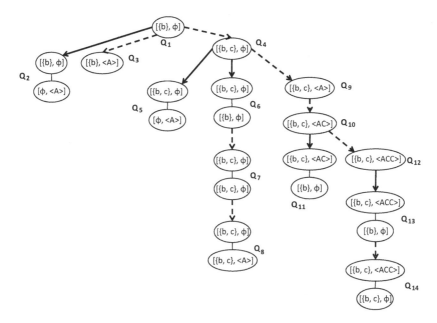

Fig. 8. A part of search space of FMG

expansion of the external structure does not be applied until the internal structure in the rightmost vertex satisfies the internal monotone constraints. In other words, the external structure will be expanded only after the internal structure in the rightmost vertex satisfies the internal monotone constraints. By employing the above enumeration strategy, we can avoid the generation of graph patterns which have non-rightmost vertices that do not satisfy the internal monotone constraints. Note that, no graph pattern satisfying the constraints can be enumerated by expanding graph patterns having non-rightmost vertices that do not satisfy the internal monotone constraints.

We will explain the pruning based on an internal monotone constraint with an example shown in Fig. 7. We assume that the given internal monotone constraint is that the length of a sequential pattern is more than 2. No pattern will be enumerated by expanding the external structure of P_{12} because the sequential pattern $\langle A \rangle$ violates the constraint. Thus, the only internal expansion will be applied to P_{12} and as a result, P_{13} will be obtained. Again, because P_{13} does not satisfy the internal monotone constraint, external expansion will not be applied. For P_{14} and its successors, both of internal and external expansions are allowed. As explained, introducing internal monotone constraints enable to reduce the number of patterns to be generated by expanding the external structure of the pattern.

On the other hand, suppose that we enumerate a pattern by expanding external structure of P_{12}. Then the pattern has a vertex that does not satisfy the internal monotone constraint. In FMG, internal expansion will be applied to the

---

Algorithm 1. FMG(σ, $\mathcal{D}$, C_{in}, C_{ex})

1: **Input**: a multi-structured graph database $\mathcal{D}$, minimum support σ, internal monotone constraint C_{in}, external monotone constraint C_{ex}.
2: **Output**: a set of frequent subgraphs $\mathcal{FG}$ that satisfies constraints.
3: $\mathcal{FG} := \emptyset$;
4: $\mathcal{SD} :=$ a set of initial patterns of $\mathcal{D}$;
5: **for all** $g \in \mathcal{SD}$ **do**
6:    call expand(g, σ, $\mathcal{D}$, $\mathcal{FG}$, C_{in}, C_{ex});
7: **end for**
8: **return** $\mathcal{FG}$;

---

rightmost vertex only. Therefore, no pattern satisfying the constraints can be obtained from such patterns. This shows that the pruning based on the internal monotone constraint does not affect on the completeness.

3.5   FMG: A Frequent Multi-structured Subgraph Pattern Miner

We show the pseudo code of FMG in Algorithms 1 and 2. In these algorithms, line 3–8 in expand/6 corresponds to the pruning based on the internal monotone constraint. If a frequent graph pattern g violates the internal monotone constraint, the external expansion will not be applied to g. On the other hand, regardless of the external monotone constraint, the external expansion will be applied only to the canonical patterns which satisfy the internal monotone constraint. The applicability of the internal closedness pruning is examined for the graph patterns which satisfy both of internal and external monotone constraints in line 10–21 in expand/6.

In Fig. 8, we show a part of search space of FMG for a database in Fig. 1 under the conditions that the minimum support σ is 2, minimum number of items in each vertex is 2, and minimum number of vertices is 2. The last two conditions are internal and external monotone constraints, respectively. In this figure, dashed lines denote the internal expansions, and solid lines indicate the external expansions. Q_1 is one of the initial patterns. Q_2 and Q_3 are not to be enumerated because of the pruning based on the internal monotone constraint. While Q_{11} can be obtained by applying external expansion of Q_{10}, FMG does not generate Q_{11} because the condition for the internal closedness pruning holds between Q_{10} and Q_{12}.

4   Experimental Results

In order to assess the effectiveness of the proposed algorithm, we implement the prototype of FMG in Java and conduct some preliminary experiments with real world datasets obtained from KEGG database. We use the pathway database of RIBOFLAVIN METABOLISM. This database includes pathways for different organisms. Three datasets of different size are prepared. The average number of vertices and edges of each dataset are shown in Table 1. The vertex of the

---

Algorithm 2. Expand(g, σ, $\mathcal{D}$, $\mathcal{FG}$, C_{in}, C_{ex})

1: **Input**: a multi-structured graph pattern g, minimum support σ, a multi-structured graph database $\mathcal{D}$, internal monotone constraint C_{in}, external monotone constraint C_{ex}.
2: **if** $sup_{\mathcal{D}}(g) \geq \sigma$ **then**
3:     **if** g violates C_{in} **then**
4:         $\mathcal{P}_{in}$:= a set of patterns obtained by expanding the internal structure of g;
5:         **for all** $p \in \mathcal{P}_{in}$ **do**
6:             call expand(p, σ, $\mathcal{D}$, $\mathcal{FG}$, C_{in}, C_{ex});
7:         **end for**
8:     **else**
9:         $\mathcal{P}_{in}$:= a set of patterns obtained by expanding the internal structure of g;
10:         $bool := false$;
11:         **if** g satisfies C_{ex} **then**
12:             $bool := true$;
13:             **for all** $p_{in} \in \mathcal{P}_{in}$ **do**
14:                 **if** $OM_{D}(g, p_{in})$ **then**
15:                     $bool := false$;
16:                 **end if**
17:             **end for**
18:             **if** $bool \wedge g$ is canonical **then**
19:                 $\mathcal{FG} := \mathcal{FG} \cup \{g\}$;
20:             **end if**
21:         **end if**
22:         $\mathcal{P}_{ex} := \emptyset$;
23:         **if** $bool \wedge g$ is canonical **then**
24:             $\mathcal{P}_{ex}$:= a set of patterns obtained by expanding the external structure of g;
25:         **end if**
26:         $\mathcal{P} := \mathcal{P}_{in} \cup \mathcal{P}_{ex}$;
27:         **for all** $p \in \mathcal{P}$ **do**
28:             call expand(p, σ, $\mathcal{D}$, $\mathcal{FG}$, C_{in}, C_{ex});
29:         **end for**
30:     **end if**
31: **end if**

---

pathway consists of enzymes, links to other pathway and compounds. Enzymes consist of an amino acid sequence and the enzyme number. Links are treated as an item. Compounds consist of a set of links to other pathway and its label. Those are treated as an item sets. As a result, we define the vertex element list as [item, number, item set, sequence].

The external and internal constraints used in the experiments are that the minimum number of vertices in the pattern is 5 and the minimum length of the sequence in a vertex is 7. All experiments were done on a PC(Intel Pentium IV, 2GHz) with 2GB of main memory running Windows XP. The experimental results are shown in Table 2.

While changing the minimum support value σ, we measured the number of patterns discovered (patterns), execution time (time), number of patterns pruned

Table 1. Three datasets

size	average vertices	average edges
50	20.24	18.78
100	21.03	19.51
150	20.29	18.67

Table 2. Experimental results with three datasets

size	σ [%]	patterns	time[sec]	closed	monotone
100	60	35	2885	5839	31110
	50	116	7256	15301	115646
	45	918	16789	41692	347786
	40	66762	49320	146993	1404278
50	50	8	4394	42140	144854
	40	26308	52610	470893	1925831
150	50	110	5106	6147	43714
	40	1077	23431	31180	296794

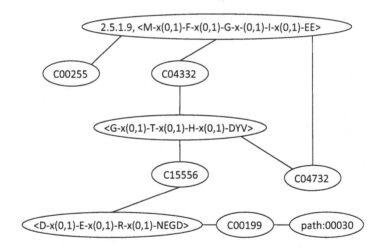

Fig. 9. An example of extracted patterns

by internal closedness pruning (closed) and number of patterns pruned by internal monotone constraints (monotone). Both the number of patterns discovered and execution time increase exponentially as the minimum support decreases. On the other hand, the number of patterns pruned by the internal closedness and the internal monotone constraint pruning are greater than the the number of patterns discovered. Not introducing the pruning techniques, patterns pruned are expanded to more specific patterns. We can easily imagine that it takes too long time to finish experiments.

We conducted another experiments by using FMG without internal monotone constraint. However, the result was not able to be obtained within 24 hours. These results indicate the effectiveness of two pruning techniques. We show an example of extracted patterns in Fig. 9. In this pattern, the vertex denoted as C00255 is riboflavin, the most important chemical compound in the riboflavin metabolism. At the same time, the sequential patterns of the enzyme are extracted. As shown, we succeeded in discovering a special kind of patterns with FMG.

5 Related Work

Recently, many graph mining algorithms have been developed and applied to several real world problems[1,2,5,6,7,9,10,11,15,16,17]. However, since most of these algorithms do not take the internal structure into account, they might fail to discover some meaningful patterns which will be found by FMG.

DAG Miner[3] differs from these traditional graph miners and it is one of the most related studies to FMG. DAG Miner extracts frequent patterns from directed acyclic graphs in which each vertex has an item sets. In DAG Miner, all of frequent item sets will be found in advance, and then, by using these item sets, a restricted form of frequent DAG patterns called pyramid patterns will be mined. In contrast to DAG miner, FMG enumerates internal and external structures simultaneously.

On the other hand, mining algorithms for complex tree-structured patterns have been proposed. FAT-miner[8] is an algorithm for discovering frequent tree patterns that consists of vertices holding a set of attributes. pFreqT[13] mines frequent subtrees in which each vertex forms a sequence. While these two miners handle the tree-structured data, the target of FMG is complex graphs. In addition, FMG permits the combination of various structured patterns in the vertex.

6 Conclusion

In this paper, we focus on the problem of frequent pattern discovery in complex graph structured databases. By combining several algorithms for mining (structured) data such as item sets, sequences and graphs, we propose an algorithm FMG for mining frequent patterns in multi-structured graphs. Through the preliminary experiments, we show the effectiveness of the proposed algorithm.

As one of the future works, we plan to exploit FMG in order to extract more meaningful patterns.

References

1. Borgelt, C.: On Canonical Forms for Frequent Graph Mining. In: Proc. of the 3rd International Workshop on Mining Graphs, pp. 1–12 (2005)
2. Borgelt, C., Berthold, M.R.: Mining molecular fragments: Finding relevant substructures of molecules. In: Proc. of the 2nd IEEE International Conference on Data Mining, pp. 51–58 (2002)

3. Chen, Y.L., Kao, H., Ko, M.: Mining DAG Patterns from DAG Databases. In: Li, Q., Wang, G., Feng, L. (eds.) WAIM 2004. LNCS, vol. 3129, pp. 579–588. Springer, Heidelberg (2004)
4. De Raedt, L., Washio, T., Kok, J.N. (eds.): Advances in Mining Graphs, Trees and Sequences, Frontiers in Artificial Intelligence and Applications, vol. 124. IOS Press, Amsterdam (2005)
5. Huan, J., Wang, W., Prins, J.: Efficient mining of frequent subgraphs in the presence of isomorphism. In: Proc. of the 3rd IEEE International Conference on Data Mining, pp. 549–552 (2003)
6. Inokuchi, A., Washio, T., Motoda, H.: An Apriori-Based Algorithm for Mining Frequent Substructures from Graph Data. In: Proc. of the 4th European Conference on Principles and Practice of Knowledge Discovery in Databases, pp. 13–23 (2000)
7. Inokuchi, A., Washio, T., Motoda, H.: Complete mining of frequent patterns from graphs: Mining graph data. Machine Learning 50, 321–354 (2003)
8. Knijf, J.D.: FAT-miner: Mining Frequent Attribute Trees. In: Proc. of the 2007 ACM symposium on Applied computing, pp. 417–422 (2007)
9. Koyuturk, M., Grama, A., Szpankowski, W.: An efficient algorithm for detecting frequent subgraphs in biological networks. Bioinformatics 20, 200–207 (2004)
10. Kuramochi, M., Karypis, G.: Frequent subgraph discovery. In: Proc. of the 1st IEEE International Conference on Data Mining, pp. 313–320 (2001)
11. Nijssen, S., Kok, J.N.: A Quickstart in Frequent Structure Mining can make a Difference. In: Proc. of 10th ACM SIGKDD International Conference on Knowledge Discovery and Data Mining, pp. 647–652 (2004)
12. Pei, J., Han, J., Mortazavi-Asl, B., Pnto, H., Chen, Q., Dayal, U., Hsu, M.: PrefixSpan: Mining Sequential Patterns Efficiently by Prefix-Projected Pattern Growth. In: Proc. of International Conference on Data Engineering, pp. 215–224 (2001)
13. Sato, I., Nakagawa, H.: Semi-structure Mining Method for Text Mining with a Chunk-Based Dependency Structure. In: Zhou, Z.-H., Li, H., Yang, Q. (eds.) PAKDD 2007. LNCS (LNAI), vol. 4426, pp. 777–784. Springer, Heidelberg (2007)
14. Uno, T., Asai, T., Uchida, Y., Arimura, H.: An Efficient Algorithm for Enumerating Closed Patterns in Transaction Databases. In: Proc. of Discovery Science (2004)
15. Washio, T., Motoda, H.: State of the art of graph-based data mining. SIGKDD Explorations 5(1), 59–68 (2003)
16. Yan, X., Han, J.: CloseGraph: Mining Closed Frequent Graph Patterns. In: Proc. of the 9th ACM SIGKDD International Conference on Knowledge Discovery and Data Mining, pp. 286–295 (2003)
17. Yan, X., Han, J.: gSpan: Graph-Based Substructure Pattern Mining. In: Proc. of the 2nd IEEE International Conference on Data Mining, pp. 721–724 (2002)

Finding Composite Episodes

Ronnie Bathoorn and Arno Siebes

Institute of Information & Computing Sciences, Utrecht University
P.O. Box 80.089, 3508TB Utrecht, The Netherlands
{ronnie,arno}@cs.uu.nl

Abstract. Mining frequent patterns is a major topic in data mining research, resulting in many seminal papers and algorithms on item set and episode discovery. The combination of these, called composite episodes, has attracted far less attention in literature, however. The main reason is that the well-known frequent pattern explosion is far worse for composite episodes than it is for item sets or episodes. Yet, there are many applications where composite episodes are required, e.g., in developmental biology were sequences containing gene activity sets over time are analyzed.

This paper introduces an effective algorithm for the discovery of a small, descriptive set of composite episodes. It builds on our earlier work employing MDL for finding such sets for item sets and episodes. This combination yields an optimization problem. For the best results the components descriptive power has to be balanced. Again, this problem is solved using MDL.

Keywords: Composite episodes, MDL.

1 Introduction

Frequent pattern mining is a major area in data mining research. Many seminal papers and algorithms have been written on the discovery of patterns such as item sets and episodes. However the combination of these two, called composite episodes, has attracted far less attention in the literature. Such composite episodes [1] are episodes of the form

$$\{A, B\} \rightarrow \{C, D\} \rightarrow \{E\}.$$

There are applications were one would like to discover frequent composite episodes. In developmental biology one has data sets that consist of time series were at each time point sets of events are registered. In one type of data, these events are the active genes at that moment in the development. In another type of data, the events are the morphological characters that occur for the first time at that moment in the development. For both types of data, frequent composite episodes would yield important insight in the development of species.

The main reason why there has been little attention to the discovery of composite episodes in the literature is that the frequent pattern explosion is worse for composite episodes than it is for both frequent item sets and for frequent episodes. In other words, the number of frequent patterns quickly explodes. For example, if $\{A, B\} \rightarrow \{C, D\}$ is frequent, then so are $\{A\}$, $\{A\} \rightarrow \{C\}$, $\{A\} \rightarrow \{D\}$, $\{A\} \rightarrow \{C, D\}$, $\{B\}$, $\{B\} \rightarrow$

Z.W. Raś, S. Tsumoto, and D. Zighed (Eds.): MCD 2007, LNAI 4944, pp. 157–168, 2008.

$\{C\}, \{B\} \rightarrow \{D\}, \{B\} \rightarrow \{C, D\}, \{A, B\} \rightarrow \{C\}$, and $\{A, B\} \rightarrow \{D\}$. So, clearly an A Priori like property holds, but the number of results will simply swamp the user.

In related work [2] so called Follow-Correlation Item set-Pairs are extracted. These are patterns of the form $< A^m, B^n >$ meaning: B likely occurs n times after A occurs m times. Patterns of this form only describe the interaction between two subsequent item sets and their complexity lies somewhere between item sets and composite episodes. And unlike our method this method does not offer a solution to restrict the number of patterns generated for low minimal support values. Other related work, item set summarization [3], does offer a method to restrict the number of item sets. However, there is no straightforward generalization to composite episodes.

In earlier work [4] we showed that MDL can be used to select interesting patterns that give a good description of the database. This paper extends on our earlier work on the use of MDL to select interesting item sets and episodes, we propose a method that reduces the number of generated patterns before it starts combining item sets into composite episodes. This reduces the number of generated patterns dramatically, while still discovering the important patterns.

Briefly the method works as follows: using a reduced set of item sets as building blocks for the patterns in the time sequences, we limit the number of possible patterns for a given dataset. MDL is used to select the item sets extracted from the data that contribute the most to the compression of that data as shown in [5]. Then the reduced set of patterns is used to encode the database after which episodes are extracted from this encoded database. Finally MDL is used again to reduce this set of episodes [6].

Simply running these two stages independently after each other, however, doesn't necessarily produce the best results. A too selective first stage will limit our abilities to select good composite sequences in the second stage. The selectivity of both stages has to be balanced for optimal results. We again use MDL to achieve this balance.

The rest of this papers is structured as follows. In Section 2 we will give some definitions of the concepts used in the rest of this paper. Section 3 introduces our 2-Fold MDL compression method used to extract composite episodes. Section 4 introduces the dataset we used in the experiments. The experiments and their results are discussed in Section 5. Section 6 contains our conclusions.

2   Composite Episode Patterns

Episodes are partially ordered sets of events that occur frequently in a time sequence. Mannila described 3 kind of episodes in [1], parallel episodes which can be seen as item sets, serial episodes and composite episodes as shown in figure 1.

The data and the composite episodes can be formalized as follows: For item sets x_i and x_j in a sequence x, we will use the notation $x_i \preceq x_j$ to denote that x_i occurs before x_j in x.

Definition 1. *Given a finite set of events* $\mathcal{I}$,

1. *A sequence* s *over* $\mathcal{I}$ *is an ordered set of item sets*

$$s = \{(is_i, i)\}_{i \in \{1,\dots,n\}},$$

in which the $is_i \subseteq \mathcal{I}$. *If* $1 \leq i \leq j \leq n$, *then* $(is_i, i) \preceq (is_j, j)$.

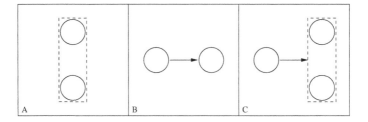

Fig. 1. (a) Item set, (b) episode, (c) composite episode

2. *An item set is is a set of events that happen together.*

$$is = (e_1, \ldots, e_j)$$

where j is the size of the item set.
3. *A composite episode ep is a sequence of item sets.*

$$ep = (is_1, \ldots, is_k)$$
$$is_i = (e_1, \ldots, e_{il})$$

where i_l is the size of the ith item set.

The database db consists of a set of sequences of item sets, i.e., it consists of composite episodes. In this database, we want to find the frequent composite episodes. To define these, we need the notion of an occurrence of a composite episode. Note that, because of our application, we do not allow gaps in occurrences.

Definition 2

1. *Let x be composite episodes and y be a sequence. Let I be the set of composite episodes and $\Phi : I \to I$ an injective mapping. x occurs in y, denoted by $x \subseteq y$, iff*
    (a) $\forall x_i \in x : x_i \subseteq \Phi(x_i)$
    (b) $\forall x_i, x_j \in x$:
        i. $\Phi(x_i) \preceq \Phi(x_j) \Leftrightarrow x_i \preceq x_j$
        ii. $\exists y_k \in y : \Phi(x_i) \preceq y_k \preceq \Phi(x_j) \Leftrightarrow$
            $\exists x_k \in x : \Phi(x_k) = y_k \land x_i \preceq x_k \preceq x_j$.
    The mapping Φ is called the occurrence.
2. *Length of an occurrence o is time interval between the first (t_s) and the last (t_e) event in the occurrence*

$$length(o) = t_e(o) - t_s(o)$$

3. *Support of an episode is the number of occurrences of the episode in the database.*

So, $(\{A, B\}, \{C\})$ occurs once in $\{A, B, C\}, \{C, D\}$, while $(\{A\}, \{B\})$ doesn't.

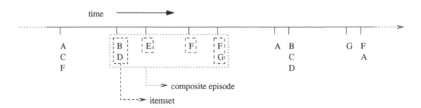

Fig. 2. Example sequence with a composite episode occurrence highlighted

2.1 Pattern Explosion

As noted before, the number of composite episodes quickly explodes. In a composite episode of length 10 containing 3 possible events we have 7 possible item sets. This leads to $\Sigma_{i=1}^{10} 7^i = 329554456$ possible composite episodes. More in general, with n events and a sequence of length k we have as number of possible composite episodes:

$$\sum_{i=1}^{k}(2^n - 1)^i$$

Clearly, if the number of frequent item sets is not the maximum, then the number of possibly frequent composite episodes also goes down.

While the growth remains exponential, the fewer (frequent) item sets we consider, the fewer composite episodes we have to consider. This is exactly the power of our approach: by dramatically reducing the number of item sets to consider, the number of of composite episodes to consider becomes manageable.

2.2 Item Set MDL

The basic building blocks of our database are the items $\mathcal{I}$, e.g., the items for sale in a shop. A transaction $t \in \mathcal{P}(\mathcal{I})$ is a set of items, e.g. representing the items a client bought at that store. A database db over $\mathcal{I}$ is a bag of transactions, e.g., the different sale transactions on a given day. An item set $I \in \mathcal{I}$ occurs in a transaction $t \in db$ iff $I \subseteq t$. The support of I in db is the number of transactions in the database in which I occurs.

We will now give a quick summary on how MDL can be used to select a small and descriptive set of item sets, using the Krimp algorithm which was introduced in [5]. This is a shortened version of the description given in [4].

The key idea of our compression based approach is the code table, a code table has item sets on the left-hand side and a code for each item set on its right-hand side. The item sets in the code table are ordered descending on 1) item set length and 2) support. The actual codes on the right-hand side are of no importance: their lengths are. To explain how these lengths are computed we first have to introduce the coding algorithm. A transaction t is encoded by Krimp by searching for the first item set c in the code table for which $c \subseteq t$. The code for c becomes part of the encoding of t. If $t \setminus c \neq \emptyset$, the algorithm continues to encode $t\ c$. Since we insist that each code table contains at least all singleton item sets, this algorithm gives a unique encoding to each (possible) transaction. The set of item sets used to encode a transaction is called its cover. Note that the coding algorithm implies that a cover consists of non-overlapping item sets. The length of the code of an item in a code table CT_i depends on the database we want to compress; the more often a code is used, the shorter it should be. To compute this code length, we encode each transaction in the database db. The frequency of an item set $c \in CT$ is the number of transactions $t \in db$ which have c in their cover. The relative frequency of $c \in CT_i$ is the probability that c is used to encode an arbitrary $t \in db$. For optimal compression of db, the higher P(c), the shorter its code should be. In fact, from information theory [7] we have the optimal code length for c as:

$$l_{CT_i}(c) = -\log(P(c|db)) = -\log\left(\frac{freq(c)}{\sum_{d\in CT_i} freq(d)}\right) \tag{1}$$

The length of the encoding of a transaction is now simply the sum of the code lengths of the item sets in its cover. Therefore the encoded size of a transaction $t \in db$ compressed using a specified code table CT_i is calculated as follows:

$$L_{CT_i}(t) = \sum_{c\in cover(t,CT_i)} l_{CT_i}(c) \tag{2}$$

The size of the encoded database is the sum of the sizes of the encoded transactions, but can also be computed from the frequencies of each of the elements in the code table:

$$L_{CT_i}(db) = \sum_{t\in db} L_{CT_i}(t) = -\sum_{c\in CT_i} freq(c)\cdot \log\left(\frac{freq(c)}{\sum_{d\in CT_i} freq(d)}\right) \tag{3}$$

Finding the Right Code Table. To find the optimal code table using MDL, we need to take into account both the compressed database size as described above as well as the size of the code table. (Otherwise, the code table could grow without limits and become even larger than the original database!) For the size of the code table, we only count those item sets that have a non-zero frequency. The size of the right-hand side column is obvious; it is simply the sum of all the different code lengths. For the size of the left-hand side column, note that the simplest valid code table consists only of the singleton item sets. This is the *standard encoding (st)* which we use to compute the size of the item sets in the left-hand side column. Hence, the size of the code table is given by:

$$L(CT) = \sum_{c\in CT:freq(c)\neq 0} l_{st}(c) + l_{CT}(c) \tag{4}$$

In [5] we defined the optimal set of (frequent) item sets as that one whose associated code table minimizes the total compressed size:

$$L(CT) + L_{CT}(db) \tag{5}$$

The algorithm starts with a valid code table (generally only the collection of singletons) and a sorted list of candidates. These candidates are assumed to be sorted descending on 1) support and 2) item set length. Each candidate item set is considered by inserting it at the right position in CT and calculating the new total compressed size. A candidate is only kept in the code table iff the resulting total size is smaller than it was before adding the candidate. For more details, please see [5].

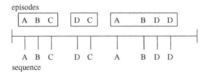

Fig. 3. Example of a sequence cover

2.3 Episode MDL

For episode mining the basic building blocks of our database db are item sets I, e.g. all the genes active at one point in time. Each transaction $t \in db$ is a sequence of item sets, e.g. a time sequence recording the activity of genes over time. An episode e occurs in a transaction t if all the item sets in e occur in t without gaps between them as described in Definition 2. Reducing a set of episodes using MDL follows the same steps as used in item set MDL. Thus we start with a codetable with two columns, it has an episode on it's left-hand side and a code for each episode on the right-hand side. The code table is used to cover all sequences in the database, an example of such a cover can be seen in Figure 3. The frequency with which the codes in the code table are used to cover all the sequences in the database determines their code size, the more a code is used the shorter its code. It is important to note that because of our application in developmental biology we do not allow overlap between the episodes in a cover, or gaps within the episodes. To determine the size of our episode code table we need to define a *standard encoding* for episodes l_{st_e} as well. As the length of an episode in the codetable we use the length of that episode as it would be when we encoded it using only episodes of length 1, this is called this episodes *standard encoding*. With this standard encoding the size of our episode code table CT_e becomes:

$$L(CT_e) = \sum_{c \in CT_e : freq(c) \neq 0} l_{st_e}(c) + l_{CT_e}(c) \qquad (6)$$

Using the episodes in our code table to encode our database leads to the following database size:

$$L_{CT_e}(db) = \sum_{t \in db} L_{CT_e}(t) = - \sum_{c \in CT_e} freq(c) \cdot \log\left(\frac{freq(c)}{\sum_{d \in CT_e} freq(d)}\right) \qquad (7)$$

More details on reducing frequent episode sets using MDL can be found in [6].

2.4 Combining Item Set and Episode MDL

In our method for finding composite episodes we are combing item set an episode MDL. First we use a set of item sets to compress our database. Then we extract episodes from the encoded database that results from the item set compression. Using MDL we select the episodes that give a good compression of our *item set encoded* database. To determine which item sets and episodes are used in the compression we have to optimize L_{total}.

$$L_{total} = L(CT_i) + L(CT_e) + L_{CT_e}(enc(CT_i, db)) \qquad (8)$$

where $enc(CT_i, db)$ is the database as encoded by item set code table CT_i. This last equation shows that to compute the total size we now need the item set code table CT_i as well as the episode code table CT_e plus the double encoded database.

3   2-Fold MDL Compression

The basis of the algorithm used to find the composite episode patterns consists of 4 steps.

BASE($data, min\_sup, max\_length$)

1   Find item sets for given min_sup
2   Compress the database using item set MDL
3   Find episodes in the encoded database with given max_length
4   Compress the database using episode MDL
5   **return** *composite episodes*

Fig. 4. Base algorithm

In the first step we use a standard FP-growth algorithm [8] to find all the item sets for a given minimal support. Which minimal support to use is the subject of the next subsection.

The set of item sets is used in the second step where MDL is used to heuristically select those item sets that give the shortest description of the database. This results in a compressed database together with a code table used to obtain the compressed database as described in Section 2.2. This code table is a set of item sets selected from all frequent item sets and the encoded database is a copy of the original database in which all occurrences of codetable elements are replaced by a code that represents this item set.

In the third step we get all frequent episodes from the encoded database that was generated in the previous step. The frequent episodes are extracted using a minimal occurrence episode discovery algorithm from [1]. Note that because we extract our episodes from the encoded database each event in the discovered episodes could now also be an item set, thus the episodes extracted are composite episodes.

And finally we use our method from Section 2.3 to get a set of episodes that give a good description of the database.

The output of our method consists of 2 codetables one from step 2 and one from step 4 together with the compressed database from step 4. Our MDL method enforces a loss-less compression of the database thus the 2 code tables can be used to decompress the database and generate the original dataset.

3.1   Compression Optimization

What is the right minimal support to use for extracting the item sets from the data? The minimal support limits the amount of episodes that could possibly be found. This can

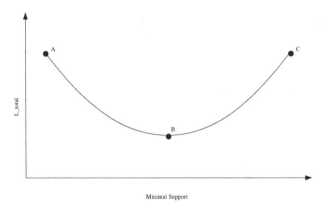

Fig. 5. Compressed Database size against minimal support values

2-FOLD($data, max\_length, start, end$)

1    $best_result = $ BASE($data, start, max\_length$)

2   **foreach** $minsup$ **in** $[start + 1..end]$

3        $cur_result = $ BASE($data, minsup, max\_length$)

4   **if** ($cur\_result.mld\_size < best\_result.mdl\_size$) **then**

5            $best\_result = cur\_result$

6   **return** $best\_result$

Fig. 6. Complete algorithm

be seen as an optimization problem where we take L_{total} (equation 8) as the value to be optimized. We are interested in finding the minimal support that results in the lowest possible value of L_{total}.

Figure 5 shows the overall compression of the database for different minimal support levels. On the x-axis we have the minimal support used in the extraction of the item sets. Changes in the compression are caused by the interaction of item sets and episodes used in the compression.

At point 'A' in the graph the minimal support is set to 1, which means that all possible item sets will be extracted from the database. As the item sets are extracted before the episodes this means there is a strong bias towards the use of large item sets. Additionally the reduced set of item sets generated with the use of MDL is used in the encoding of our database before we proceed with the extraction of episodes. This will lower the probability of finding long episodes as they have to be build up of large item sets. As all item sets used in the encoding of our database are substituted by a single code this makes it impossible to use subsets of these item sets.

Increasing the minimal support will lower the number and size of the found item sets and will increase the possibility of longer episodes being used in the compression. After reaching a certain threshold no item sets are used anymore and the compression will be based solely on episodes. This point is reached at 'C'.

Because of this interaction between item sets and episodes, we expect the best compression of our database somewhere near point 'B' in the graph. This changes our problem of finding the best minimal support for our algorithm to an optimization problem where we optimize the compression of the database by varying the minimal support. Our base algorithm can be extended by putting it inside a loop that runs the method for all the minimal support values we are interested in. So now the entire algorithm becomes as can be seen in Figure 6.

Computing this optimal solution comes at the prize of having to do one run of the composite episode extraction for each minimal support value. But as these runs do not depend on each other they can be run on different processors or different computers all together. Making this algorithm well suited for parallel computation and cutting the runtime down to the runtime of one run of the composite episode extraction algorithm.

4   The Data

For our experiments we use two datasets from the biological domain. The first dataset contains time sequences containing developmental information of 5 different species. In these time sequences the timing of the activity of 28 different genes are recorded. There are large differences in the time sequences in length as well as in the number of times the events are present in each. More background information on the biology involved in the analysis of developmental sequences can be found in [9]. The second dataset contains time sequences of 24 mammals. It records the start of 116 different morphological characters such as the forming of the optic lens.

Table 1. Dataset description

dataset	#sequences	#events
gene activity	5 species	28 genes
morphological characters	24 species	116 characters

The datasets currently produced by the biologists are so small that the codetable is very large in relation to the database hampering the compression. As the biologists are working on producing bigger datasets in the future, we used the following method to test our method on a bigger datasets. The gene activity dataset was enlarged by combining time sequences of two randomly chosen species in a new time sequence by adding a reversed copy of the second time sequence to the back of the first. This recombination is used to preserve the types of patterns that can be found in the data but increases the number of patterns found as well as their frequency. The artificial dataset contains 10 of these combined time sequences, doubling the number of time sequences as well as the average length of these sequences.

5   Experimental Results

The first experiment was done on the original gene activity dataset and composite episodes were extracted for minimal support values ranging from 3 to 21. Figure 7 shows

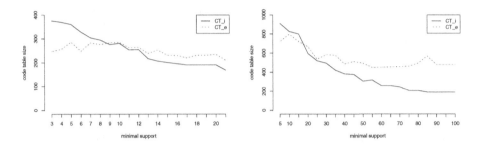

Fig. 7. Item set code table (ct_i) and episode code table (ct_e) sizes for the original (left) and the artificial (right) dataset

the size of the code tables for the different minimal support values. Here we can see that for lower minimal support levels the item set codetable is bigger than the episode code table and this is the other way around when the minimal support is increased.

In our experiments on the morphological characters dataset the algorithm was run multiple times for 7 different minimal support values ranging from 2 to 8. Where the minimal support is the minimal support for the item set extraction. The item sets are extracted using the implementation of fp-growth taken from [10]. The episodes were extracted using 3 different maximal episode lengths, 25, 35 and 45. It is important to note that the end result of our method is a set of composite episodes, the compression values in this experiment are only used to select the best minimal support for the item set discovery.

Figure 8 (left) shows the compressed database size as a function of the minimal support of the item set extraction for two maximal episode lengths. The compressed database size shown in the figure is the sum of the item set codetable, the episode codetable and the size of the encoded database. The figure looks very similar to figure 5 which was what we predicted based on the interaction of the item sets and the episodes. In figure 8 we can see that we reach the best compression for a minimal support of 4. The same experiment was done on the artificially enlarged dataset. With item sets being extracted for 20 different minimal support values ranging from 5 to 100. The episodes are extracted using 2 different maximal episode lengths, 250 and 350. The results are also shown in Figure 8 (right) we can see that we reach the best compression for a minimal support of 40.

To give an indication of the amount of reduction reached in the total number of patterns generated, only between 0.002% and 6.667% of the frequent item sets were selected by MDL. The reduction decreased for higher minimal support. As we showed in Section 2.1 this gives a tremendous reduction in the possible composite episodes. For the frequent episodes only between 0.33% and 1,7% of the episodes were selected by MDL as being interesting. Here the reduction was better for higher minimal support, due to the interaction between the item sets and the episodes in the total compression.

For the original dataset 2-Fold started with a set of 58 item sets from which it constructed 18 composite episodes. An example of such a composite episode: $\{hoxc10\} \rightarrow \{hoxd11, hoxd12\}$ which describes the temporal collinearity of hox genes that Biologists already know from their experiments.

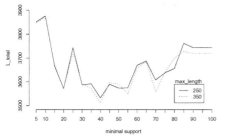 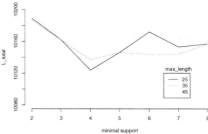

Fig. 8. Compressed Database size against minimal support values for the gene activity dataset (left) and the morphological dataset (right)

6   Conclusions and Future Work

In this paper we show that it is possible to mine for the descriptive composite episodes from data. The 2-Fold algorithm uses MDL to keep the combinatorial explosion of potential patterns under control. 2-Fold uses MDL in three different ways. Firstly to mine for descriptive item sets. Secondly to mine for descriptive episodes. Thirdly to balance the first two, to ensure the discovery of descriptive composite episodes. The experiments show first of all that MDL performs well in all three of its tasks. The number of composite episodes discovered is small enough that experts can still verify them. Moreover, the validity of the results we discovered has been verified by domain experts.

Acknowledgment

We would like to thank Matthijs van Leeuwen & Jilles Vreeken for their short introduction on item set MDL.

References

1. Mannila, H., Toivonen, H., Verkamo, A.I.: Discovery of frequent episodes in event sequences. Data Mining and Knowledge Discovery 1, 259–289 (1997)
2. Zhang, S., Zhang, J., Zhu, X., Huang, Z.: Identifying follow-correlation itemset-pairs. In: ICDM 2006: Proceedings of the Sixth International Conference on Data Mining, pp. 765–774. IEEE Computer Society, Washington (2006)
3. Wang, C., Parthasarathy, S.: Summarizing itemset patterns using probabilistic models. In: KDD 2006: Proceedings of the 12th ACM SIGKDD international conference on Knowledge discovery and data mining, pp. 730–735. ACM Press, New York (2006)
4. van Leeuwen, M., Vreeken, J., Siebes, A.: Compression picks item sets that matter. In: Fürnkranz, J., Scheffer, T., Spiliopoulou, M. (eds.) PKDD 2006. LNCS (LNAI), vol. 4213, pp. 585–592. Springer, Heidelberg (2006)
5. Siebes, A., Vreeken, J., van Leeuwen, M.: Itemsets that compress. In: SIAM 2006: Proceedings of the SIAM Conference on Data Mining, Maryland, USA, pp. 393–404 (2006)

6. Bathoorn, R., Koopman, A., Siebes, A.: Reducing the frequent pattern set. In: Tsumoto, S., Clifton, C., Zhong, N., Wu, X., Liu, J., Wah, B., Cheung, Y.M. (eds.) ICDM 2006: Proceedings of the 6th International Conference on Data Mining - Workshops, ICDM workshops, vol. 6, pp. 55–59. IEEE Computer Society, Los Alamitos (2006)
7. Grünwald, P.: A tutorial introduction to the minimum description length principle. In: Advances in Minimum Description Length, MIT Press, Cambridge (2005)
8. Han, J., Pei, J., Yin, Y.: Mining frequent patterns without candidate generation. In: Chen, W., Naughton, J., Bernstein, P.A. (eds.) 2000 ACM SIGMOD Intl. Conference on Management of Data, 05 2000, pp. 1–12. ACM Press, New York (2000)
9. Welten, M.C.M., Verbeek, F.J., Meijer, A.H., Richardson, M.K.: Gene expression and digit homology in the chicken embryo wing. Evolution & Development 7, 18–28 (2005)
10. Rácz, B., Bodon, F., Schmidt-Thieme, L.: On benchmarking frequent itemset mining algorithms. In: Proceedings of the 1st International Workshop on Open Source Data Mining, in conjunction with ACM SIGKDD (2005)

Ordinal Classification with Decision Rules

Krzysztof Dembczyński[1], Wojciech Kotłowski[1], and Roman Słowiński[1,2]

[1] Institute of Computing Science, Poznań University of Technology,
60-965 Poznań, Poland
{kdembczynski,wkotlowski,rslowinski}@cs.put.poznan.pl
[2] Institute for Systems Research, Polish Academy of Sciences, 01-447 Warsaw, Poland

Abstract. We consider the problem of ordinal classification, in which a value set of the decision attribute (output, dependent variable) is finite and ordered. This problem shares some characteristics of multi-class classification and regression, however, in contrast to the former, the order between class labels cannot be neglected, and, in the contrast to the latter, the scale of the decision attribute is not cardinal. In the paper, following the theoretical framework for ordinal classification, we introduce two algorithms based on gradient descent approach for learning ensemble of base classifiers being decision rules. The learning is performed by greedy minimization of so-called threshold loss, using a forward stagewise additive modeling. Experimental results are given that demonstrate the usefulness of the approach.

1 Introduction

In the prediction problem, the aim is to predict the unknown value of an attribute y (called *decision attribute, output* or *dependent variable*) of an object using known joint values of other attributes (called *condition attributes, predictors,* or *independent variables*) $\mathbf{x} = (x_1, x_2, \ldots, x_n)$. In the *ordinal classification*, it is assumed that $y = \{r_1, \ldots, r_K\}$, with r_k, $k \in \mathcal{K} = \{1, \ldots, K\}$, being K distinct and ordered class labels $r_K \succ r_{K-1} \succ \ldots \succ r_1$, where $\succ$ denotes the ordering relation between labels. Let us assume in the following, without loss of generality, that $r_k = k$. This problem shares some characteristics of multi-class classification and regression. A value set of y is finite, but in contrast to the multi-class classification, the order between class labels cannot be neglected. The values of y are ordered, but in contrast to regression, the scale of y is not cardinal. Such a setting of the prediction problem is very common in real applications. For example, in recommender systems, users are often asked to evaluate items on five value scale (see Netflix Prize problem [16]). Another example is the problem of email classification to ordered groups, like: "very important", "important", "normal", and "later".

The problem of ordinal classification is often solved by multi-class classification or regression methods. In recent years, however, some new approaches tailored for ordinal classification were introduced [13,6,7,18,17,3,14,15]. In this paper, we take first a closer look at the nature of ordinal classification. Later

Z.W. Raś, S. Tsumoto, and D. Zighed (Eds.): MCD 2007, LNAI 4944, pp. 169–181, 2008.
© Springer-Verlag Berlin Heidelberg 2008

on, we introduce two novel algorithms based on gradient descent approach for learning ensemble of base classifiers. The learning is performed by greedy minimization of so-called threshold loss [17] using a forward stagewise additive modeling [12]. As a base classifier, we have chosen single decision rule which is a logical expression having the form: *if [conditions], then [decision]*. This choice is motivated by simplicity and ease in interpretation of decision rule models. Recently, one can observe a growing interest in decision rule models for classification purposes (e.g. such algorithms like SLIPPER [5], LRI [19], RuleFit [11], ensemble of decision rules [1,2]).

Finally, we report experimental results that demonstrate the usefulness of the proposed approach for ordinal classification. In particular our approach is competitive to traditional regression and multi-class classification methods, and also to existing ordinal classification methods.

2   Statistical Framework for Ordinal Classification

Similarly to classification and regression, the task is to find a function $F(\mathbf{x})$ that predicts accurately an ordered label of y. The optimal prediction function (or Bayes optimal decision) is given by:

$$F^*(\mathbf{x}) = \arg\min_{F(\mathbf{x})} E_{y\mathbf{x}} L(y, F(\mathbf{x})) \tag{1}$$

where the expected value $E_{y\mathbf{x}}$ is over joint distribution of all variables $P(y, \mathbf{x})$ for the data to be predicted. $L(y, F(\mathbf{x}))$ is a loss or cost for predicting $F(\mathbf{x})$ when the actual value is y. $E_{y\mathbf{x}} L(y, F(\mathbf{x}))$ is called *prediction risk* or *expected loss*. Since $P(y, \mathbf{x})$ is generally unknown, the learning procedure uses only a set of training examples $\{y_i, \mathbf{x}_i\}_1^N$ to construct $F(\mathbf{x})$ to be the best possible approximation of $F^*(\mathbf{x})$. Usually, it is performed by minimization of *empirical risk*:

$$R_e = \frac{1}{N} \sum_{i=1}^{N} L(y_i, F(\mathbf{x}_i)).$$

Let us remind that the typical loss function in binary classification (for which $y \in \{-1, 1\}$) is 0-1 loss:

$$L_{0-1}(y, F(\mathbf{x})) = \begin{cases} 0 & \text{if } y = F(\mathbf{x}), \\ 1 & \text{if } y \neq F(\mathbf{x}), \end{cases} \tag{2}$$

and in regression (for which $y \in \mathbb{R}$), it is squared-error loss:

$$L_{se}(y, F(\mathbf{x})) = (y - F(\mathbf{x}))^2. \tag{3}$$

One of the important properties of the loss function is a form of prediction function minimizing the expected risk $F^*(\mathbf{x})$, sometimes called *population minimizer* [12]. In other words, it is an answer to a question: what does a minimization of expected loss estimate on a population level? Let us remind that the population minimizers for 0-1 loss and squared-error loss are, respectively:

$$F^*(\mathbf{x}) = \mathrm{sgn}\left(\Pr(y = 1|\mathbf{x}) - 0.5\right), \qquad F^*(\mathbf{x}) = E_{y|\mathbf{x}}(y).$$

Table 1. Commonly used loss functions and their population minimizers

Loss function	Notation	$L(y, F(\mathbf{x}))$	$F^*(\mathbf{x})$
Binary classification, $y \in \{-1, 1\}$:			
Exponential loss	L_{exp}	$\exp(-y \cdot F(\mathbf{x}))$	$\frac{1}{2} \log \frac{\Pr(y=1|\mathbf{x})}{\Pr(y=-1|\mathbf{x})}$
Deviance	L_{dev}	$\log(1 + \exp(-2 \cdot y \cdot F(\mathbf{x})))$	$\frac{1}{2} \log \frac{\Pr(y=1|\mathbf{x})}{\Pr(y=-1|\mathbf{x})}$
Regression, $y \in \mathbb{R}$:			
Least absolute deviance	L_{lad}	$|y - F(\mathbf{x})|$	$\text{median}_{y|\mathbf{x}}(y)$

Apart from 0-1 and squared error loss, some other important loss functions are considered. Their definitions and population minimizers are given in Table 1.

In ordinal classification, one minimizes prediction risk based on the $K \times K$ loss matrix:

$$L_{K \times K}(y, F(\mathbf{x})) = [l_{ij}]_{K \times K} \tag{4}$$

where $y, F(\mathbf{x}) \in \mathcal{K}$, and $i = y$, $j = F(\mathbf{x})$. The only constraints that (4) must satisfy in ordinal classification problem are the following, $l_{ii} = 0, \forall i$, $l_{ik} \geq l_{ij}, \forall\, k \geq j \geq i$, and $l_{ik} \geq l_{ij}, \forall\, k \leq j \leq i$. Observe that for

$$l_{ij} = 1, \quad \text{if } i \neq j, \tag{5}$$

loss matrix (4) boils down to the 0-1 loss for ordinary multi-class classification problem. One can also simulate typical regression loss functions, such as least absolute deviance and squared-error, by taking:

$$l_{ij} = |i - j|, \tag{6}$$
$$l_{ij} = (i - j)^2, \tag{7}$$

respectively. It is interesting to see, what are the population minimizers of the loss matrices (5)-(7). Let us observe that we deal here with the multinomial distribution of y, and let us denote $\Pr(y = k|\mathbf{x})$ by $p_k(\mathbf{x})$. The population minimizer is then defined as:

$$F^*(\mathbf{x}) = \arg \min_{F(\mathbf{x})} \sum_{k=1}^{K} p_k(\mathbf{x}) \cdot L_{K \times K}(k, F(\mathbf{x})). \tag{8}$$

For loss matrices (5)-(7) we obtain, respectively:

$$F^*(\mathbf{x}) = \arg \max_{k \in \mathcal{K}} p_k(\mathbf{x}), \tag{9}$$

$$F^*(\mathbf{x}) = \text{median}_{p_k(\mathbf{x})}(y) = \text{median}_{y|\mathbf{x}}(y), \tag{10}$$

$$F^*(\mathbf{x}) = \sum_{k=1}^{K} k \cdot p_k(\mathbf{x}) = E_{y|\mathbf{x}}(y). \tag{11}$$

In (11) it is assumed that the range of $F(\mathbf{x})$ is a set of real values.

The interesting corollary from the above is that in order to solve ordinal classification problem one can use any multi-class classification method that estimates $p_k(\mathbf{x})$, $k \in \mathcal{K}$. This can be, for example, logistic regression or gradient boosting machine [9]. A final decision is then computed according to (8) with respect to chosen loss matrix. For (5)-(7) this can be done by computing mode, median or average over y with respect to estimated $p_k(\mathbf{x})$, respectively. For loss matrix entries defined by (7) one can use any regression method that aims at estimating $E_{y|\mathbf{x}}(y)$. We refer to such an approach as *simple ordinal classifier*.

Let us notice that multi-class classification problem is often solved as K (one class against $K - 1$ classes) or $K \times (K - 1)$ (one class against one class) binary problems. However, taking into account the order on y, we can solve the ordinal classification by solving $K - 1$ binary classification problems. In the k-th ($k = 1, \ldots, K - 1$) binary problem, objects for which $y \leq k$ are labeled as $y' = -1$ and objects for which $y > k$ are labeled as $y' = 1$. Such an approach has been used in [6].

The ordinal classification problem can also be formulated from a value function perspective. Let us assume that there exists a latent value function that maps objects to scalar values. The ordered classes correspond to contiguous intervals on a range of this function. In order to define K intervals, one needs $K + 1$ thresholds: $\theta_0 = -\infty < \theta_1 < \ldots < \theta_{K-1} < \theta_K = \infty$. Thus k-th class is determined by $(\theta_{k-1}, \theta_k]$. The aim is to find a function $F(\mathbf{x})$ that is possibly close to any monotone transformation of the latent value function and to estimate thresholds $\{\theta_k\}_1^{K-1}$. Then, instead of the loss matrix (4) one can use immediate-threshold or all-threshold loss [17] defined respectively as:

$$L^{imm}(y, F(\mathbf{x})) = L(1, F(\mathbf{x}) - \theta_{y-1}) + L(-1, F(\mathbf{x}) - \theta_y), \qquad (12)$$

$$L^{all}(y, F(\mathbf{x})) = \sum_{k=1}^{y-1} L(1, F(\mathbf{x}) - \theta_k) + \sum_{k=y}^{K-1} L(-1, F(\mathbf{x}) - \theta_k). \qquad (13)$$

In the above, $L(y, f)$ is one of the standard binary classification loss functions. When using exponential or deviance loss, (12) and (13) become continuous and convex functions that are easy to minimize.

There is, however, a problem with interpretation what does minimization of expected threshold loss estimate. Only in the case when 0-1 loss is chosen as the basis of (12) and (13), the population minimizer has a nice interpretable form. For (12), we have:

$$F^*(\mathbf{x}) = \arg\min_{F(\mathbf{x})} \sum_{k=1}^{K} p_k(\mathbf{x}) \cdot L_{0-1}^{imm}(k, F(\mathbf{x})) = \arg\max_{k \in \mathcal{K}} p_k(\mathbf{x}), \qquad (14)$$

and for (13), we have:

$$F^*(\mathbf{x}) = \arg\min_{F(\mathbf{x})} \sum_{k=1}^{K} p_k(\mathbf{x}) \cdot L_{0-1}^{all}(k, F(\mathbf{x})) = \text{median}_{y|\mathbf{x}}(y). \qquad (15)$$

An interesting theoretical result is obtained in [15], where (12) and (13) are used in derivation of the upper bound of generalization error for any loss matrix (4).

Threshold loss functions were already considered in building classifiers. In [17] the classifier was learned by conjugate gradient descent. Among different base loss functions, also deviance was used. In [18,3,15], a generalization of SVM (support vector machines) was derived. The algorithm based on AdaBoost [8] was proposed in [15]. In the next section, we present two algorithms based on forward stagewise additive modeling. The first one is an alternative boosting formulation for threshold loss functions. The second one is an extension of the gradient boosting machine [9].

Let us remark at the end of our theoretical considerations that (13) can also be formulated as a specific case of so-called rank loss [13,7,4]:

$$L_{rank}\big(y_1, y_2, F(\mathbf{x}_1), F(\mathbf{x}_2)\big) = L\big(\text{sgn}(y_1 - y_2), F(\mathbf{x}_1) - F(\mathbf{x}_2)\big). \quad (16)$$

This loss function requires that all objects are compared pairwise. Assuming that thresholds $\{\theta_k\}_1^{K-1}$ are values of $F(\mathbf{x})$ for some virtual objects/profiles and all other objects are compared only with these virtual profiles, one obtains (13). Rank loss was used in [13] to introduce a generalization of SVM for ordinal classification problems, and in [7], an extension of AdaBoost for ranking problems was presented. The drawback of this approach is the complexity of empirical risk minimization defined by rank loss that grows quadratically with the problem size (number of training examples). For this reason we do not use this approach in our study.

3   Ensemble of Decision Rules for Ordinal Classification

The introduced algorithms generating an ensemble of ordinal decision rules are based on forward stagewise additive modeling [12]. The decision rule being the base classifier is a logical expression having the form: *if [conditions], then [decision]*. If an object satisfies conditions of the rule, then the suggested decision is taken. Otherwise no action is performed. By conditions we mean a conjunction of expressions of the form $x_j \in S$, where S is a value subset of j-th attribute, $j \in \{1, \ldots, n\}$. Denoting set of conditions by Φ and decision by α, the decision rule can be equivalently defined as:

$$r(\mathbf{x}, \mathbf{c}) = \begin{cases} \alpha & \text{if } \mathbf{x} \in cov(\Phi), \\ 0 & \text{if } \mathbf{x} \notin cov(\Phi), \end{cases} \quad (17)$$

where $\mathbf{c} = (\Phi, \alpha)$ is a set of parameters. Objects that satisfy Φ are denoted by $cov(\Phi)$ and referred to as cover of conditions Φ.

The general scheme of the algorithm is presented as Algorithm 1. In this procedure, $F_m(\mathbf{x})$ is a real function being a linear combination of m decision rules $r(\mathbf{x}, \mathbf{c})$, $\{\theta_k\}_1^{K-1}$ are thresholds and M is a number of rules to be generated. $L^{all}(y_i, F(\mathbf{x}))$ is an all-threshold loss function. The algorithm starts with $F_0(\mathbf{x}) = 0$ and $\{\theta_k\}_1^{K-1} = 0$. In each iteration of the algorithm, function $F_{m-1}(\mathbf{x})$ is

---

Algorithm 1. Ensemble of ordinal decision rules

input : set of training examples $\{y_i, \mathbf{x}_i\}_1^N$,
            M – number of decision rules to be generated.
output: ensemble of decision rules $\{r_m(\mathbf{x})\}_1^M$,
            thresholds $\{\theta_k\}_1^{K-1}$.
$F_0(\mathbf{x}) := 0; \{\theta_{k0}\}_1^{K-1} := 0;$
for $m = 1$ *to* M **do**
  $\quad (\mathbf{c}, \{\theta_k\}_1^{K-1}) := \arg\min_{(\mathbf{c}, \{\theta_k\}_1^{K-1})} \sum_{i=1}^N L^{all}(y_i, F_{m-1}(\mathbf{x}_i) + r(\mathbf{x}_i, \mathbf{c}));$
  $\quad r_m(\mathbf{x}, \mathbf{c}) := r(\mathbf{x}, \mathbf{c});$
  $\quad \{\theta_{km}\}_1^{K-1} := \{\theta_k\}_1^{K-1};$
  $\quad F_m(\mathbf{x}) := F_{m-1}(\mathbf{x}) + r_m(\mathbf{x}, \mathbf{c});$
end
$ensemble = \{r_m(\mathbf{x}, \mathbf{c})\}_1^M; thresholds = \{\theta_{kM}\}_1^{K-1};$

---

augmented by one additional rule $r_m(\mathbf{x}, \mathbf{c})$. A single rule is built by sequential addition of new conditions to Φ and computation of α. This is done in view of minimizing

$$L_m = \sum_{i=1}^N L^{all}(y_i, F_{m-1}(\mathbf{x}_i) + r(\mathbf{x}_i, \mathbf{c}))$$

$$= \sum_{\mathbf{x}_i \in cov(\Phi)} \left(\sum_{k=1}^{y_i-1} L(1, F_{m-1}(\mathbf{x}_i) + \alpha - \theta_k) + \sum_{k=y_i}^{K-1} L(-1, F(\mathbf{x}_i)_{m-1} + \alpha - \theta_k) \right)$$

$$+ \sum_{\mathbf{x}_i \notin cov(\Phi)} \left(\sum_{k=1}^{y_i-1} L(1, F_{m-1}(\mathbf{x}_i) - \theta_k) + \sum_{k=y_i}^{K-1} L(-1, F(\mathbf{x}_i)_{m-1} - \theta_k) \right) \quad (18)$$

with respect to Φ, α and $\{\theta_k\}_1^{K-1}$. A single rule is built until L_m cannot be decreased.

Ordinal classification decision is computed according to:

$$F(\mathbf{x}) = \sum_{k=1}^K k \cdot I \left(\sum_{m=1}^M r_m(\mathbf{x}, \mathbf{c}) \in [\theta_{k-1}, \theta_k) \right), \quad (19)$$

where $I(a)$ is an indicator function, i.e. if a is true then $I(a) = 1$, otherwise $I(a) = 0$. Some other approaches are also possible. For example, in experiments we have used a procedure that assigns intermediate values between class labels in order to minimize squared error.

In the following two subsections, we give details of two introduced algorithms.

3.1   Ordinal Decision Rules Based on Exponential Boosting (ORDER-E)

The algorithm described in this subsection can be treated as generalization of AdaBoost [8] with decision rules as base classifiers. In each iteration of the algorithm, a strictly convex function (18) defined using the exponential loss L_{exp} is minimized with respect to parameters Φ, α and $\{\theta_k\}_1^{K-1}$. In iteration m, it is easy to compute the following auxiliary values that depend only on $F_{m-1}(\mathbf{x})$ and Φ:

$$A_{km} = \sum_{\mathbf{x}_i \in cov(\Phi)} I(y_i > k)e^{-F_{m-1}(\mathbf{x}_i)} \quad B_{km} = \sum_{\mathbf{x}_i \in cov(\Phi)} I(y_i \leq k)e^{F_{m-1}(\mathbf{x}_i)}$$

$$C_{km} = \sum_{\mathbf{x}_i \notin cov(\Phi)} I(y_i > k)e^{-F_{m-1}(\mathbf{x}_i)} \quad D_{km} = \sum_{\mathbf{x}_i \notin cov(\Phi)} I(y_i \leq k)e^{F_{m-1}(\mathbf{x}_i)}$$

These values are then used in computation of the parameters. The optimal values for thresholds $\{\theta_k\}_1^{K-1}$ are obtained by setting the derivative to zero:

$$\frac{\partial L_m}{\partial \theta_k} = 0 \Leftrightarrow \theta_k = \frac{1}{2} \log \frac{B_k \cdot \exp(\alpha) + D_k}{A_k \exp(-\alpha) + C_k}, \tag{20}$$

where parameter α is still to be determined. Putting (20) into (18), we obtain the formula for L_m:

$$L_m = 2 \sum_{k=1}^{K-1} \sqrt{(B_k \cdot \exp(\alpha) + D_k)(A_k \cdot \exp(-\alpha) + C_k)}. \tag{21}$$

which now depends only on single parameter α. The optimal value of α can be obtained by solving

$$\frac{\partial L_m}{\partial \alpha} = 0 \Leftrightarrow \sum_{k=1}^{K-1} \frac{B_k \cdot C_k \cdot \exp(\alpha) - A_k \cdot D_k \cdot \exp(-\alpha)}{\sqrt{(B_k \cdot \exp(\alpha) + D_k)(A_k \cdot \exp(-\alpha) + C_k)}} = 0 \tag{22}$$

There is, however, no simple and fast exact solution to (22). That is why we approximate α by a single Newton-Raphson step:

$$\alpha := \alpha_0 - \nu \cdot \frac{\partial L_m}{\partial \alpha} \cdot \left(\frac{\partial^2 L_m}{\partial^2 \alpha}\right)^{-1}\Bigg|_{\alpha=\alpha_0} \tag{23}$$

computed around zero, i.e. $\alpha_0 = 0$. Summarizing, a set of conditions Φ is chosen which minimizes (21) with α given by (23). One can notice the absence of thresholds in the formula for L_m (21). Indeed, thresholds are necessary only for further classification and can be determined once, at the end of induction procedure. However, L_m (21) is not additive anymore, i.e. it is not the sum of losses of objects due to implicit dependence between objects through the (hidden) thresholds values.

Another boosting scheme for ordinal classification has been proposed in [14]. Similar loss function has been used, although expressed in terms of margins (therefore called "left-right margins" and "all-margins" instead of "immediate-thresholds" and "all-thresholds"). The difference is that in [14] optimization over parameters is performed sequentially. First, a base learner is fitted with $\alpha = 1$.

Then, the optimal value of α is obtained, using thresholds values from previous iterations. Finally, the thresholds are updated. In section 4, we compared this boosting strategy with our methods, showing that such a sequential optimization does not work well with decision rule as a base learner.

3.2 Ordinal Decision Rules Based on Gradient Boosting (ORDER-G)

The second algorithm is an extension of the gradient boosting machine [9]. Here, the goal is to minimize (18) defined by deviance loss L_{dev}. Φ is determined by searching for regression rule that fits pseudoresponses $\tilde{y}_i$ being negative gradients:

$$\tilde{y}_i = -\frac{\partial L_{dev}^{all}(y_i, F(\mathbf{x}_i))}{\partial F(\mathbf{x}_i)}\bigg|_{F(\mathbf{x}_i)=F_{m-1}(\mathbf{x}_i)} \tag{24}$$

with $\{\theta_{km-1}\}_1^{K-1}$ determined in iteration $m-1$. The regression rule is fit by minimization of the squared-error loss:

$$\sum_{\mathbf{x}_i \in cov(\Phi)} (\tilde{y}_i - F_{m-1}(\mathbf{x}_i) - \tilde{\alpha})^2 + \sum_{\mathbf{x}_i \notin cov(\Phi)} (\tilde{y}_i - F_{m-1}(\mathbf{x}_i))^2. \tag{25}$$

The minimum of (25) is reached for

$$\tilde{\alpha} = \sum_{\mathbf{x}_i \in cov(\Phi)} (\tilde{y}_i - F_{m-1}(\mathbf{x}_i)) / \sum_{\mathbf{x}_i \in cov(\Phi)} 1. \tag{26}$$

The optimal value for α is obtained by setting

$$\frac{\partial L_m}{\partial \alpha} = 0$$

with Φ already determined in previous step. However, since this equation has no closed-form solution, the value of α is then approximated by a single Newton-Raphson step, as in (23). Finally, $\{\theta_{km}\}_1^{K-1}$ are determined by

$$\frac{\partial L_m}{\partial \theta_{km}} = 0.$$

Once again, since there is no closed-form solution, θ_{km} is approximated by a single Newton-Raphson step,

$$\theta_{km} = \theta_{km-1} - \frac{\partial L_m}{\partial \theta_{km}} \cdot \left(\frac{\partial^2 L_m}{\partial^2 \theta_{km}}\right)^{-1}\bigg|_{\theta_{km}=\theta_{km-1}},$$

with Φ and α previously determined.

Notice that the scheme presented here is valid not only for L_{dev}, but for any other convex, differentiable loss function used as a base loss function in (18).

4   Experimental Results

We performed two experiments. Our aim was to compare simple ordinal classifiers, ordinal decision rules and approaches introduced in [3,14]. We also wanted to check, how the introduced approaches works on Netflix Prize dataset [16]. As a comparison criteria we chose zero-one error (ZOE), mean absolute error (MEA) and root mean squared error (RMSE). The former two were used in referred papers. RMSE was chosen because of Netflix Prize rules.

The simple ordinal classifiers were based on logistic regression, LogitBoost [10,9] with decision stumps, linear regression and additive regression [9]. Implementations of these methods were taken from Weka package [20]. In the case of logistic regression and LogitBoost, decisions were computed according to the analysis given in section 2. In order to minimize, ZOE, MAE and RMSE a final decision was computed as a mode, median or average over the distribution given by these methods, respectively. We used three ordinal rule ensembles. The first one is based on ORBoost-All scheme introduced in [14]. The other two are ORDER-E and ORDER-G introduced in this paper. In this case, a final decision was computed according to (19) in order to minimize ZOE and MAE. For minimization of RMSE, we have assumed that the ensemble constructs $F_M(\mathbf{x})$ which is a monotone transformation of a value function defined on an interval $[1,5] \subseteq \mathbb{R}$. In classification procedure, values of $F_M(\mathbf{x})$ are mapped to $[1,5] \subseteq \mathbb{R}$ by:

$$F(\mathbf{x}) = \sum_{k=1}^{K} \left(k + \frac{F_M(\mathbf{x}) - (\theta_k + \theta_{k-1})/2}{\theta_k - \theta_{k-1}} \right) \cdot I(F_M(\mathbf{x}) \in [\theta_{k-1}, \theta_k)),$$

where $\theta_0 = \theta_1 - 2 \cdot (\theta_2 - \theta_1)$ and $\theta_K = \theta_{K-1} + 2 \cdot (\theta_{K-1} - \theta_{K-2})$. These methods were compared with SVM with explicit constraints and SVM with implicit constraints introduced in [3] and with ORBoost-LR and ORBoost-All with perceptron and sigmoid base classifiers introduced in [14].

In the first experiment we used the same datasets and settings as in [3,14] in order to compare the algorithms. These datasets were discretized by equal-frequency bins from some metric regression datasets. We used the same $K = 10$, the same "training/test" partition ratio, and also averaged the results over 20 trials. We report in Table 2 the means and standard errors for ZOE and MEA as it was done in the referred papers. In the last column of the table we put the best result found in [3,14] for a given dataset. The optimal parameters for simple ordinal classifiers and ordinal rule ensembles were obtained in 5 trials without changing all other settings.

Second experiment was performed on Netflix Prize dataset [16]. We chose 10 first movies from the list of Netflix movies, which have been evaluated by at least 10 000 and at most 30 000 users. Three types of error (ZOE, MEA and RMSE) were calculated. We compared here only simple ordinal classifiers with ORDER-E and ORDER-G. Classifiers were learned on Netflix-training dataset and tested on Netflix-probe dataset (all evaluations from probe dataset were

Table 2. Experimental results on datasets used in [3,14]. The same data preprocessing is used that enables comparison of the results. In the last column, the best results obtained by [1]SVM with explicit constraints [3], [2]SVM with implicit constraints [3], [3]ORBoost-LR [14], and [4]ORBoost-All [14] are reported. Two types of error are considered (zero-one and mean-absolute). Best results are marked in bold among all compared methods and among methods introduced in this paper.

Zero-one error (ZOE)					
Dataset	Logistic Regression	LogitBoost with DS	ORBoost-All with Rules	ORDER-E	ORDER-G	Best result from [3,14]
Pyrim.	**0.754±0.017**	0.773±0.018	0.852±0.011	**0.754±0.019**	0.779±0.018	**0.719±0.066**[2]
CPU	0.648±0.009	0.587±0.012	0.722±0.011	0.594±0.014	**0.562±0.009**	0.605±0.010[4]
Boston	0.615±0.007	0.581±0.007	0.653±0.008	**0.560±0.006**	0.581±0.007	**0.549±0.007**[3]
Abal.	**0.678±0.002**	0.694±0.002	0.761±0.003	0.710±0.002	0.712±0.002	0.716±0.002[3]
Bank	**0.679±0.001**	0.693±0.001	0.852±0.002	0.754±0.001	0.759±0.001	0.744±0.005[1]
Comp.	0.489±0.001	0.494±0.001	0.593±0.002	**0.476±0.002**	0.479±0.001	**0.462±0.001**[1]
Calif.	0.665±0.001	**0.606±0.001**	0.773±0.002	0.631±0.001	0.609±0.001	**0.605±0.001**[3]
Census	0.707±0.001	**0.665±0.001**	0.793±0.002	0.691±0.001	0.687±0.001	0.694±0.001[3]
Mean absolute error (MAE)					
Dataset	Logistic Regression	LogitBoost with DS	ORBoost-All with Rules	ORDER-E	ORDER-G	Best result from [3,14]
Pyrim.	1.665±0.056	1.754±0.050	1.858±0.074	**1.306±0.041**	1.356±0.063	**1.294±0.046**[2]
CPU	0.934±0.021	0.905±0.025	1.164±0.026	0.878±0.027	**0.843±0.022**	0.889±0.019[4]
Boston	0.903±0.013	0.908±0.017	1.068±0.017	**0.813±0.010**	0.828±0.014	**0.747±0.011**[2]
Abal.	**1.202±0.003**	1.272±0.003	1.520±0.008	1.257±0.002	1.281±0.004	1.361±0.003[2]
Bank	**1.445±0.003**	1.568±0.003	2.183±0.005	1.605±0.005	1.611±0.004	**1.393±0.002**[2]
Comp.	0.628±0.002	0.619±0.002	0.930±0.005	**0.583±0.002**	0.588±0.002	0.596±0.002[2]
Calif.	1.130±0.004	0.957±0.001	1.646±0.007	0.955±0.003	**0.897±0.002**	0.942±0.002[4]
Census	1.432±0.003	1.172±0.002	1.669±0.006	**1.152±0.002**	1.166±0.002	1.198±0.002[4]

removed from training dataset). Ratings on 100 movies, selected in the same way for each movie, were used as condition attributes. For each method, we tuned its parameters to optimize its performance, using 10% of training set as a validation set; to avoid favouring methods with more parameters, for each algorithm we performed the same number of tuning trials. The results are shown in Table 3.

The results from both experiments indicate that ensembles of ordinal decision rules are competitive to other methods used in the experiment:

- From the first experiment, one can conclude that ORBoost strategy does not work well with decision rule as a base learner, and that simple ordinal classifiers and ordinal decision rules perform comparably to approaches introduced in [3,14].
- The second experiment shows that especially ORDER-E outperforms other methods in RMSE for most of the movies and in MAE for half of the movies. However, this method was the slowest between all tested algorithms. ORDER-G is much more faster than ORDER-E, but it obtained moderate results.
- In both experiments logistic regression and LogitBoost perform well. It is clear that these algorithms achieved the best results with respect to ZOE. The reason is that they can be tailored to multi-classification problem with zero-one loss, while ordinal decision rules cannot.

Table 3. Experimental results on 10 movies from Netflix Prize data set. Three types of error are considered (zero-one, mean-absolute and root mean squared). For each movie, best results are marked in bold.

Zero-one error (ZOE)					
Movie #	Linear Regression	Additive Regression	Logistic Regression	LogitBoost with DS	ORDER-E	ORDER-G
8	0.761	0.753	0.753	**0.714**	0.740	0.752
18	0.547	0.540	0.517	**0.493**	0.557	0.577
58	0.519	0.496	0.490	**0.487**	0.513	0.496
77	0.596	0.602	0.583	**0.580**	0.599	0.605
83	0.486	0.486	0.483	**0.398**	0.462	0.450
97	0.607	0.607	0.591	**0.389**	0.436	0.544
108	0.610	0.602	0.599	**0.593**	0.613	0.596
111	0.563	0.561	0.567	**0.555**	0.572	0.563
118	0.594	0.596	0.532	0.524	**0.511**	0.551
148	0.602	0.610	0.593	0.536	**0.522**	0.573
Mean absolute error (MAE)					
Movie #	Linear Regression	Additive Regression	Logistic Regression	LogitBoost with DS	ORDER-E	ORDER-G
8	1.133	1.135	1.115	1.087	**1.013**	1.018
18	0.645	0.651	**0.583**	0.587	0.603	0.613
58	0.679	0.663	0.566	**0.543**	0.558	0.560
77	0.831	0.839	0.803	0.781	**0.737**	0.755
83	0.608	0.614	0.519	**0.448**	0.500	0.502
97	0.754	0.752	0.701	**0.530**	0.537	0.654
108	0.777	0.776	**0.739**	**0.739**	0.768	**0.739**
111	0.749	0.766	0.720	0.715	**0.693**	0.705
118	0.720	0.734	0.626	0.630	**0.596**	0.658
148	0.747	0.735	0.688	0.626	**0.604**	0.659
Root mean squared error (RMSE)					
Movie #	Linear Regression	Additive Regression	Logistic Regression	LogitBoost with DS	ORDER-E	ORDER-G
8	1.332	1.328	1.317	1.314	**1.268**	1.299
18	0.828	0.836	**0.809**	0.856	0.832	0.826
58	0.852	0.847	0.839	**0.805**	0.808	0.817
77	1.067	1.056	1.056	1.015	**0.999**	1.043
83	0.775	0.772	0.737	0.740	**0.729**	0.735
97	0.968	0.970	0.874	0.865	**0.835**	0.857
108	0.984	0.993	**0.969**	0.979	0.970	0.989
111	0.985	0.992	0.970	0.971	**0.967**	0.986
118	0.895	0.928	0.862	0.860	**0.836**	0.873
148	0.924	0.910	0.900	0.863	**0.838**	0.893

- It is worth noticing, that regression algorithms resulted in poor accuracy in many cases.
- We have observed during the experiment that ORDER-E and ORDER-G are sensitive to parameters setting. We plan to work on some simple method for parameters selection.

5   Conclusions

From the theoretical analysis, it follows that ordinal classification problem can be solved by different approaches. In our opinion, there is still a lot to do in order to establish a theoretic framework for ordinal classification. In this paper, we introduced a decision rule induction algorithm based on forward stagewise additive modeling that utilizes the notion of threshold loss function. The experiment indicates that ordinal decision rules are quite promising. They are competitive to traditional regression and multi-class classification methods, and also to existing ordinal classification methods. Let us remark that the algorithm can also be used for other base classifiers like decision trees instead of decision rules. In this paper, we remained with rules because of their simplicity in interpretation. It is also interesting that such a simple classifier works so well as a part of the ensemble.

References

1. Błaszczyński, J., Dembczyński, K., Kotłowski, W., Słowiński, R., Szeląg, M.: Ensembles of Decision Rules. Foundations of Computing and Decision Sciences 31, 221–232 (2006)
2. Błaszczyński, J., Dembczyński, K., Kotłowski, W., Słowiński, R., Szeląg, M.: Ensembles of Decision Rules for Solving Binary Classification Problems in the Presence of Missing Values. In: Greco, S., Hata, Y., Hirano, S., Inuiguchi, M., Miyamoto, S., Nguyen, H.S., Słowiński, R. (eds.) RSCTC 2006. LNCS (LNAI), vol. 4259, pp. 224–234. Springer, Heidelberg (2006)
3. Chu, W., Keerthi, S.S.: New approaches to support vector ordinal regression. In: Proc. of 22nd International Conference on Machine Learning, pp. 321–328 (2005)
4. Clémençon, S., Lugosi, G., Vayatis, N.: Ranking and empirical minimization of U-statistics (to appear)
5. Cohen, W., Singer, Y.: A simple, fast, and effective rule learner. In: Proc. of 16th National Conference on Artificial Intelligence, pp. 335–342 (1999)
6. Frank, E., Hall, M.: A simple approach to ordinal classification. In: Flach, P.A., De Raedt, L. (eds.) ECML 2001. LNCS (LNAI), vol. 2167, pp. 145–157. Springer, Heidelberg (2001)
7. Freund, Y., Iyer, R., Schapire, R., Singer, Y.: An efficient boosting algorithm for combining preferences. J. of Machine Learning Research 4, 933–969 (2003)
8. Freund, Y., Schapire, R.: A decision-theoretic generalization of on-line learning and an application to boosting. J. of Computer and System Sciences 55(1), 119–139 (1997)
9. Friedman, J.: Greedy function approximation: A gradient boosting machine. The Annals of Statistics 29(5), 1189–1232 (2001)
10. Friedman, J., Hastie, T., Tibshirani, R.: Additive logistic regression: a statistical view of boosting. Annals of Statistics 28(2), 337–407 (2000)
11. Friedman, J., Popescu, B.: Predictive learning via rule ensembles. Research report, Dept. of Statistics, Stanford University (2005)
12. Hastie, T., Tibshirani, R., Friedman, J.: Elements of Statistical Learning: Data Mining, Inference, and Prediction. Springer, Heidelberg (2003)
13. Herbrich, R., Graepel, T., Obermayer, K.: Regression models for ordinal data: A machine learning approach. Technical report TR-99/03, TU Berlin (1999)
14. Lin, H.T., Li, L.: Large-margin thresholded ensembles for ordinal regression: Theory and practice. In: Balcázar, J.L., Long, P.M., Stephan, F. (eds.) ALT 2006. LNCS (LNAI), vol. 4264, pp. 319–333. Springer, Heidelberg (2006)
15. Lin, H.T., Li, L.: Ordinal regression by extended binary classifications. Advances in Neural Information Processing Systems 19, 865–872 (2007)
16. Netflix prize, http://www.netflixprize.com

17. Rennie, J., Srebro, N.: Loss functions for preference levels: Regression with discrete ordered labels. In: Proc. of the IJCAI Multidisciplinary Workshop on Advances in Preference Handling (2005)
18. Shashua, A., Levin, A.: Ranking with large margin principle: Two approaches. Advances in Neural Information Processing Systems 15 (2003)
19. Weiss, S., Indurkhya, N.: Lightweight rule induction. In: Proc. of 17th International Conference on Machine Learning, pp. 1135–1142 (2000)
20. Witten, I., Frank, E.: Data Mining: Practical machine learning tools and techniques, 2nd edn. Morgan Kaufmann, San Francisco (2005)

Data Mining of Multi-categorized Data

Akinori Abe[1,2], Norihiro Hagita[1,3], Michiko Furutani[1],
Yoshiyuki Furutani[1], and Rumiko Matsuoka[1]

[1] International Research and Educational Institute for Integrated Medical Science
(IREIIMS), Tokyo Women's Medical University
8-1 Kawada-cho, Shinjuku-ku, Tokyo 162-8666 Japan
[2] ATR Knowledge Science Laboratories
[3] ATR Intelligent Robotics and Communication Laboratories
[4] 2-2-2, Hikaridai, Seika-cho, Soraku-gun, Kyoto 619-0288 Japan
ave@ultimaVI.arc.net.my, hagita@atr.jp,
{michi,yoshi,rumiko}@imcir.twmu.ac.jp

Abstract. At the International Research and Educational Institute for
Integrated Medical Sciences (IREIIMS) project, we are collecting complete medical data sets to determine relationships between medical data
and health status. Since the data include many items which will be categorized differently, it is not easy to generate useful rule sets. Sometimes
rare rule combinations are ignored and thus we cannot determine the
health status correctly. In this paper, we analyze the features of such
complex data, point out the merit of categorized data mining and propose categorized rule generation and health status determination by using combined rule sets.

1 Introduction

Medical science and clinical diagnosis and treatment has been progressing rapidly
in recent years with each field becoming more specialized and independent. As
a result, cooperation and communication among researchers in the two fields
has decreased which has led to problems between both communities, not only in
terms of medical research but also with regard to clinical treatment. Therefore,
an integrated and cooperative approach to research between medical researchers
and biologists is needed. Furthermore, we are living in a changing and quite
complex society, so important knowledge is always being updated and becoming more complex. Therefore, integrated and cooperative research needs to be
extended to include engineering, cultural science, and sociology.

As for medical research, the integration of conventional (Western) and uncon-
ventional (Eastern) medical research, which should be fundamentally the same
but in fact are quite different, has been suggested.

With this situation in mind, we propose a framework called Cyber Integrated
Medical Infrastructure (CIMI) [Abe et al., 2007a] which is a framework of integrated management of clinical data on computer networks consisting of a
database, a knowledge base, and an inference and learning component, which

Z.W. Raś, S. Tsumoto, and D. Zighed (Eds.): MCD 2007, LNAI 4944, pp. 182–195, 2008.

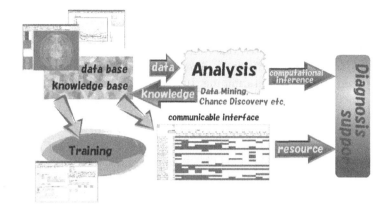

Fig. 1. Cyber Integrated Medical Infrastructure

are connected to each other in the network (Fig.1). In this framework, medical information (e.g. clinical data) is analyzed or data mined to build a knowledge base for the prediction of all possible diseases and to support medical diagnosis.

For medical data mining, several techniques such as Inductive Logic Programming (ILP), statistical methods, decision tree learning, Rough Sets and KeyGraph have been applied (e.g. [Ichise and Numao, 2005], [Ohsawa, 2003] and [Tsumoto, 2004]) and acceptable results have been obtained. We have also applied C4.5 [Quinlan, 1993] to medical data and obtained acceptable results. However, we used incomplete medical data sets which lack many parts of the data, since the data were collected during clinical examination. To save costs, physicians do not collect unnecessary data. For instance, if certain data are not related to the patient's situation, physicians will not collect them. If parts of the data are missing, even if we can collect many data sets, some of them are ignored during simple data mining procedures. To prevent this situation, we need to supplement missing data sets. However, it is difficult to automatically supplement the missing data sets. In fact, to supplement missing data sets, Ichise proposed a non-linear supplemental method [Ichise and Numao, 2003], but when we collected data from various patients it was difficult to guess relationship among the data, so we gave up to the idea of introducing such a supplemental method. Instead, we introduced a boosting method which estimates the distribution of original data sets from incomplete data sets and increases data by adding Gaussian noise [Abe et al., 2004]. We obtained results with robustness but we could not guarantee the results. In addition, when we used data sets collected in clinical inspections, we could only collect a small number of incomplete data sets. Therefore, in the International Research and Educational Institute for Integrated Medical Science (IREIIMS) project, we decided to collect complete medical data sets. Currently, we could have collected only 1800 medical data sets, though they will be increasing by the scheduled medical data collections. Even if we collect considerable size of complete data sets, we still have additional problems. The data include various types of data. That is, they contain data, for instance, of persons with lung cancer, those with stomach cancer etc. It is sometimes hazardous to use such mixed and complex data to perform data mining. In [Abe et al., 2007a], we pointed out that if we deal with multiple categorized (mixed) data, it is rather difficult to discover hidden or potential knowledge and we proposed integrated data mining. In [Abe et al., 2007b], we proposed an interface for medical diagnosis support which helps the user to discover hidden factors for the results. In this study, we introduce the interface to help to discover rare, hidden or potential data relationships.

In this paper, we analyze the collected medical data consisting of multiple categorized items then propose categorized rule generation (data mining) and a health level[1] determination method by applying the combined rule sets. In Section 2, we briefly describe the features of the collected medical data. In Section 3, we analyze (data mine) the collected data by C4.5 and apply generated rule sets to the medical data to determine the patients' situations. In section 4, we analyze the data mined results to point out the limitation of simple data mining of the collected data. In section 5, we suggest several strategies to deal with complex data that enable better health level determination.

2   Features of the Collected Medical Data

In this section, we analyze and describe features of the medical data collected in the IREIIMS project.

To construct the database in CIMI, we are now collecting various types of medical data, such as those obtained in blood and urine tests. We have collected medical data from about 1800 persons (For certain persons, the data were collected more than once.) and more than 130 items are included in the medical data of each person. Item sets in the data are, for instance, total protein, albumin, serum protein fraction-α1-globulin, Na, K, Ferritin, total acid phosphatase, urobilinogen, urine acetone, mycoplasma pneumoniae antibody, cellular immunity, immunosuppressive acidic protein, Sialyl Le X-i antigen, and urine β2-microglobulin. In addition, health levels are assigned by physicians resulting from the medical data and by clinical interviews. Health levels that express the health status of patients are defined based on *Tumor stages* [Kobayashi and Kawakubo, 1994] and modified by Matsuoka. Categorization of the health levels is shown in Fig. 2 ("%" represents a typical distribution ratio of persons in each level.). Persons at levels I and II can be regarded as being healthy, but those at levels III, IV, and V can possibly develop cancer. In [Kobayashi and Kawakubo, 1994], level III is defined as the stage before the shift to preclininal cancer, level IV is defined as conventional stage 0 cancer (G0), and level V is defined as conventional stages 1–4 cancer (G1–G4).

As shown in Fig. 2, Kobayashi categorized health levels into 5 categories. For more detailed analysis, Matsuoka categorized health levels into 8 categories which are 1, 2, 3, 4a, 4b, 4c, 5a, and 5b, since levels 4 and 5 include many clients' data. Table 1 shows the distribution ratio of health levels of the collected

---

[1] Health level is explained in section 2.

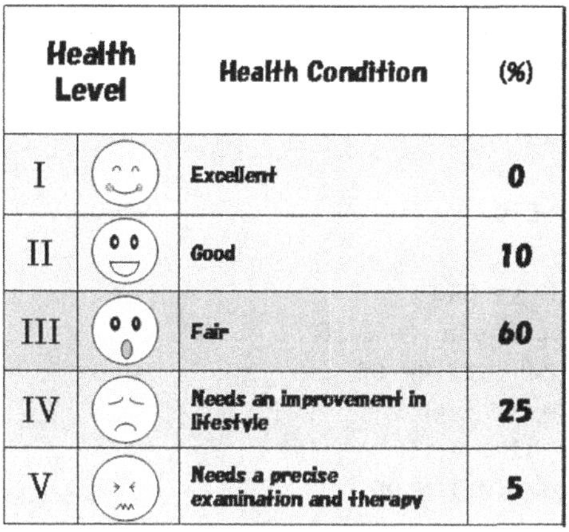

Health Level		Health Condition	(%)
I	☺	Excellent	0
II	😀	Good	10
III	😮	Fair	60
IV	☹	Needs an improvement in lifestyle	25
V	😖	Needs a precise examination and therapy	5

Fig. 2. Health levels

Table 1. Health levels

health level	1	2	3	4a	4b	4c	5a	5b
ratio (%)	0.0	0.0	3.09	17.23	46.99	19.22	10.77	2.71
ratio (%)	0.0	0.0	3.09	83.44			13.48	

data. The distribution ratio is quite different from that shown in Fig. 2, as we are currently collecting data from office workers (aged 40 to 50 years old) but not from students or younger persons. Accordingly the data distribution shifts to level 5 and 80% of the data are assigned to level 4. This imbalance and distribution might influence the data mining results. However, in this study, we did not apply any adjustments as we have no idea or models for proper adjustment. Adjustments will be proposed after analysis of the data sets.

3   Analysis of the Data

In this section, we analyze the collected medical data. First, we simply apply C4.5 to obtain relationships between the health levels and medical data sets. Then we apply the obtained relationships to medical data to estimate the patient's health situation. If we have the actual health level information, we can determine whether obtained rule sets are good or not. In addition, we can obtain the features of medical data sets.

3.1   Data Analysis by C4.5

First to determine the features of the data, we simply applied C4.5 to the collected data. To check the effect of data size, we analyzed both 1200 and 1500 medical data sets that were chronologically[2] extracted from 1800 medical data sets. Both results are shown below.

- 1200 medical data sets
  ```
  ICTP <= 5.8
  | TK activity <= 5.4
  | | CEA <= 4.1
  | | | EBV-VCA-IgG <= 640
  | | | | γ-seminoprotein <= 2.15
  | | | | | Chloride (Cl) <= 96
  | | | | | | CK <= 82 : 4b
  | | | | | | CK > 82 : 4c
  | | | | | Chloride (Cl) > 96
  ...
  ```

- 1500 medical data sets
  ```
  TK activity <= 5.4
  | ICTP <= 5.8
  | | CYFRA <= 2.1
  | | | γ-seminoprotein <= 2.1
  | | | | EBV-VCA-IgG <= 640
  | | | | | Chloride (Cl) <= 96
  | | | | | | B-Cell(CD20) <= 22 : 4b
  | | | | | | B-Cell(CD20) > 22 : 4c
  | | | | | Chloride (Cl) > 96
  | | | | | | CEA <= 4.2
  ...
  ```

Since the data set size will not be large enough for general data mining, there are some differences between the results of 1200 medical data sets and 1500 medical data sets. If we can collect more medical data sets, we will be able to obtain more stable results. Nevertheless, even from current data, acceptable results can be obtained. If we focus on the first few lines, they are almost the same.

3.2   Applying the Result to Determine Health Levels

Next, we applied the obtained rule sets (decision trees) to the rest of the collected medical data to determine the health levels. For the results from 1200 medical data sets, we can estimate the health levels of about 600 (=1800−1200) persons' health levels. For the result from 1500 medical data sets, we can estimate the health levels of about 300 (=1800−1500) persons' health levels. A series of nodes in a decision tree is used to determine the health level. This is a first trial to use the data mined rule sets for medical diagnosis. Therefore, we do not have

---

[2] "Chronological" extraction is performed because we aim to use generated rule sets to determine health levels. It is natural to use previous data sets to generate models for future estimation.

Table 2. Health level estimation

Difference	from 1200 data Correct ratio (%)	from 1500 data Correct ratio (%)
−3	1.4	1.1
−2	6.3	3.7
−1	19.2	23.1
0	42.4	38.4
+1	23.9	25.4
+2	4.8	6.7
+3	1.8	1.1
+4	0.2	0.0

any model and currently the combination of multiple decision tree clusters is not considered. We adopt a very simple strategy to follow a decision tree from the root point (top of a decision tree) to a leaf. Table 2 shows accuracies of the results. For difference, we mean that if the estimated health level is 4b and the actual level is 4c, then the difference is −1. If the estimated health level is 4b and the actual level is 3, the difference is +2. Of course, if the estimated and actual health levels are the same, the difference is 0.

An exact estimation (0 estimation) ratio is about 40%. Generally, this is not a good result. However, if we regard both +1 and −1 as correct estimations, the ratio becomes about 85% which is usually regarded as a good result. Even for us, it is sometimes rather difficult to distinguish level 4a from 4b, so it might be acceptable to extend the correct estimation to +1 and −1. In addition, we could not find any superiority due to the size of the data sets. In fact, as for an exact estimation, rules generated from 1200 data sets are better than those generated from 1500 data sets. The difference in number is only 300, so it might be difficult to find a superiority due to the size of the data sets. From the accuracy ratio, in medical situations, it might be difficult to use the generated rule sets as they are. However, with a certain modification or improvement, they can help to determine the health levels of patients during medical diagnosis.

After obtaining the results, for us, it is more interesting and significant to find the reasons for incorrect estimations. From these reasons, we can propose a proper strategy for reducing incorrect estimations. In the next section, we analyze the results in detail and try to find the reasons for incorrect estimation.

4 Analysis of Results

In this section, we analyze the reasons for incorrect estimations by using the interface proposed in [Abe et al., 2007b] which can deal with data interactively. Figure 3 shows a result (decision tree) obtained in the web browser interface. In the browser, the left tab shows ID lists of a person such as 1186 and the right tab shows a decision tree. In the decision tree, "White blood cell differentiation:Neutro > 66.1: 5a(2/1/0)" can be interpreted as "... and if

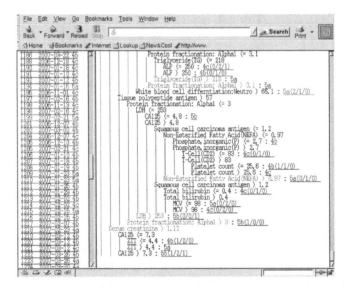

Fig. 3. Analysis result shown in the proposed interface

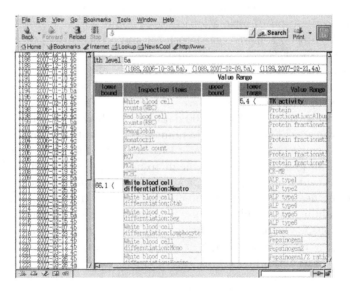

Fig. 4. Health level estimation result

White blood cell differentiation:Neutro > 66.1 then the health level is 5a."
In addition, the generated rule set (series of nodes) can explain three persons of which two explanations are correct and one is incorrect. When the user clicks the link point "5a(2/1/0)," another browser appears (Fig. 4). In the browser, <1199,2007-02-21,4a> shows that an estimation of ID 1199 is incorrect

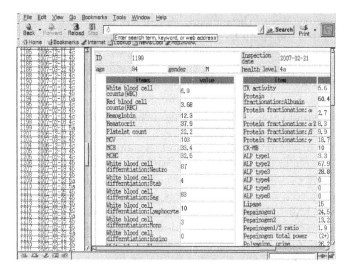

Fig. 5. Data of ID:1199

(blue colored and the assigned level is 4a; the actual level is 5a). When the user clicks the link point "<1199,2007-02-21,4a>," another browser appears (Fig. 5). In this case, even if we review the medical data, if we are not physicians we cannot determine whether the person is in level 4a or 5a. However we can ask physicians for reasons or confirmation. In contrast, we also found a case where, in spite of the assigned health level being 4b, the estimated level was 3. Similarly to the above case, we checked the data to find that NSE is 7.8 but it was not considered during the estimation. Actually in the decision tree the following pattern appears but in the estimation process, the system cannot refer to the rule. That is, the pattern does not appear in the inference path to determine the health levels.

```
. . . .
|  |  |  NSE > 7.2 . . . . .
|  |  |  |  Lipase <= 20
|  |  |  |  Albumin <= 4.3: 4a
|  |  |  |  Albumin > 4.3: . . . .
|  |  |  |  |  Na <= 139: 3
|  |  |  |  |  Na > 139
|  |  |  |  |  |  Protein fractionation: γ <= 17.5
|  |  |  |  |  |  |  Acid Phosphatase (ACP) <= 8.3 : 4b
. . . .
```

In fact, as shown in Table 1, only 3% are in level 3. Thus the number of examples is too small to generate proper rules or models, so it will be necessary to collect more data in level 3. In fact, none of the persons in health level 3 have been estimated correctly. Cases where the an estimation of the health level is 4a may be satisfactory, but some cases are estimated as 4b or 4c which cannot be regarded as correct. From these cases, we can say that the number of examples in health level 3 is too small to generate a proper model.

First, for this type of phenomenon, we assumed that a rare case might be present in level 4x. That is, some of the items that would place a person in level 4x are rare. However, as shown above, the decision tree includes rules of NSE for level 4x. Accordingly, since a combination of NSE and certain factors for level 4 are rare, the system cannot determine the health levels correctly by using such rules.

As suggested in section 2, imbalance and the non-standard distribution of data play a negative role in modeling. In fact, we intentionally remove any imbalance in the data, that is, we reduce the data according to the number of persons in level 3. The data contain only 300 samples (all levels have about 50 data), so the number is too small to generate a proper model. The result is not satisfactory as a diagnosis system but shows that balanced data will generate better models for medical diagnosis. If we can collect many data, we will be able to modify them. However currently this is not possible, so to overcome this problem, it is necessary to apply or develop another type of modeling.

5   Proposal for the Complex Data Treatment

In the previous section, we analyzed the collected medical data and showed several problems with the data mining of the collected data and the application of generated rules to determine health levels. We found at least three features of the collected medical data as shown below:

1) The collected data showed imbalance and did not follow the standard distribution.
2) Proper models might not be generated for health level 3 due to the small number of examples in this level.
3) Parts of the generated rules cannot be referred to in health level determination.

Due to the features of the collected medical data, a simple data mining method cannot be applied to them to obtain a satisfying result. As for 3), for estimating health levels, currently our medical diagnosis system can only follow a series of nodes in a decision tree. In fact, a decision tree has many points of division, so that after a certain division of the tree, the system cannot refer to rule sets on the other decision tree clusters that might be necessary for proper health level determination. Thus the data contains several problems for data mining, but currently we cannot correct or improve the data future. Therefore, it is necessary to develop methods that can remove or relief the above problems.

In the below, we show two methods to overcome the above problems.

5.1   Majority Rule Criterion Application Strategy

We previously proposed boosting method which tried to improve the distribution of the original incomplete data sets and increases data by adding Gaussian noise

Table 3. Results: majority rule criterion application strategy

hit rate	average of 5 results	25.52%
hit and near miss rate	average of 5 results	68.28%
hit rate	majority rule criterion applied	39.93%
hit and near miss rate	majority rule criterion applied	86.19%

[Abe et al., 2004]. However we cannot guarantee the result by data mining the pseudo-data. Since now we have complete medical data set, it would be better to adopt another type of data treatment.

To solve the problem (1), it is necessary to remove or relief harmful effects of imbalance of data. For that, one of simple methods is to divide whole data into several data sets and data mine each data sets. Then, generated multiple results are compared each other to take major results into consideration during health level determination. Our assumption is that by dividing data set, part of imbalance data will become minority of the whole data sets.

For instance, when we obtain 3 level 4a results and 2 level 4b results, the estimated health level is 4a. When we obtain 1 level 3 result, 2 level 4a results, and 2 level 4b results, the estimated health level is 4a. On the other hand, when we obtain 1 level 4c result, 2 level 4a results, and 2 level 4b results, the estimated health level is 4b. For the special case where the distribution is the same, we will use the middle level. For instance, when we obtain 2 level 4a, 1 level 4b, and 2 level 4c, the estimated health level is 4b. For a simplification, we currently regard the distance between neighbour levels is the same. That is, the distance between level 3 and 4a and that between level 4a and 4b is the same. Of course, strictly speaking, the distance between level 3 and 4a and that between level 4a and 4b should be different. Because in general, level 3 and 4 are quite different in the medical sense. This treatment might cause another problem, but in this paper we ignore this type of problem.

In fact, we currently have only 1800 data sets. If we divide the data into 5 data sets plus another data set for an estimation, each data set has only 281 data, which seem rather small number. Therefore, results might be miserable, but our main aim is to check whether majority rule criterion application strategy can solve the problem (1) or not. We obtained the result shown in Table 3.

Since currently we have rather small number of data, and we adopt a non-sensitive level selection strategy, the above results are not satisfactory, but we can say that even a simple method that follows majority rule criterion can work well. Anyway, it is necessary to collect as many data as possible to obtain better and satisfactory results. However, this method cannot solve the problems (2) and (3) even when we can collect sufficient number of data. It is necessary to introduce another or additional method.

5.2   Categorized Rule Generation and Application Strategy

To solve problems (2) and (3), in [Abe et al., 2007a], we proposed integrated data mining that categorizes the medical data into multiple categories, to discover relationships between the items in each category and the health levels, and integrates the results. In addition, we discovered an order of influential power of each category which controls the results. In the followings, we show the actual results of categorized data mining.

If we apply a rule generated from all the data, to person ID1035, for instance, the estimated health level is 4b, though the assigned health level is 4c. When we check the medical data of ID1035, we find that the LDH (Lactate dehydrogenase) value is rather higher than the reference value. In fact, "LDH is an intracellular enzyme from particularly in the kidney, heart, skeletal muscle, brain, liver and lungs. Increases are usually found in cellular death and/or leakage from the cell or in some cases it can be useful in confirming myocardial or pulmonary infarction (only in relation to other tests). Decreased levels of the enzyme may be seen in cases of malnutrition, hypoglycemia, adrenal exhaustion or low tissue or organ activity" (from The Danish Hepatitis C web site (`http://home3.inet.tele.dk/omni/alttest.htm`)). As for our standard classification, LDH is categorized into the liver, pancreas, and kidney test data group.

Accordingly, when we apply a rule generated from the liver, pancreas, and kidney test data, the estimated health level becomes 4c. The reason is that if we apply a rule generated from all the data, the effect of LDH cannot be considered. Because LDH does not appear in the decision tree. However, if we apply a rule generated from the liver, pancreas, and kidney test data, the effect of LDH can be considered during computational medical diagnosis. As shown above, since LDH is categorized into the liver, pancreas, and kidney test data, if we generate a rule only from these data, any influence from tumour markers[3] will be cancelled and rules including the effect of LDH can appear. In fact, we can obtain a part of a decision tree which includes LDH as shown below:

```
...
LDH > 241
| ALP type3 <= 52.4
| | Protein fractionation: α2 <= 9.7
| | | Lipase <= 5
| | | | Blood urea nitrogen <= 12 : 4b
| | | | Blood urea nitrogen > 12 : 5b
| | | Lipase > 5
| | | | AST(GOT) <= 20 : 4b
| | | | AST(GOT) > 20
.....
```

Thus for these type of problems, a combination of multiple rules will work well. For a person ID1035, even a single rule set that is generated from data of a single category can work well. Of course, in general,we cannot estimate the health level correctly by only using a rule set generated by categorized data mining, we cannot estimate health level correctly. In fact, correct estimation ratios are fewer

---

[3] In [Abe et al., 2007a], we discovered that tumour marker is the most influential factor.

when obtained in this manner than those obtained by rules generated from all the data.

On the contrary, for NSE (neuron-specific enolase) which is categorized to a tumour marker group, as pointed out in section 4, since an effect of NSE cannot be considered, an estimated health level becomes 3, though the assigned health level is 4b. In fact, as shown in the part of decision tree in section 4, NSE appears in the decision tree generated from all the medical data.

It might be better to apply the similar strategy to the above, but since NSE is categorized as a tumour marker member, we cannot apply a similar strategy. Of course the generated model itself might not be sufficiently proper, but it is necessary to introduce the other strategy to overcome such problems.

In fact, we categorized the medical data sets according to the standard classification including

1) liver, pancreas, and kidney test data,
2) metabolic function test data,
3) general urine test data,
4) blood and immunity test data, and
5) tumor markers.

For the case of ID1035, categorized rule generation and application strategy worked well. However, if we take the case of NSE into consideration, it might be necessary to introduce another classification or a more complicated or detailed classification. Zheng proposed committee learning [Zheng and Webb, 1998] which divide data set into several parts and perform data mining for each divided data set and generates a result after comparison of each result. The classification strategy is different from categorized data mining and needs many data sets as shown in majority rule criterion, but if we can collect sufficient number of medical data, it will be better to introduce committee learning. In fact, as shown above, although we have only 1800 medical data, we apply a simple type of committee learning that is majority rule criterion applied learning to the medical data to obtain better results than an ordinal learning strategy. Anyhow, we need to generate rules with properly categorized data sets and apply generated rules with a proper combination. Then we will be able to deal with complex or mixed data.

5.3   Health Level Estimation According to the Patient's Situation

We proposed two types of data mining strategies in the above. However, physicians usually do different type of "data mining" during medical examinations. When they assign a health level to a patient, they will focus on a part of the medical data according to the patients situation or clinical interview. Thus they do not take all the data into account. Since some of the data are not related to the patient's situation, physicians usually ignore unnecessary data. For a better estimation of health levels, it might be necessary to prepare or install such an intuitional reasoning as physicians do. That is, during the health level estimation procedure, the system should focus on proper data clusters, and apply rules related to the data clusters.

As shown above, we conducted categorized data mining and applied the generated rule sets to determine health levels. Currently we have not discovered general models for generation and applying rules. However, as shown above, we discovered several case that can estimate health levels correctly. Therefore, it is necessary to develop an automatic categorization method that can properly categorize medical data. Simple Principal Component Analysis could not properly categorize the medical data sets. In fact, if we know the patient's situation, we can focus on the data category relating to the patient's organ which is the source of the health problem. In the future, we will construct a data categorization model and a rule set combination model by analyzing data sets and physicians' health level determination models.

Finally, we can also say that the combination of categorized learning and committee learning will generate better result in data mining.

6 Conclusions

In this paper, we discussed the treatment of complex medical data which are collected in the International Research and Educational Institute for Integrated Medical Science (IREIIMS) project. Our main aim is to determine relationships between health levels and medical data. By applying C4.5, we could obtained acceptable results, but for even better results, we suggested and introduced several strategies including a majority rule criterion application strategy, and a categorized data mining and combined rule application strategy. We have not discovered general models for categorization and combination. However, we point out that a general model can be obtained by referring to physicians' determination or diagnosis patterns. Thus we need to discover more strong relationships between medical data and health status.

For relationship or association, Agrawal proposed an association rule that represents relationships between items in databases [Agrawal et al., 1993]. The association rule is frequently used when analyzing POS data to discover tendencies of users' shopping patterns (basket analysis). However, from the analysis, we can only discover frequently co-occurring patterns. Also, relational data mining has recently been proposed [Džroski and Lavrač, 2001]. This paradigm also discovers relationships between items in a (relational) database by using ILP techniques. Their approaches are important for complex data mining. However our major aim is not to discover relationships between each category but to determine an effective classification for data mining of complex data. Nevertheless, their concept can be introduced to discover relationships.

Finally, we emphasize that our approach is based on the concept of chance discovery [Ohsawa and McBurney, 2003]. A rare relationship that cannot be extracted by simple C4.5 application can play a significant role in a proper health level determination. For instance, due to imbalance distribution of health levels, it is rather difficult to obtain a proper model for health level 3. In the situation, combination of the NSE and certain factors for level 4 might be rare in our data set. Accordingly, our main aim is to discover such rare and significant relationships that can be used for accurate health level determination.

Acknowledgments

This research was supported in part by the Program for Promoting the Establishment of Strategic Research Centers, Special Coordination Funds for Promoting Science and Technology, Ministry of Education, Culture, Sports, Science and Technology (Japan). We thank Mr. Ken Chang (NTT-AT) for supporting to develop analysis tools.

References

[Abe et al., 2004] Abe, A., Naya, F., Kogure, K., Hagita, N.: Rule Acquisition from small and heterogeneous data set, Technical Report of JSAI, SIG-KBS-A304-32, pp. 189–194 (2004) (in Japanese)

[Abe et al., 2007a] Abe, A., Hagita, N., Furutani, M., Furutani, Y., Matsuoka, R.: Possibility of Integrated Data Mining of Clinical Data. Data Science Journal 6 (Supplement), S104–S115 (2007)

[Abe et al., 2007b] Abe, A., Hagita, N., Furutani, M., Furutani, Y., Matsuoka, R.: An interface for medical diagnosis support. In: Apolloni, B., Howlett, R.J., Jain, L. (eds.) KES 2007, Part II. LNCS (LNAI), vol. 4693, pp. 909–916. Springer, Heidelberg (2007)

[Agrawal et al., 1993] Agrawal, R., Imielinski, T., Swami, A.: Mining association rules between sets of items in large databases. In: Proc. of ACM SIGMOD Int'l Conf. on Management of Data, pp. 207–216 (1993)

[Džroski and Lavrač, 2001] Džroski, S., Lavrač, N. (eds.): Relational Data Mining. Springer, Heidelberg (2001)

[Ichise and Numao, 2003] Ichise, R., Numao, M.: A Graph-based Approach for Temporal Relationship Mining, Technical Report of JSAI, SIG-FAI-A301, pp. 121–126 (2003)

[Ichise and Numao, 2005] Ichise, R., Numao, M.: First-Order Rule Mining by Using Graphs Created from Temporal Medical Data. In: Tsumoto, S., Yamaguchi, T., Numao, M., Motoda, H. (eds.) AM 2003. LNCS (LNAI), vol. 3430, pp. 112–125. Springer, Heidelberg (2005)

[Kobayashi and Kawakubo, 1994] Kobayashi, T., Kawakubo, T.: Prospective Investigation of Tumor Markers and Risk Assessment in Early Cancer Screening. Cancer 73(7), 1946–1953 (1994)

[Ohsawa, 2003] Ohsawa, Y., Okazaki, N., Matsumura, N.: A Scenario Development on Hepatics B and C, Technical Report of JSAI, SIG-KBS-A301, pp. 177–182 (2003)

[Ohsawa and McBurney, 2003] Osawa, Y., McBurney, P. (eds.): Chance Discovery. Springer, Heidelberg (2003)

[Tsumoto, 2004] Tsumoto, S.: Mining Diagnostic Rules from Clinical Databases Using Rough Sets and Medical Diagnostic Model. Information Sciences 162(2), 65–80 (2004)

[Quinlan, 1993] Quinlan, J.R.: C4.5: Programs for Machine Learning. Morgan Kaufman, San Francisco (1993)

[Zheng and Webb, 1998] Zheng, Z., Webb, G.I.: Stochastic Attribute Selection Committees. In: Antoniou, G., Slaney, J.K. (eds.) AI 1998. LNCS, vol. 1502, pp. 321–332. Springer, Heidelberg (1998)

ARAS: Action Rules Discovery Based on Agglomerative Strategy

Zbigniew W. Raś[1,3], Elżbieta Wyrzykowska[2], and Hanna Wasyluk[4]

[1] Univ. of North Carolina, Dept. of Comp. Science, Charlotte, N.C. 28223, USA
ras@uncc.edu
[2] Univ. of Information Technology and Management, ul. Newelska, Warsaw, Poland
ewyrzyko@wit.edu.pl
[3] Polish-Japanese Institute of Information Technology,
ul. Koszykowa 86, 02-008 Warsaw, Poland
ras@pjwstk.edu.pl
[4] Medical Center for Postgraduate Education,
ul. Marymoncka 99, 01-815 Warsaw, Poland
hwasyluk@cmkp.edu.pl

Abstract. Action rules can be seen as logical terms describing knowledge about possible actions associated with objects which is hidden in a decision system. Classical strategy for discovering them from a database requires prior extraction of classification rules which next are evaluated pair by pair with a goal to build a strategy of action based on condition features in order to get a desired effect on a decision feature. An actionable strategy is represented as a term $r = [(\omega) \wedge (\alpha \rightarrow \beta)] \Rightarrow [\phi \rightarrow \psi]$, where ω, α, β, ϕ, and ψ are descriptions of objects or events. The term r states that when the fixed condition ω is satisfied and the changeable behavior $(\alpha \rightarrow \beta)$ occurs in objects represented as tuples from a database so does the expectation $(\phi \rightarrow \psi)$. This paper proposes a new strategy, called $ARAS$, for constructing action rules with the main module resembling $LERS$ [6]. $ARAS$ system is more simple than $DEAR$ and its time complexity is also lower.

1 Introduction

Finding useful rules is an important task of a knowledge discovery process. Most researchers focus on techniques for generating patterns from a data set such as classification rules, association rules...etc. They assume that it is user's responsibility to analyze these patterns in order to infer solutions for specific problems within a given domain. The classical knowledge discovery algorithms have the potential to identify enormous number of significant patterns from data. Therefore, people are overwhelmed by a large number of uninteresting patterns and it is very difficult for a human being to analyze them in order to form timely solutions. Therefore, a significant need exists for a new generation of techniques and tools with the ability to assist users in analyzing a large number of rules for a useful knowledge.

There are two aspects of interestingness of rules that have been studied in data mining literature, objective and subjective measures [1], [12]. Objective

Z.W. Raś, S. Tsumoto, and D. Zighed (Eds.): MCD 2007, LNAI 4944, pp. 196–208, 2008.

measures are data-driven and domain-independent. Generally, they evaluate the rules based on their quality and similarity between them. Subjective measures, including unexpectedness, novelty and actionability, are user-driven and domain-dependent.

For example, classification rules found from a bank's data are very useful to describe who is a good client (whom to offer some additional services) and who is a bad client (whom to watch carefully to minimize the bank loses). However, if bank managers hope to improve their understanding of customers and seek specific actions to improve services, mere classification rules will not be convincing for them. Therefore, we can use the classification rules to build a strategy of action based on condition features in order to get a desired effect on a decision feature [9]. Going back to the bank example, the strategy of action would consist of modifying some condition features in order to improve their understanding of customers and then improve services.

Action rules, introduced in [9] and investigated further in [14], [16], [11], are constructed from certain pairs of classification rules. Interventions, defined in [5], are conceptually very similar to action rules.

The process of constructing action rules from pairs of classification rules is not only unnecessarily expensive but also gives too much freedom in constructing their classification parts. In [11] it was shown that action rules do not have to be built from pairs of classification rules and that single classification rules are sufficient to achieve the same goal. However, the paper only proposed a theoretical lattice-theory type framework without giving any detailed algorithm for action rules construction. In this paper we propose a very simple *LERS*-type algorithm for constructing action rules from a single classification rule. *LERS* is a classical example of a bottom-up strategy which constructs rules with a conditional part of the length $k+1$ after all rules with a conditional part of the length k have been constructed. Relations representing rules produced by *LERS* are marked. System *ARAS* assumes that *LERS* is used to extract classification rules. This way *ARAS* instead of verifying the validity of certain relations only has to check if these relations are marked by *LERS*. The same, by using *LERS* as the pre-processing module for *ARAS*, the overall complexity of the algorithm is decreased.

2   Action Rules

In paper [9], the notion of an action rule was introduced. The main idea was to generate, from a database, special type of rules which basically form a hint to users showing a way to re-classify objects with respect to values of some distinguished attribute (called a decision attribute).

We start with a definition of an information system given in [8].

By an information system we mean a pair $S = (U, A)$, where:

- U is a nonempty, finite set of objects (object identifiers),
- A is a nonempty, finite set of attributes (partial functions) i.e. $a : U \to V_a$ for $a \in A$, where V_a is called the domain of a.

We often write (a, v) instead of v, assuming that $v \in V_a$. Information systems can be used to model decision tables. In any decision table together with the set of attributes a partition of that set into conditions and decisions is given. Additionally, we assume that the set of conditions is partitioned into stable and flexible [9]. Attribute $a \in A$ is called stable for the set U, if its values assigned to objects from U can not change in time. Otherwise, it is called flexible. *"Date of Birth"* is an example of a stable attribute. *"Interest rate"* on any customer account is an example of a flexible attribute. For simplicity reason, we will consider decision tables with only one decision. We adopt the following definition of a decision table:

By a decision table we mean an information system $S = (U, A_1 \cup A_2 \cup \{d\})$, where $d \notin A_1 \cup A_2$ is a distinguished attribute called decision. Additionally, it is assumed that d is a total function. The elements of A_1 are called stable attributes, whereas the elements of $A_2 \cup \{d\}$ are called flexible. Our goal is to suggest changes in values of attributes in A_2 for some objects from U so the values of the attribute d for these objects may change as well. A formal expression describing such a property is called an action rule [9], [14].

Table 1. Two classification rules extracted from S

Stable	Flexible	Stable	Flexible	Stable	Flexible	Decision
A	B	C	E	G	H	D
a_1	b_1	c_1	e_1			d_1
a_1	b_2			g_2	h_2	d_2

To construct an action rule [14], let us assume that two classification rules, each one referring to a different decision class, are considered. We assume here that these two rules have to be equal on their stable attributes, if they are both defined on them. We use Table 1 to clarify the process of action rule construction. Here, *"Stable"* means stable attribute and *"Flexible"* means flexible one. In a standard representation, these two classification rules have a form:

$$r_1 = [(a_1 \wedge b_1 \wedge c_1 \wedge e_1) \rightarrow d_1], \ r_2 = [(a_1 \wedge b_2 \wedge g_2 \wedge h_2) \rightarrow d_2].$$

Assume now that object x supports rule r_1 which means that it is classified as d_1. In order to reclassify x to a class d_2, we need to change not only the value of B from b_1 to b_2 but also to assume that $G(x) = g_2$ and that the value H for object x has to be changed to h_2. This is the meaning of the (r_1, r_2)-action rule defined by the expression below:

$$r = [[a_1 \wedge g_2 \wedge (B, b_1 \rightarrow b_2) \wedge (H, \rightarrow h_2)] \rightarrow (D, d_1 \rightarrow d_2)].$$

The term $[a_1 \wedge g_2]$ is called the header of the action rule. Assume now that by $Sup(t)$ we mean the number of tuples having property t. By the support of

(r_1, r_2)-action rule (given above) we mean: $Sup[a_1 \wedge g_2 \wedge b_1 \wedge d_1]$. By the confidence $Conf(r)$ of (r_1, r_2)-action rule r (given above) we mean (see [14], [15]):

$$[Sup[a_1 \wedge g_2 \wedge b_1 \wedge d_1]/Sup[a_1 \wedge g_2 \wedge b_1]] \cdot [Sup[a_1 \wedge b_2 \wedge c_1 \wedge d_2]/Sup[a_1 \wedge b_2 \wedge c_1]].$$

Assume now that $S = (U, A_1 \cup A_2 \cup \{d\})$ is decision system, where $A_1 = \{a, b\}$ are stable attributes, $A_2 = \{c, e, f\}$ are flexible attributes, and d is the decision. For a generality reason, we take an incomplete decision system. It is represented as Table 2. Our goal is to re-classify objects in S from $(d, 2)$ to $(d, 1)$. Additionally, we assume that $Dom(a) = \{2, 3, 10\}$, $Dom(b) = \{2, 3, 4, 5\}$, and the null value is represented as -1. We will follow optimistic approach in the process of action rules discovery, which means that the Null value is interpreted as the disjunction of all possible attribute values in the corresponding domain.

Table 2. Incomplete Decision System S

Stable a	Stable b	Flexible c	Flexible e	Flexible f	Decision d
2	−1	−1	7	8	1
2	5	4	6	8	1
−1	−1	4	9	4	2
10	4	5	8	7	2
2	2	5	−1	9	3
2	2	4	7	6	3
−1	2	4	7	−1	2
2	−1	−1	6	8	3
3	2	4	6	8	2
3	3	5	7	4	2
3	3	5	6	2	3
2	5	4	9	4	1

Now, we present the preprocessing step for action rules discovery. We start with our incomplete decision system S as the root of the Reduction Tree (see Fig. 1). The next step is to split S into sub-tables taking an attribute with the minimal number of distinct values as the splitting one. In our example, we chose attribute a. Because the 3rd and the 7th tuple in S contain null values in column a, we move them both to all three newly created sub-tables. This process is recursively continued for all stable attributes. Sub-tables corresponding to outgoing edges from the root node which are labelled by $a = 10$, $a = 3$ are removed because they do not contain decision value 1. Any remaining node in

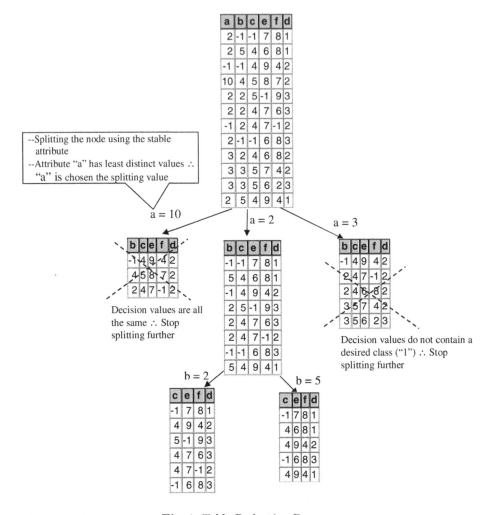

Fig. 1. Table Reduction Process

the resulting tree can be used for discovering action rules. Clearly, if node n is used to construct action rules, then its children are not used for that purpose.

3   ARAS: Algorithm for Discovering Action Rules

This section covers only complete information systems. For an incomplete information system, we can use $ERID$ [3] to discover classification rules. Their syntax is the same as the syntax of rules discovered from a complete system.

Let us assume that $S = (U, A_1 \cup A_2 \cup \{d\})$ is a complete decision system, where $d \notin A_1 \cup A_2$ is a distinguished attribute called the decision. The elements of A_1 are stable conditions, whereas the elements of $A_2 \cup \{d\}$ are flexible. Assume that $d_1 \in V_d$ and $x \in U$. We say that x is a d_1-object if $d(x) = d_1$. We also assume that $\{a_1, a_2, \ldots, a_p\} \subseteq A_1$, $\{a_{p+1}, a_{p+2}, \ldots a_n\} = A_1 - \{a_1, a_2, \ldots, a_p\}$, $\{b_1, b_2, \ldots, b_q\} \subseteq A_2$, $a_{[i,j]}$ denotes a value of attribute a_i, $b_{[i,j]}$ denotes a value of attribute b_i, for any i, j and that

$$r = [[a_{[1,1]} \wedge a_{[2,1]} \wedge \cdots \wedge a_{[p,1]} \wedge b_{[1,1]} \wedge b_{[2,1]} \wedge \ldots \wedge b_{[q,1]}] \longrightarrow d_1]$$

is a classification rule extracted from S supporting some d_1-objects in S. Class d_1 is a preferable class and our goal is to reclassify d_2-objects into d_1 class, where $d_2 \in V_d$.

By an action rule schema $r_{[d_2 \longrightarrow d_1]}$ associated with r and the above reclassification task $(d, d_2 \longrightarrow d_1)$ we mean the following expression:

$$r_{[d_2 \longrightarrow d_1]} = [[a_{[1,1]} \wedge a_{[2,1]} \wedge \cdots \wedge a_{[p,1]} \wedge (b_1, \longrightarrow b_{[1,1]})$$
$$\wedge (b_2, \longrightarrow b_{[2,1]}) \wedge \ldots \wedge (b_q, \longrightarrow b_{[q,1]})] \longrightarrow (d, d_2 \longrightarrow d_1)]$$

In a similar way, by an action rule schema $r_{[\longrightarrow d_1]}$ associated with r and the reclassification task $(d, \longrightarrow d_1)$ we mean the following expression:

$$r_{[\longrightarrow d_1]} = [[a_{[1,1]} \wedge a_{[2,1]} \wedge \cdots \wedge a_{[p,1]} \wedge (b_1, \longrightarrow b_{[1,1]})$$
$$\wedge (b_2, \longrightarrow b_{[2,1]}) \wedge \ldots \wedge (b_q, \longrightarrow b_{[q,1]})] \longrightarrow (d, \longrightarrow d_1)]$$

The term $[a_{[1,1]} \wedge a_{[2,1]} \wedge \cdots \wedge a_{[p,1]}]$, built from values of stable attributes, is called the header of $r_{[d_2 \longrightarrow d_1]}$ and its values can not be changed. It is denoted by $h[r_{[d_2 \longrightarrow d_1]}]$.

The support set of the action rule schema $r_{[d_2 \longrightarrow d_1]}$ is defined as $Sup(r_{[d_2 \rightarrow d_1]})$ $= \{x \in U : (a_1(x) = a_{[1,1]}) \wedge (a_2(x) = a_{[2,1]}) \wedge \cdots \wedge (a_p(x) = a_{[p,1]}) \wedge (d(x) = d_2)\}$.

Now, we outline $ARAS$ strategy for generating the set **AR** of action rules from the action rule schema $r_{[d_2 \rightarrow d_1]}$.

Assume that:

- $V_{a_{p+1}} = \{a_{[p+1,1]}, a_{[p+1,2]}, \ldots, a_{[p+1,J(1)]}\}$
- $V_{a_{p+2}} = \{a_{[p+2,1]}, a_{[p+2,2]}, \ldots, a_{[p+2,J(2)]}\}$
- ...
- $V_{a_{p+n}} = \{a_{[p+n,1]}, a_{[p+n,2]}, \ldots, a_{[p+n,J(n)]}\}$
- $V_{b_1} = \{b_{[1,1]}, b_{[1,2]}, \ldots, b_{[1,J(n+1)]}\}$
- $V_{b_2} = \{b_{[2,1]}, b_{[2,2]}, \ldots, b_{[2,J(n+2)]}\}$
- ....
- $V_{b_q} = \{b_{[q,1]}, b_{[q,2]}, \ldots, b_{[q,J(n+q)]}\}$

To simplify the presentation of the algorithm we assume that:

- $c_k = a_{p+k}$ and $c_{[k,i]} = a_{[p+k,i]}$, for $1 \leq i \leq J(k)$, $1 \leq k \leq n$,
- $c_{n+m} = b_m$ and $c_{[n+m,i]} = b_{[m,i]}$, for $1 \leq i \leq J(n+m)$, $1 \leq m \leq q$.

For simplicity reason, we use $U_{[r,d_2]}$ to denote $Sup(r_{[d_2 \rightarrow d_1]})$. We assume that the term $c_{[i_1,j_1]} \wedge c_{[i_2,j_2]} \wedge \cdots \wedge c_{[i_r,j_r]}$ is denoted by $[c_{[i_k,j_k]}]_{k \in \{1,2,\ldots,r\}}$, where all $i_1, i_2, \ldots, i_r$ are distinct integers and $j_p \leq J(i_p)$, $1 \leq p \leq r$. Following **LERS** notation [6], we also assume that $t^\star$ denotes the set of all objects in S having property t.

Algorithm $AR(r, d_2)$
    i:=1
  while $i \leq n + q$ **do**
   begin
   j:=2; m:=1
   while $j < J(i)$ **do**
    begin
    if $[h[r_{[d_2 \longrightarrow d_1]}] \wedge c_{(i,j)}]^\star \subseteq U_{[r,d_2]} \wedge c_i \in A_2$ **then**
     begin
     mark$[c_{(i,j)}]$;
     output Action Rule
      $[[h[r_{[d_2 \longrightarrow d_1]}] \wedge (c_i, c_{(i,j)} \longrightarrow c_{(i,1)})] \longrightarrow [d, d_2 \longrightarrow d_1]]$
     end
    if $[h[r_{[d_2 \longrightarrow d_1]}] \wedge c_{(i,j)}]^\star \subseteq U_{[r,d_2]} \wedge c_i \in A_1$ **then**
     begin
     mark$[c_{(i,j)}]$;
     output Action Rule
      $[[h[r_{[d_2 \longrightarrow d_1]}] \wedge (c_i, c_{(i,j)})] \longrightarrow [d, d_2 \longrightarrow d_1]]$
     end
    j:=j+1
    end
   end
   $I_k := \{i_k\}$;
   (where i_k - index randomly chosen from $\{2, 3, \ldots, q + n\}$).
   for all $j_k \leq J(i_k)$ **do** $[c_{(i_k, j_k)}]_{i_k \in I_k} := c(i_k, j_k)$;
   for all i, j **such that** both sets $[c_{(i_k, j_k)}]_{i_k \in I_k}$, $c_{(i,j)}$ are not marked and
$i \in I_k$
   do
    begin
    if $[[h[r_{[d_2 \longrightarrow d_1]}] \wedge [c_{(i_k, j_k)}]_{i_k \in I_k} \wedge c_{(i,j)}]]^\star \subseteq U_{[r,d_2]} \wedge c_i \in A_2$ **then**
     begin
     mark $[[c_{(i_k, j_k)}]_{i_k \in I_k} \wedge c_{(i,j)}]$;
     output Action Rule
      $[[h[r_{[d_2 \longrightarrow d_1]}] \wedge [c_{(i_k, j_k)}]_{i_k \in I_k} \wedge (c_i, c_{(i,j)} \longrightarrow c_{(i,1)})] \longrightarrow [d, d_2 \longrightarrow d_1]]$
     end
    if $[[h[r_{[d_2 \longrightarrow d_1]}] \wedge [c_{(i_k, j_k)}]_{i_k \in I_k} \wedge c_{(i,j)}]]^\star \subseteq U_{[r,d_2]} \wedge c_i \in A_1$ **then**
     begin
     mark $[[c_{(i_k, j_k)}]_{i_k \in I_k} \wedge c_{(i,j)}]$;
     output Action Rule
      $[[h[r_{[d_2 \longrightarrow d_1]}] \wedge [c_{(i_k, j_k)}]_{i_k \in I_k} \wedge (c_i, c_{(i,j)})] \longrightarrow [d, d_2 \longrightarrow d_1]]$
     end
    else
     begin
     $I_k := I_k \cup \{i\}$; $[c_{(i_k, j_k)}]_{i_k \in I_k} := [c_{(i_k, j_k)}]_{i_k \in I_k} \wedge c_{(i,j)}$
     end

The complexity of $ARAS$ is lower than the complexity of $DEAR$ system discovering action rules. The justification here is quite simple. $DEAR$ system [14] groups classification rules into clusters of non-conflicting rules and then takes all possible pairs of classification rules within each cluster and tries to build action rules from them. $ARAS$ algorithm is treating each classification rule describing the target decision value as a seed and grabs other classification rules describing non-target decision values to form a cluster and then it builds decision rules automatically from them. Rules grabbed into a seed are only compared with that seed. So, the number of pairs of rules which have to be checked, in comparison to $DEAR$ is greatly reduced. Another advantage of the current strategy is that the module generating action rules in $ARAS$ only checks if a mark is assigned by $LERS$ to the relation $[h[r_{[d_2 \longrightarrow d_1]}] \wedge c_{(i,j)}]^\star \subseteq U_{[r,d_2]}$ instead of checking its validity.

The confidence of generated action rules depends on the number of remaining objects supporting them. Also, if $Conf(r) \neq 1$, then some objects in S satisfying the description $[a_{1,1} \wedge a_{2,1} \wedge \cdots \wedge a_{p,1} \wedge b_{1,1} \wedge b_{2,1} \wedge \ldots \wedge b_{q,1}]$ are classified as d_2. According to the rule $r_{[d_2 \rightarrow d_1]}$ they should be classified as d_1 which means that the confidence of $r_{[d_2 \rightarrow d_1]}$ will get also decreased.

If $Sup(r_{[d_2 \rightarrow d_1]}) = \emptyset$, then $r_{[d_2 \rightarrow d_1]}$ can not be used for reclassification of objects. Similarly, $r_{[\rightarrow d_1]}$ can not be used for reclassification, if $Sup(r_{[d_2 \rightarrow d_1]}) = \emptyset$, for each d_2 where $d_2 \neq d_1$. From the point of view of actionability, such rules are not interesting.

Let $Sup(r_{[\rightarrow d_1]}) = \bigcup \{Sup(r_{[d_2 \rightarrow d_1]}) : (d_2 \in V_d) \wedge (d_2 \neq d_1)\}$ and $Sup(R_{[\rightarrow d_1]}) = \bigcup \{Sup(r_{[\rightarrow d_1]}) : r \in R(d_1)\}$, where $R(d_1)$ is the set of all classification rules extracted from S which are defining d_1. So, $Sup(R_S) = \bigcup \{Sup(R_{[\rightarrow d_1]}) : d_1 \in V_d\}$ contains all objects in S which potentially can be reclassified.

Assume now that $U(d_1) = \{x \in U : d(x) \neq d_1\}$. Objects in the set $B(d_1) = [U(d_1) - Sup(R_{[\rightarrow d_1]})]$ can not be reclassified to the class d_1 and they are called d_1-resistant [11].

Let $B(\neg d_1) = \bigcap \{B(d_i) : (d_i \in V_d) \wedge (d_i \neq d_1)\}$. Clearly $B(\neg d_1)$ represents the set of d_1-objects which can not be reclassified. They are called d_1-stable. Similarly, the set $B_d = \bigcup \{B(\neg d_i) : d_i \in V_d\}$ represents objects in U which can not be reclassified to any decision class. All these objects are called d-stable. In order to show how to find them, the notion of a confidence of an action rule is needed.

Let $r_{[d_2 \rightarrow d_1]}$, $r'_{[d_2 \rightarrow d_3]}$ are two action rules extracted from S. We say that these rules are p-equivalent ($\simeq$), if the condition given below holds for every $b_i \in A_1 \cup A_2$:

if r/b_i, r'/b_i are both defined, then $r/b_i = r'/b_i$.

Now, we explain how to calculate the confidence of $r_{[d_2 \rightarrow d_1]}$. Let us take d_2-object $x \in Sup(r_{[d_2 \rightarrow d_1]})$. We say that x positively supports $r_{[d_2 \rightarrow d_1]}$ if there is no classification rule r' extracted from S and describing $d_3 \in V_d$, $d_3 \neq d_1$, which is p-equivalent to r, such that $x \in Sup(r'_{[d_2 \rightarrow d_3]})$. The corresponding subset of $Sup(r_{[d_2 \rightarrow d_1]})$ is denoted by $Sup^+(r_{[d_2 \rightarrow d_1]})$. Otherwise, we say that

x negatively supports $r_{[d_2 \to d_1]}$. The corresponding subset of $Sup(r_{[d_2 \to d_1]})$ is denoted by $Sup^-(r_{[d_2 \to d_1]})$.

By the confidence of $r_{[d_2 \to d_1]}$ in S we mean [11]:

$$Conf(r_{[d_2 \to d_1]}) = [card[Sup^+(r_{[d_2 \to d_1]})]/card[Sup(r_{[d_2 \to d_1]})]] \cdot conf(r).$$

Now, if we assume that $Sup^+(r_{[\to d_1]}) = \bigcup\{Sup^+(r_{[d_2 \to d_1]}) : (d_2 \in V_d) \wedge (d_2 \neq d_1)\}$, then by the confidence of $r_{[\to d_1]}$ in S we mean:

$$Conf(r_{[\to d_1]}) = [card[Sup^+(r_{[\to d_1]})]/card[Sup(r_{[\to d_1]})]] \cdot conf(r).$$

It can be easily shown that the definition of support and confidence of action rules given in Section 3 is equivalent to the definition of support and confidence given in Section 2.

4   An Example

Let us assume that the decision system $S = (U, \{A_1 \cup A_2 \cup \{d\}\})$, where $U = \{x_1, x_2, x_3, x_4, x_5, x_6, x_7, x_8\}$, is represented by Table 3 [11]. The set $A_1 = \{a, b, c\}$ contains stable attributes and $A_2 = \{e, f, g\}$ contains flexible attributes. System $LERS$ [6] is used to extract classification rules.

Table 3. Decision System

U	a	b	c	e	f	g	d
x_1	a_1	b_1	c_1	e_1	f_2	g_1	d_1
x_2	a_2	b_1	c_2	e_2	f_2	g_2	d_3
x_3	a_3	b_1	c_1	e_2	f_2	g_3	d_2
x_4	a_1	b_1	c_2	e_2	f_2	g_1	d_2
x_5	a_1	b_2	c_1	e_3	f_2	g_1	d_2
x_6	a_2	b_1	c_1	e_2	f_3	g_1	d_2
x_7	a_2	b_3	c_2	e_2	f_2	g_2	d_2
x_8	a_2	b_1	c_1	e_3	f_2	g_3	d_2

We are interested in reclassifying d_2-objects either to class d_1 or d_3. Four certain classification rules describing d_1, d_3 can be discovered by $LERS$ in the decision system S. They are given below:

$$r1 = [b_1 \wedge c_1 \wedge f_2 \wedge g_1] \to d_1, \quad r2 = [a_2 \wedge b_1 \wedge e_2 \wedge f_2] \to d_3,$$
$$r3 = e_1 \to d_1, \quad r4 = [b_1 \wedge g_2] \longrightarrow d_3.$$

It can be shown that $R_{[d, \to d_1]} = \{r1, r3\}$ and $R_{[d, \to d_3]} = \{r2, r4\}$. Action rule schemas associated with $r1$, $r2$, $r3$, $r4$ and the reclassification task either $(d, d_2 \to d_1)$ or $(d, d_2 \to d_3)$ are:

$$r1_{[d_2 \rightarrow d_1]} = [b_1 \wedge c_1 \wedge (f, \rightarrow f_2) \wedge (g, \rightarrow g_1)] \rightarrow (d, d_2 \rightarrow d_1),$$
$$r2_{[d_2 \rightarrow d_3]} = [a_2 \wedge b_1 \wedge (e, \rightarrow e_2) \wedge (f, \rightarrow f_2)] \rightarrow (d, d_2 \rightarrow d_3),$$
$$r3_{[d_2 \rightarrow d_1]} = [(e, \rightarrow e_1)] \rightarrow (d, d_2 \rightarrow d_1),$$
$$r4_{[d_2 \rightarrow d_3]} = [b_1 \wedge (g, \rightarrow g_2)] \rightarrow (d, d_2 \rightarrow d_3).$$

We can also show that $U_{[r1,d_2]} = Sup(r1_{[d_2 \rightarrow d_1]}) = \{x_3, x_6, x_8\}$, $U_{[r2,d_2]} = Sup(r2_{[d_2 \rightarrow d_3]}) = \{x_6, x_8\}$, $U_{[r3,d_2]} = Sup(r3_{[d_2 \rightarrow d_1]}) = \{x_3, x_4, x_5, x_6, x_7, x_8\}$, $U_{[r4,d_2]} = Sup(r4_{[d_2 \rightarrow d_3]}) = \{x_3, x_4, x_6, x_8\}$.

Following $AR(r1, d_2)$ algorithm we get: $[b_1 \wedge c_1 \wedge a_1]^\star = \{x_1\} \not\subseteq U_{[r1,d_2]}$, $[b_1 \wedge c_1 \wedge a_2]^\star = \{x_6, x_8\} \subseteq U_{[r1,d_2]}$, $[b_1 \wedge c_1 \wedge f_3]^\star = \{x_6\} \subseteq U_{[r1,d_2]}$, $[b_1 \wedge c_1 \wedge g_2]^\star = \{x_2, x_7\} \not\subseteq U_{[r1,d_2]}$, $[b_1 \wedge c_1 \wedge g_3]^\star = \{x_3, x_8\} \subseteq U_{[r1,d_2]}$.

It will generate two action rules:

$$[b_1 \wedge c_1 \wedge (f, f_3 \rightarrow f_2) \wedge (g, \rightarrow g_1)] \rightarrow (d, d_2 \rightarrow d_1),$$
$$[b_1 \wedge c_1 \wedge (f, \rightarrow f_2) \wedge (g, g_3 \rightarrow g_1)] \rightarrow (d, d_2 \rightarrow d_1).$$

In a similar way we construct action rules from the remaining three action rule schemas.

The action rules discovery process, presented above, is called *ARAS* and it consists of two main modules. For its further clarification, we use another example which has no connection with Table 3. The first module extracts all classification rules from S following *LERS strategy*. Assuming that d is the decision attribute and user is interested in re-classifying objects from its value d_1 to d_2, we treat the rules defining d_1 as seeds and build clusters around them.

For instance, if $A_1 = \{a, b, g\}$ are stable attributes, $A_2 = \{c, e, h\}$ are flexible in $S = (U, A_1 \cup A_2 \cup \{d\})$, and $r = [[a_1 \wedge b_1 \wedge c_1 \wedge e_1] \rightarrow d_1]$ is a classification rule in S, where $V_a = \{a_1, a_2, a_3\}$, $V_b = \{b_1, b_2, b_3\}$, $V_c = \{c_1, c_2, c_3\}$, $V_e = \{e_1, e_2, e_3\}$, $V_g = \{g_1, g_2, g_3\}$, $V_h = \{h_1, h_2, h_3\}$, then we remove from S all tuples containing values $a_2, a_3, b_2, b_3, c_1, e_1$ and we use again *LERS* to extract rules from this subsystem. Each rule defining d_2 is used jointly with r to construct an action rule. The validation step of each of the set-inclusion relations, in the second module of *ARAS*, is replaced by checking if the corresponding term was marked by *LERS* in the first module of *ARAS*.

5   Mining Database HEPAR

As the application domain for our research we have chosen the HEPAR system built in collaboration between the Institute of Biocybernetics and Biomedical Engineering of the Polish Academy of Sciences and physicians at the Medical Center of Postgraduate Education in Warsaw, Poland [2], [7]. HEPAR was designed for gathering and processing clinical data on patients with liver disorders. Its integral part is a database created in 1990 and thoroughly maintained since then at the Clinical Department of Gastroenterology and Metabolizm in Warsaw, Poland. It contains information about 758 patients described by 106 attributes (including 31 laboratory tests with values discretized to: below normal, normal, above normal). It has 14 stable attributes. Two laboratory tests are invasive:

HBsAg [in tissue] and HBcAg [in tissue]. The decision attribute has 7 values: I (acute hepatitis), IIa (subacute hepatitis [types B and C]), IIb (subacute hepatitis [alcohol-abuse]), IIIa (chronic hepatitis [curable]), IIIb (chronic hepatitis [non-curable]), IV (cirrhosis-hepatitis), V (liver-cancer).

The diagnosis of liver disease depends on a combination of patient's history, physical examinations, laboratory tests, radiological tests, and frequently a liver biopsy. Blood tests play an important role in the diagnosis of liver diseases. However, their results should be analyzed along with the patient's history and physical examination. The most common radiological examinations used in the assessment of liver diseases are ultrasound and sonography. Ultrasound is a good test for the detection of liver masses, assessment of bile ducts, and detection of gallstones presence. However, it does not detect the degree of inflammation or fibrosis of the liver. Ultrasound is a noninvasive procedure and there are no risks associated with it. Liver biopsy enables the doctor to examine how much inflammation and how much scarring has occurred. Liver biopsy is an example of invasive procedure that carries certain risks to the patient. Therefore, despite of the importance of its results to the diagnosis, clinicians try to avoid biopsy as often as possible. However, liver biopsy is often the only way to establish correct diagnosis in patients with chronic liver disorders.

A medical treatment is naturally associated with re-classification of patients from one decision class into another one. In our research we are mainly interested in the re-classification of patients from the class IIb into class I and from the class IIIa into class I but without referring to any invasive tests results in action rules.

Database HEPAR has many missing values so we decided to remove all its attributes with more than 90% of null values assuming that these attributes are not related to invasive tests. Also, subjective attributes (like *history of alcohol abuse*) and cammong performed basic medical tests have been removed. Finally, we used classical null value imputation techniques to make the resulting database complete.

The next step of our strategy is to apply RSES software [13] to find d-reducts. The set R = {m, n, q, u, y, aa, ah, ai, am, an, aw, bb, bg, bm, by, cj, cm} is one of them and it does not contain invasive tests. The description of its values is given below:

- m - Bleeding
- n - Subjaundice symptoms
- q - Eructation
- u - Obstruction
- y - Weight loss
- aa - Smoking
- ah - History of viral hepatitis (stable)
- ai - Surgeries in the past (stable)
- am - History of hospitalization (stable)
- an - Jaundice in pregnancy
- aw - Erythematous dermatitis

− bb - Cysts
− bg - Sharp liver edge (stable)
− bm - Blood cell plaque
− by - Alkaline phosphatase
− cj - Prothrombin index
− cm - Total cholesterol
− dd Decision attribute

Two action rules discovered by $ARAS$ from the database reduced to d-reduct R are given below.

$[(ah = 1) \wedge (ai = 2) \wedge (am = 2) \wedge (bg = 1)] \wedge [(cm = 1) \wedge (aw = 1) \wedge$
$(u, \rightarrow 1) \wedge (bb{=}1) \wedge (aa = 1) \wedge (q, \rightarrow 1) \wedge (m, \rightarrow 1) \wedge (n{=}1) \wedge (bm, \rightarrow down) \wedge$
$(y = 1) \wedge (by, norm \rightarrow up)] \Longrightarrow (dd, IIIa \rightarrow I)$

$[(ah = 1) \wedge (ai = 2) \wedge (am = 2) \wedge (bg = 1)] \wedge [(cm = 1) \wedge (aw = 1) \wedge$
$(u, \rightarrow 1) \wedge (bb{=}1) \wedge (aa = 1) \wedge (q, \rightarrow 1) \wedge (m, \rightarrow 1) \wedge (n{=}1) \wedge (bm, \rightarrow down) \wedge$
$(y = 1) \wedge (by, norm \rightarrow down)] \Longrightarrow (dd, IIIa \rightarrow I)$

Both action rules are applicable to patients with no history of viral hepatitis but with a history of surgery and hospitalization. Sharp liver edge has to be normal, no subjaundice symptoms, total cholesterol, erythematous dermatitis, and weight have to be normal, no cysts, and patient can not smoke.

For this class of patients, the action rule says that:

By getting rid of obstruction, eructation, bleeding, by decreasing the blood cell plaque and by changing the level of alkaline phosphatase we should be able to reclassify the patient from class IIIa to class I. Medical doctor should decide if the alkaline phosphatase level needs to be decreased or increased. Attribute values of total cholesterol, weight, and smoking have to remain unchanged.

6   Conclusion

System $ARAS$ differs from the tree-based strategies for action rules discovery (for instance from $DEAR$ [14]) because clusters generated by its second module are formed around target classification rules. An action rule can be constructed in $ARAS$ from two classification rules only if both of them belong to the same cluster and one of them is a target classification rule. So, the complexity of the second module of $ARAS$ is $0(k \cdot n)$, where n is the number of classification rules extracted by $LERS$ and k is the number of clusters. The time complexity of the second module of $DEAR$ is equal to $0(n \cdot n)$, where n is the same as in $ARAS$. The first module of $ARAS$ is the same as the first module of $DEAR$, so their complexities are the same.

Acknowledgements

This research was partially supported by the National Science Foundation under grant IIS-0414815.

References

1. Adomavicius, G., Tuzhilin, A.: Discovery of actionable patterns in databases: the action hierarchy approach. In: Proceedings of KDD 1997 Conference, Newport Beach, CA, AAAI Press, Menlo Park (1997)
2. Bobrowski, L.: HEPAR: Computer system for diagnosis support and data analysis. In: Prace IBIB 31, Institute of Biocybernetics and Biomedical Engeneering, Polish Academy of Sciences, Warsaw, Poland (1992)
3. Dardzinska, A., Ras, Z.W.: Extracting Rules from Incomplete Decision Systems: System ERID. In: Lin, T.Y., Ohsuga, S., Liau, C.J., Hu, X. (eds.) Foundations and Novel Approaches in Data Mining, Advances in Soft Computing, vol. 9, pp. 143–154. Springer, Heidelberg (2006)
4. Hilderman, R.J., Hamilton, H.J.: Knowledge Discovery and Measures of Interest. Kluwer, Dordrecht (2001)
5. Greco, S., Matarazzo, B., Pappalardo, N., Slowiński, R.: Measuring expected effects of interventions based on decision rules. Journal of Experimental and Theoretical Artificial Intelligence 17(1-2) (2005)
6. Grzymala-Busse, J.: A new version of the rule induction system LERS. Fundamenta Informaticae 31(1), 27–39 (1997)
7. Onińsko, A., Druzdzel, M., Wasyluk, H.: Extension of the HEPAR II model to multiple-disorder diagnosis. In: Intelligent Information Systems, Advances in Soft Computing, pp. 303–313. Springer, Heidelberg (2000)
8. Pawlak, Z.: Information systems - theoretical foundations. Information Systems Journal 6, 205–218 (1991)
9. Raś, Z., Wieczorkowska, A.: Action Rules: how to increase profit of a company. In: Zighed, A.D.A., Komorowski, J., Żytkow, J.M. (eds.) PKDD 2000. LNCS (LNAI), vol. 1910, pp. 587–592. Springer, Heidelberg (2000)
10. Raś, Z.W., Tzacheva, A., Tsay, L.-S.: Action rules. In: Wang, J. (ed.) Encyclopedia of Data Warehousing and Mining, pp. 1–5. Idea Group Inc. (2005)
11. Raś, Z.W., Dardzińska, A.: Action rules discovery, a new simplified strategy. In: Esposito, F., Raś, Z.W., Malerba, D., Semeraro, G. (eds.) ISMIS 2006. LNCS (LNAI), vol. 4203, pp. 445–453. Springer, Heidelberg (2006)
12. Silberschatz, A., Tuzhilin, A.: On subjective measures of interestingness in knowledge discovery. In: Proceedings of KDD 1995 Conference, AAAI Press, Menlo Park (1995)
13. RSES - http://logic.mimuw.edu.pl/~rses/
14. Tsay, L.-S., Raś, Z.W.: Action rules discovery system DEAR, method and experiments. Journal of Experimental and Theoretical Artificial Intelligence 17(1-2), 119–128 (2005)
15. Tsay, L.-S., Raś, Z.W.: Action rules discovery system DEAR3. In: Esposito, F., Raś, Z.W., Malerba, D., Semeraro, G. (eds.) ISMIS 2006. LNCS (LNAI), vol. 4203, pp. 483–492. Springer, Heidelberg (2006)
16. Tzacheva, A., Raś, Z.W.: Action rules mining. International Journal of Intelligent Systems 20(7), 719–736 (2005)

Learning to Order: A Relational Approach

Donato Malerba and Michelangelo Ceci

Dipartimento di Informatica, Università degli Studi di Bari
via Orabona, 4 - 70126 Bari - Italy
{malerba,ceci}@di.uniba.it

Abstract. In some applications it is necessary to sort a set of elements according to an order relationship which is not known *a priori*. In these cases, a training set of ordered elements is often available, from which the order relationship can be automatically learned. In this work, it is assumed that the correct succession of elements in a training sequence (or chain) is given, so that it is possible to induce the definition of two predicates, $first/1$ and $succ/2$, which are then used to establish an ordering relationship. A peculiarity of this work is the relational representation of training data which allows various relationships between ordered elements to be expressed in addition to the ordering relationship. Therefore, an ILP learning algorithm is applied to induce the definitions of the two predicates. Two methods are reported for the identification of either single chains or multiple chains on new objects. They have been applied to the problem of learning the reading order of layout components extracted from document images. Experimental results show the effectiveness of the proposed solution.

1 Introduction

Many applications require sorting a set of elements according to either a partial or a total ordering relationship. The problem can be efficiently solved by applying sorting algorithms when the ordering relationship is known *a priori*, but there are cases in which no definition of ordering relationship is available due to several difficulties in formalizing one. A prominent example is represented by preference functions, which indicate whether one element should be ranked before another. A preference function is typically subjective and difficult to elicit, although it can be relatively easy to collect instances of ordered elements from which the preference function can be automatically induced. Another example is the reading order of layout components in a page. The rule of thumb "Western-style documents are usually read top-bottom and left-right" is an ambiguous statement, which is not appropriate for many newspapers and magazines. Also in this case, it is easy to collect examples of correct reading sequences from which the reading order rules specific to a class of documents can be learned.

The problem of learning how to construct an ordering, given a collection of instances of ordered elements, has been faced by Cohen *et al* [6] who propose a two-stage approach. In the first stage (*learning*), a binary preference function $PREF(u, v)$ is learned, which indicates how certain it is that u should be

Z.W. Raś, S. Tsumoto, and D. Zighed (Eds.): MCD 2007, LNAI 4944, pp. 209–223, 2008.
© Springer-Verlag Berlin Heidelberg 2008

ranked before v. In the second stage (*sorting*), new instances are ordered so as to maximize the agreement with the learned preference function. The function $PREF(u, v)$ is a linear combination of primitive preference functions, each of which is associated with a ranking expert, and the learning process consists in defining a weight for each primitive preference function on the basis of training sets of ordered pairs (u, v). Kamishima and Akaho [13] propose a naive Bayes approach to estimate $PREF(u, v)$ and compare two alternative optimality criteria (sum and product of values taken by PREF) to be maximized in the sorting stage.

In both works, an ordered pair (u, v) in the training set is interpreted as v is ranked above u, and the learning task aims to induce a definition of an ordering which is consistent with input ordered pairs. However, in many applications the ordered pair (u, v) in the training set can be interpreted as "v is the successor of u", in which case the learning task can be slightly different, namely learning the definition of the *successor* relationship between elements. Once the successor relationship is learned, an ordering relationship can be established.

In this paper, we follow this approach to learning how to order elements. Given both positive and negative instances of two predicates, namely $first/1$ and $succ/2$, we first induce the definitions of "first element" and "successor element" in a sequence (or *chain*) of elements, and then we apply these definitions to a new set of elements, in order to reconstruct a possible partial or total ordering of these elements. The main advantage of this approach is that it can also be applied to those tasks characterized by the following properties: a) not all elements have to be ordered - only those involved in a direct succession relationship; b) different sequences can be defined on subsets of elements. Two examples of these tasks are the detection of the reading order between document layout components [2] and the design of workflows from process logs [20].

The task considered in this paper is *predictive* and differs from another *descriptive* task reported in the works [19], where the problem is discovering fragments of order, and [10], where the problem is that of describing a set of sequences by a single (or a set of) partial orders occurring in the sequences.

Another important difference with respect to related works is the representation of training data. In all previous works, the ordering relationship is the only one considered between elements, which are represented as rows of a single table, whose columns correspond to attributes of the elements. However, in several applications this representation is quite restrictive. For instance, in reading order detection some spatial relationships can be defined between layout components and the reading order can actually depend on these spatial relationships (e.g., the next layout component is 'below' the one currently read). To consider these additional relationships we resort to a *relational* representation with several tables which describe possibly different types of elements and the various relationships between them. Therefore, a peculiarity of this work is that training data are *complex objects* and that ordered elements are *basic components* of these complex objects.

The relational representation of complex objects requires the application of inductive logic programming (ILP) [23,14,24] methods in order to induce a definition of the predicates $first/1$ and $succ/2$. In this work, we resort to the application of the ILP system ATRE [17] to learn a logical theory which defines the two predicates and is then used to reconstruct a partial or total ordering in new structured complex objects.

The paper is organized as follows. The problem of learning to order objects is formally defined in Section 2. The machine learning system ATRE, applied to the problem of learning the logical theory, is introduced in Section 3, while the application of the learned theory in order to reconstruct a partial order relationship, is reported in Section 4. Finally, the application to the document image processing domain is illustrated in Section 5, where experimental results are also reported and commented.

2   Problem Definition

In order to formalize the learning problem, some useful definition are necessary.

Definition 1 (*Partial Order [11]*). *Let A be a set of basic components of a complex object, a partial order P over A is a relation $P \in A \times A$ such that P is*

1. *reflexive $\forall s \in A \Rightarrow (s, s) \in P$*
2. *antisymmetric $\forall s_1, s_2 \in A$: $(s_1, s_2) \in P \wedge (s_2, s_1) \in P \Leftrightarrow s_1 = s_2$*
3. *transitive $\forall s_1, s_2, s_3 \in A$: $(s_1, s_2) \in P \wedge (s_2, s_3) \in P \Rightarrow (s_1, s_3) \in P$*

When P satisfies the antisymmetric, the transitive and the irreflexive ($\forall s \in A \Rightarrow (s, s) \notin P$) properties, it is called a weak partial order over A.

Definition 2 (*Total Order*). *Let A be a set of basic components of a complex object, a partial order T over the set A is a total order iff $\forall s_1, s_2 \in A$: $(s_1, s_2) \in T \vee (s_2, s_1) \in T$*

Definition 3 (*Complete chain, Chain reduction*). *Let A be a set of basic components of a complex object, let D be a weak partial order over A, let $B = \{a \in A | (\exists b \in A \text{ s.t. } (a, b) \in D \vee (b, a) \in D)\}$ be the subset of elements in A related to any element in A itself. If $D \cup \{(a, a) | a \in B\}$ is a total order over B then D is a complete chain over A.*

Furthermore, $C = \{(a, b) \in D | \neg \exists c \in A \text{ s.t. } (a, c) \in D \wedge (c, b) \in D\}$ is the reduction of the chain D over A.

Example 1. Let $A = \{a, b, c, d, e\}$. $D = \{(a, b), (a, c), (a, d), (b, c), (b, d), (c, d)\}$ is a complete chain over A, then $C = \{(a, b), (b, c), (c, d)\}$ is its reduction.

Indeed, for our purposes it is equivalent to deal with complete chains or their reduction. Henceforth, for the sake of simplicity, the term *chain* will denote the reduction of a complete chain. By resorting to definitions above, it is possible to formalize the ordering induction problem as follows:

Given

- A description $DesTO_i$ in the language L of the set of n training complex objects $TrainingObjs = \{TP_i \in \Pi | i = 1..n\}$ (where Π is the set of complex objects).
- A description $DesTC_i$ in the language L of the set TC_i of chains (over $TP_i \in TrainingObjs$) for each $TP_i \in TrainingObjs$.

Find: An intensional definition T in the language L of a chain over a generic compex object $O \in \Pi$ such that T is complete and consistent with respect to all training chains descriptions $DesTC_i$, $i = 1..n$.

In this problem definition, we refer to the intensional definition T as a first order logic theory, that is, a set of first order definite clauses [16]. The fact that T is complete and consistent with respect to all training chains descriptions can be formally described as follows:

Definition 4 (Completeness and Consistency). *Let*

- *T be a logic theory describing chains instances expressed in the language L.*
- *E^+ be the set of positive examples for the chains instances.*
  $(E^+ = \bigcup_{i=1..n}(\bigcup_{TC \in TC_i} TC))$.
- *E^- be the set of negative examples for the chains instances.*
  $(E^- = \bigcup_{i=1..n}(TP_i \times TP_i)/E^+)$.
- *$DesE^+$ be the description of E^+ in L.*
- *$DesE^-$ be the description of E^- in L.*

then T is complete and consistent with respect to all training chains descriptions iff $T \models DesE^+ \wedge T \not\models DesE^-$

This formalization of the problem permits to represent and identify distinct orderings on the same complex object and allows to avoid to include in the ordering basic components that should not be included.

3   Learning the Logical Theory with ATRE

ATRE is an ILP system that can learn recursive theories from examples. The learning problem solved by ATRE can be formulated as follows:

Given

- a set of *concepts* $K_1, K_2, \ldots, K_r$ to be learned
- a set of *observations* O described in a language $\mathcal{L}_O$
- a *background theory* BK described in a language $\mathcal{L}_{BK}$,
- a *language* of hypotheses $\mathcal{L}_H$ which defines the space of hypotheses S_H
- a user's *preference criterion* PC

Find

A logical theory $T \in S_H$, which defines the concepts K_1, K_2, ..., K_r, such that T is complete and consistent with respect to O and satisfies the preference criterion *PC*.

The *completeness* property holds when the theory T explains all observations in O of the r concepts K_1, K_2, ..., K_r, while the *consistency* property holds when the theory T explains no counter-example in O of any concept C_i. The satisfaction of these properties guarantees the correctness of the induced theory, with respect to the given observations O. The selection of the "best" theory is made on the basis of an inductive bias embedded in some heuristic function or expressed by the user of the learning system (preference criterion).

In the context of the ordering induction problem, we identified two concepts to be learned, namely $first/1$ and $succ/2$. The former refers to the the first basic component of a chain, while the latter refers to the relation *successor* between two basic components in a chain. By combining the two concepts it is possible to identify a partial ordering of basic components of a complex object.

Both the language of hypotheses $\mathcal{L}_H$ and the language of background knowledge $\mathcal{L}_{BK}$ are limited to linked, range-restricted definite clauses [7], whose literals can be of the two distinct forms:

$$f(t_1, \ldots, t_n) = \text{Value} \quad \text{(simple literal)}$$
$$f(t_1, \ldots, t_n) \in \text{Range} \quad \text{(set literal)},$$

where f and g are function symbols called *descriptors*, t_i's are *terms* (constants or variables) and *Range* is a closed interval of possible values taken by f.

Observations are represented as ground multiple-head clauses, called *objects*, which have a conjunction of simple literals in the head. Each object is associated with a unique object identifier (OID). The notion of multiple-head clauses in ATRE adapts the notion of *interpretation*, which is common to many relational data mining systems for efficiency reasons [8]. ATRE distinguishes objects from *examples*, which are described as pairs $\langle L, OID \rangle$, where L is a literal in the head of the object identified by *OID*. Examples can be either *positive* or *negative*.

At the high-level ATRE implements a *sequential covering* algorithm [22]. A recursive theory T is built iteratively, starting from an empty theory T_0, and then adding a new clause at each iteration. In this way we obtain a sequence of theories:

$$T_0 = \emptyset, T_1, \ldots, T_i, T_{i+1}, \ldots, T_n = T$$

such that $T_{i+1} = T_i \cup \{C\}$ for some clause C and $LHM(T_i) \subseteq LHM(T_{i+1})$, where $LHM(T)$ denotes the least Herbrand model of a theory T [16]. Let $pos(LHM(T))$ and $neg(LHM(T))$ be the number of positive and negative examples in $LHM(T)$, respectively. If we guarantee that:

1. $pos(LHM(T_i)) < pos(LHM(T_{i+1}))$, for each $i \in \{0, 1, \ldots, n-1\}$ and
2. $neg(LHM(T_i)) = 0$, for each $i \in \{0, 1, \ldots, n\}$,

then, after a finite number of iterations, a theory T, which is complete and consistent, is built. The first condition is guaranteed by selecting a positive example (or *seed*) $e^+ \notin LHM(T_i)$ of a concept K_j to be learned, and then by looking for a clause C, if any, such that $e^+ \in LHM(T_i \cup \{C\})$ (i.e., $pos(LHM(T_i \cup \{C\})) > pos(LHM(T_i)))$. The second condition is more difficult to guarantee since the addition of a locally consistent clause C to a theory T_i does not preserve consistency of $T_i \cup \{C\}$ (non-monotonicity of the normal ILP setting). The approach followed in ATRE consists of simple syntactic changes in T_i, which eventually creates new *layers* [17].

The automated discovery of dependencies between concepts K_1, K_2, ..., K_r is based on a variant of the sequential covering learning strategy, which is traditionally adopted by single concept learning systems that generate clauses with the same literal in the head at each iteration. In multiple concept learning, clauses generated at each iteration may refer to different concepts. In addition, the body of the clause generated at the i-th iteration may involve any concept K_1, K_2, ..., K_r for which at least a clause has been added to the theory partially learned in previous iterations. In this way, dependencies between concepts can be generated.

At each iteration of the main loop of the sequential covering algorithm, clauses for distinct concepts are generated, and then one of them is picked. Since the generation of a clause depends on a seed, several seeds have to be chosen (if any, at least one seed per concept to be learned). Therefore, the search space is a forest of as many search-trees as the number of chosen seeds. Each search-tree is rooted with a unit clause and ordered by the generalization model adopted in ATRE (*generalized implication* [17]). The forest can be processed in parallel by as many concurrent tasks as the number of search-trees. Each task traverses the specialization hierarchy top-down through a sequential covering strategy, but synchronizes traversal with the other tasks at each level. Search proceeds toward deeper and deeper levels of the specialization hierarchies until at least a user-defined number of consistent clauses is found. Task synchronization is performed after that all "relevant" clauses at the same depth have been examined. A supervisor task decides whether the search should carry on or not on the basis of the results returned by the concurrent tasks. When the search is stopped, the supervisor selects the "best" consistent clause according to the user's preference criterion. This *separate-and-parallel-conquer* search strategy provides us with a solution to the problem of interleaving the induction of distinct concept definitions.

4   Application of the Learned Logical Theory

Once the logical theory has been learned, they can be applied to new complex objects in order to generate a set of ground atoms such as: $\{first(0) = true, succ(0, 1) = true, \ldots, succ(4, 3) = true, \ldots\}$ which can be used to reconstruct chains of (possibly logically labelled) basic components. In our approach, we propose two different solutions: 1) Identification of multiple chains of basic components. 2) Identification of a single chain of basic components.

By applying the logical theory learned by ATRE, it is possible to identify:

- A *directed* graph $G = \langle V, E \rangle^1$ where V is the set of nodes representing all the basic components of a complex object and $E = \{(b_1, b_2) \in V^2 | succ(b_1, b_2) = true\}$
- A list of initial nodes $I = \{b \in V | first(b) = true\}$
- A list of final nodes $F = \{b_2 \in V - I | \forall b_1 \in V \ (b_1, b_2) \notin E\}$.

Both approaches make use of G, I and T in order to identify chains.

Multiple Chains Identification. This approach aims at identifying a (possibly empty) set of chains over the set of basic components of a complex object. It is based on two steps, the first of which aims at identifying the heads (first elements) of the possible chains, that is the set

$$Heads = I \cup \{b_1 \in V | \ \exists b_2 \in V \ (b_1, b_2) \in E \ \wedge \forall b_0 \in V \ (b_0, b_1) \notin E\}$$

This set contains both nodes for which $first$ is true and nodes which occur as a first argument in a true $succ$ atom and do not occur as a second argument in any true $succ$ atom.

Once the set $Heads$ has been identified, it is necessary to reconstruct the distinct chains. Intuitively, each chain is the list of nodes forming a path in G which begins with a node in $Heads$ and ends with a node without outgoing edges. Formally, an extracted chain $C \subseteq E$ is defined as follows:

$$C = \{(b_1, b_2), (b_2, b_3), \ldots, (b_k, b_{k+1})\}$$

such that

1. $b_1 \in Heads$,
2. $b_{k+1} \in F$,
3. $\forall i = 1..k : (b_i, b_{i+1}) \in E$
4. $\forall i = 1..k \forall j = i + 1..k + 1 \ b_i \neq b_j$.

The last condition imposes that the same node cannot appear more than once in the same chain In order to avoid cyclic paths.

Single Chain Identification. The result of the second approach is a single chain. Following the proposal reported in [6], we aim at iteratively evaluating the most promising node to be appended to the resulting chain. More formally, let $PREF_G : V \times V \to \{0, 1\}$ be a preference function defined as follows:

$$PREF_G(b_1, b_2) = \begin{cases} 1 \text{ if } b_1 = b_2 \text{ or a path connecting } b_1 \text{ and } b_2 \text{ exists in } G \\ 0 \text{ otherwise} \end{cases}$$

Let $\mu : V \to \mathbb{N}$ be the function defined as follows:

$$\mu(L, G, I, b) = countConnections(L, G, I, b) + outGoing(V/L, b)$$
$$-inComing(V/L, b)$$

---

[1] G is not a direct acyclic graph (dag) since it could also contain cycles.

where

- $G = \langle V, E \rangle$ is the ordered graph
- L is a list of *distinct* nodes in G
- $b \in V/L$ is a candidate node
- $countConnections(L, G, I, b) = |\{d \in L \cup I | PREF_G(d, b) = 1\}|$ counts the number of nodes in $L \cup I$ from which b is reachable.
- $outGoing(V/L, b) = |\{d \in V/L | PREF_G(b, d) = 1\}|$ counts the number of nodes in V/L reachable from b.
- $inComing(V/L, b) = |\{d \in V/L | PREF_G(d, b) = 1\}|$ counts the number of nodes in V/L from which b is reachable.

Algorithm 1 fully specifies the method for the single chain identification. The rationale is that at each step a node is added to the final chain. Such a node is that for which μ is the highest. Higher values of μ are given to nodes which can be reached from I, as well as from other nodes already added to the chain, and have a high (low) number of outgoing (incoming) paths to (from) nodes in V/L. Indeed, the algorithm returns an ordered list of nodes which could be straightforwardly transformed into a chain.

Algorithm 1. Single chain identification algorithm

1: **findChain** $(G =< V, E >, I)$ *Output:* **L: chain of nodes**
2: L← ∅;
3: **repeat**
4:     $best\_mu \leftarrow -\infty$;
5:     **for all** $b \in V/L$ **do**
6:         $cc \leftarrow countConnections(L, G, I, b)$;
7:         $inC \leftarrow incoming(V/L, b)$; $outG \leftarrow outGoing(V/L, b)$;
8:         **if** $((cc \neq 0)$ OR $(inC \neq 0)$ OR $(outG \neq 0))$ **then**
9:             $\mu \leftarrow cc + outG - inC$;
10:            **if** $best\_mu < \mu$ **then**
11:                $best\_b \leftarrow b$; $best\_mu \leftarrow \mu$;
12:            **end if**
13:        **end if**
14:    **end for**
15:    **if** $(best\_mu <> -\infty)$ **then**
16:        $L.add(best\_b)$;
17:    **end if**
18: **until** $best\_mu = -\infty$
19: return L

---

5 The Application: Learning Reading Order of Layout Components

In this paper, we investigate an application to the document image understanding problem. More specifically, we are interested in determining the reading order of layout components in each page of a multi-page document. Indeed, the spatial order in which the information appears in a paper document may have more to do with optimizing the print process than with reflecting the logical order of the information contained. Determining the correct reading order can be a crucial problem for several applications. By following the reading order recognized in a document image, it is possible to cluster together text regions labeled with the same logical label into the same textual component (e.g., "introduction", "results", "method" of a scientific paper). In this way, the reconstruction of a single textual component is supported and advanced techniques for text processing can be subsequently applied. For instance, information extraction methods may be applied locally to reconstructed textual components. Moreover, retrieval of document images on the basis of their textual contents is more effectively supported.

Several works on reading order detection have already been reported in the literature [26][12][21][25][1] [4]. A common aspect of all methods is that they strongly depend on the specific domain and are not "reusable" when the classes of documents or the task at hand change. There is no work, to the best of our knowledge, that handles the reading order problem by resorting to machine learning techniques, which can generate the required knowledge from a set of training layout structures whose correct reading order has been provided by the user. In previous works on document image understanding, we investigated the application of machine learning techniques to several knowledge-based document image processing tasks, such as classification of blocks [3], automatic global layout analysis correction [18], classification of documents into a set of pre-defined classes and logical labelling. Following this mainstream of research, herein we consider the problem of learning the reading order.

In this context the limitations posed by the single table assumption are quite restrictive for at least two reasons. First, layout components cannot be realistically considered independent observations, because their spatial arrangement is mutually constrained by formatting rules typically used in document editing. Second, spatial relationships between a layout component and a variable number of other components in its neighborhood cannot be properly represented by a fixed number of attributes in a table. Even more so, the representation of properties of the other components in the neighborhood, because different layout components may have different properties (e.g., the property "brightness" is appropriate for half-tone images, but not for textual components). Since the single-table assumption limits the representation of relationships (spatial or non) between examples, it also prevents the discovery of this kind of pattern, which can be very useful in document image mining.

For these reasons, the ILP approach proposed in this paper seems to be appropriate for the task at hand. In ATRE, training observations are represented by ground multiple-head clauses [15], called *objects*, which have a conjunction of simple literals in the head. The head of an object contains positive and negative examples for the concepts to be learned, while the body contains the description of layout components on the basis of geometrical features (e.g. width, height) and topological relations (e.g. vertical and horizontal alignments) existing among

Fig. 1. A document page: the input reading order chain. Sequential numbers indicate the reading order.

blocks, the type of the content (e.g. text, horizontal line, image) and the logic type of a block (e.g. title or authors of a scientific paper). Terms of literals in objects can only be constants, where different constants represent distinct layout components within a page. An example of object description generated for the document page in Figure 1 is the following:

$object('tpami17\_1\text{-}13', [class(p) = tpami,$
$first(0) = true, first(1) = false, ...$
$succ(0, 1) = true, succ(1, 2) = true, ..., succ(7, 8) = true, succ(2, 10) = false, ...],$
$[part\_of(p, 0) = true, ...,$
$height(0) = 83, height(1) = 11, ..., width(0) = 514, width(1) = 207, ...,$
$type\_of(0) = text, ..., type\_of(11) = hor\_line,$
$title(0) = true, author(1) = true, affiliation(2) = true, .., undefined(16) = true,$
$x\_pos\_centre(0) = 300, x\_pos\_centre(1) = 299, ...,$
$y\_pos\_centre(0) = 132, y\_pos\_centre(1) = 192, ...,$
$on\_top(9, 0) = true, on\_top(15, 0) = true, ..., to\_right(6, 8) = true, ...$
$alignment(16, 8) = only\_right\_col, alignment(17, 5) = only\_left\_col, ...$
$class(p) = tpami, page(p) = first]).$

The constant p denotes the whole page while the remaining integer constants (0, 1, ..., 17) identify distinct layout components. In this example, the block number 0 corresponds to the first block to read ($first(0) = true$), it is a textual component ($type\_of(0) = text$) and it is logically labelled as 'title' ($title(0) = true$). Block number 1 (immediately) follows block 0 in the reading order ($succ(0,1) = true$), it is a textual component and it includes information on the authors of the paper ($author(1) = true$).

As explained in the previous sections, ATRE learns a logical theory T defining the concepts $first/1$ and $succ/2$ such that T is complete and consistent with respect to the examples. This means that it is necessary to represent both positive and negative examples and the representation of negative examples for the concept $succ/2$ poses some feasibility problems due to their quadratic growth. In order to reduce the number of negative examples, we resort to sampling techniques. In our case, we sampled negative examples by limiting their number to 1000% of the number of positive examples. This way, it is possible to simplify the learning stage and to have a logical theory which is less specialized and avoids overfitting.

In order to evaluate the applicability of the proposed approach to reading order identification, we considered a set of multi-page articles published in an international journal. In particular, we considered twenty-four papers, published as either regular or short articles, in the IEEE Transactions on Pattern Analysis and Machine Intelligence (TPAMI), in the January and February issues of 1996. Each paper is a multi-page document; therefore, we processed 211 document images. Each document page corresponds to a 24bit TIFF color image.

Initially, document images are pre-processed in order to segment them, perform layout analysis, identify the membership class and map the layout structure of each page into the logical structure. Training examples are then generated by manually specifying the reading order. In all, 211 positive examples and 3,263 negative examples for the concept $first/1$ and 1,418 positive examples and 15,518 negative examples for the concept $succ/2$ are generated. The average number of layout components in training chains is about 8.0.

We evaluated the performance of the proposed approach by means of a six-fold cross-validation: the dataset is first divided into six folds of equal size and then, for every fold, the learner is trained on the remaining folds and tested on it.

For each learning problem, statistics on precision and recall of the learned logical theory are recorded. In order to evaluate the ordering returned by the proposed approach, we resort to metrics used in information retrieval to evaluate the returned ranking of results [9]. Herein we consider the metrics valid for partial orders evaluation.

In particular, we consider the *normalized Spearman footrule distance* which, given two complete lists L and L_1 on a set S (that is, L and L_1 are two different permutations without repetition of all the elements in S), is defined as follows:

$$F(L, L_1) = \frac{\sum_{b \in S} abs(pos(L, b) - pos(L_1, b))}{|S|^2 / 2} \tag{1}$$

where the function $pos(L, b)$ returns the position of the element b in the ordered list L. $F(L, L_1)$ is always between 0 and 1, where 0 means that L and L_1 are exactly the same, while 1 means that the two lists are completely in the opposite order. In other terms, the lower the $F(L, L_1)$, the better.

The *induced footrule distance* $F(L|_{L_1}, L_1)$ is the variant of $F(L, L_1)$ computed by projecting the list L on L_1: it is used when L_1 is a sublist of L.

The normalized Spearman footrule distance can be straightforwardly generalized to the case of several lists:

$$F(L, L_1, \ldots, L_k) = 1/k \sum_{i=1\ldots k} F(L, L_i).$$

In order to consider partial (and not total) orders, we resorted to a variant called (*induced normalized footrule distance*):

$$F(L, L_1, \ldots, L_k) = 1/k \sum_{i=1\ldots k} F(L|_{L_i}, L_i)$$

Since this measure does not take into account the length of single lists, we also adopted the *normalized scaled footrule distance*:

$$F'(L, L_1) = \frac{\sum_{b \in S} abs(pos(L, b)/|L| - pos(L_1, b)/|L_1|)}{|L_1|/2} \tag{2}$$

Also in this case it is possible to extend the measure to the case of multiple lists:
$F'(L, L_1, \ldots, L_k) = 1/k \sum_{i=1\ldots k} F'(L|_{L_i}, L_i)$.

In this study, we apply such distance measures to chains. In particular, both FD= $F(L|_{L_1}, L_1)$ and SFD=$F'(L|_{L_1}, L_1)$ are used in the evaluation of single chain identification, while IFD=$F(L, L_1, \ldots, L_k)$ and ISFD=$F'(L, L_1, \ldots, L_k)$ are used in the evaluation of multiple chains identification.

Results reported in Table 1 show that the system has a precision of about 65% and a recall greater than 75%. Moreover, there is no significant difference in terms of recall between the two concepts, while precision is higher for the *succ* concept. This is mainly due to the specificity of the clauses learned for the concept *first*: clauses learned for the concept *first* cover (on average) fewer positive examples than clauses learned for the concept *succ* (see Table 2). We can conclude that the concept *first* appears to be more complex to learn than the concept *succ*, probably because of the lower number of training examples (one per page).

Experimental results concerning the reconstruction of single/multiple chains are reported in Table 3. We recall that the lower the distance value the better the reconstruction of the original chain(s). By comparing results in terms of

Table 1. Precision and Recall results shown per concept to be learned

Concept	first/1		succ/2		Overall	
	Precision %	Recall%	Precision%	Recall%	Precision%	Recall%
FOLD1	75.00	50.00	76.90	64.10	76.60	61.80
FOLD2	66.70	63.20	74.10	65.20	73.00	64.90
FOLD3	74.30	78.80	81.00	66.10	80.10	67.40
FOLD4	69.40	71.40	67.80	56.30	68.00	58.20
FOLD5	66.70	66.70	78.40	68.70	76.80	68.40
FOLD6	71.00	61.10	79.40	62.90	78.20	62.60
AVG	70.52	65.20	76.27	63.88	75.45	63.88

Table 2. Number of learned clauses per positive examples

Concept	$first\_to\_read/1$		$succ\_in\_reading/2$	
	No of clauses	Training POS exs	No of clauses	Training POS exs
FOLD1	42	175	162	1226
FOLD2	46	173	145	1194
FOLD3	42	178	149	1141
FOLD4	42	176	114	1171
FOLD5	40	178	166	1185
FOLD6	41	175	177	1173
AVG coverage	4.17		7.77	

Table 3. Reading order reconstruction results

Concept	Multiple chains		Single chain	
	AVG. IFD%	AVG. ISFD%	AVG. FD%	AVG. SFD%
FOLD1	13.18	21.12	47.33	10.17
FOLD2	10.98	18.51	46.32	8.13
FOLD3	1.31	26.91	47.32	17.63
FOLD4	1.32	24.00	49.96	14.51
FOLD5	0.90	22.50	49.31	10.60
FOLD6	0.90	27.65	54.38	12.97
AVG	4.76	23.45	49.10	12.33

the *footrule distance* measure (IFD vs FD), we note that the reconstruction of multiple chains shows better results than the reconstruction of single chains. Indeed, this result does not take into account the length of the lists. When considering the length of the lists (ISFD vs. SFD) the situation is completely different and the reconstruction of single chains outperforms the reconstruction of multiple chains.

Some examples of clauses learned by ATRE are reported below:

1. $first(X1) = true \leftarrow x\_pos\_centre(X1) \in [55..177]$,
   $y\_pos\_centre(X1) \in [60..121], height(X1) \in [98..138]$.
2. $first(X1) = true \leftarrow title(X1) = true, \ x\_pos\_centre(X1) \in [293..341]$,
   $succ(X1, X2) = true$.
3. $succ(X2, X1) = true \leftarrow affiliation(X1) = true, \ author(X2) = true$,
   $height(X1) \in [45..124]$.
4. $succ(X2, X1) = true \leftarrow alignment(X1, X3) = both\_columns$,
   $on\_top(X2, X3) = true, succ(X1, X3) = true, \ height(X1) \in [10..15]$

They show that ATRE is particularly indicated for the task at hand since it is able to identify dependencies among concepts to be learned or even recursion.

6   Conclusions

In this paper, we present an ILP approach to the problem of inducing a partial ordering between basic components of complex objects. The proposed solution is based on learning a logical theory which defines two predicates $first/1$ and $succ/2$. The learned theory should be able to express dependencies between the two target predicates. For this reason we used the learning system ATRE which is able to learn mutually recursive predicate definitions. In the recognition phase, learned predicate definitions are used to reconstruct reading order chains according two different modalities: single vs. multiple chains identification.

The proposed approach can be applied to several application domains. In this paper, it has been applied to a real-world problem, namely detecting the reading order between layout components extracted from images of multi-page documents. Results prove that learned logical theories are quite accurate and that the reconstruction phase significantly depends on the application at hand. In particular, if the user is interested in reconstructing the actual chain (e.g. text reconstruction for rendering purposes), the best solution is in the identification of single chains. On the contrary, when the user is interested in recomposing text such that sequential components are correctly linked (e.g. in information extraction), the most promising solution is the identification of multiple chains.

For future work we intend to compare the logic-based approach proposed in this paper with a probabilistic-based approach reported in [5]. Moreover, we plan to extend our empirical investigation to other application domains as well as to synthetically generated datasets.

Acknowledgments

This work is in partial fulfillment of the research objectives of the project "D.A.M.A." (Document Acquisition, Management and Archiving). The authors also wish to thank Lynn Rudd for her help in reading the manuscript.

References

1. Aiello, M., Monz, C., Todoran, L., Worring, M.: Document understanding for a broad class of documents. International Journal on Document Analysis and Recognition-IJDAR 5(1), 1–16 (2002)
2. Aiello, M., Smeulders, A.: Bidimensional relations for reading order detection. In: Proceedings of Joint Conference on Information Science (2003)
3. Altamura, O., Esposito, F., Malerba, D.: Transforming paper documents into XML format with WISDOM++. International Journal on Document Analysis and Recognition-IJDAR 4(1), 2–17 (2001)
4. Breuel, T.M.: High performance document layout analysis. In: Proceedings of the 2003 Symposium on Document Image Understanding (SDIUT 2003) (2003)
5. Ceci, M., Berardi, M., Porcelli, G., Malerba, D.: A data mining approach to reading order detection. In: ICDAR 2007: 9th International Conference on Document Analysis and Recognition, pp. 924–928 (2007)
6. Cohen, W.W., Schapire, R.E., Singer, Y.: Learning to order things. Journal of Artificial Intelligence Research (JAIR) 10, 243–270 (1999)
7. De Raedt, L.: Interactive Theory Revision. Academic Press, London (1992)
8. Džeroski, S., Lavrač, N.: Relational Data Mining. Springer, Berlin (2001)
9. Dwork, C., Kumar, R., Naor, M., Sivakumar, D.: Rank aggregation methods for the web. In: WWW 2001: Proceedings of the 10th international conference on World Wide Web, pp. 613–622. ACM Press, New York (2001)

10. Gionis, A., Kujala, T., Mannila, H.: Fragments of order. In: KDD 2003: Proceedings of the ninth ACM SIGKDD international conference on Knowledge discovery and data mining, pp. 129–136. ACM Press, New York (2003)
11. Grimaldi, R.P.: Discrete and Combinatorial Mathematics, an Applied Introduction, 3rd edn. Addison Wesley, Reading (1994)
12. Ishitani, Y.: Document transformation system from papers to XML data based on Pivot XML document method. In: ICDAR 2003: 7th International Conference on Document Analysis and Recognition, p. 250. IEEE Computer Society, Los Alamitos (2003)
13. Kamishima, T., Akaho, S.: Learning from order examples. In: Proceedings of the 2nd IEEE International Conference on Data Mining, pp. 645–648 (2002)
14. Lavrač, N., Džeroski, S.: Inductive Logic Programming: techniques and applications. Ellis Horwood, Chichester (1994)
15. Levi, G., Sirovich, F.: Generalized and/or graphs. Artificial Intelligence 7(3), 243–259 (1976)
16. Lloyd, J.W.: Foundations of Logic Programming, 2nd edn. Springer, Berlin (1987)
17. Malerba, D.: Learning recursive theories in the normal ILP setting. Fundamenta Informaticae 57(1), 39–77 (2003)
18. Malerba, D., Esposito, F., Altamura, O., Ceci, M., Berardi, M.: Correcting the document layout: A machine learning approach. In: ICDAR 2003: 7th International Conference on Document Analysis and Recognition, p. 97 (2003)
19. Mannila, H., Meek, C.: Global partial orders from sequential data. In: KDD 2000: Proceedings of the sixth ACM SIGKDD international conference on Knowledge discovery and data mining, pp. 161–168. ACM Press, New York (2000)
20. Maruster, L., Weijters, A., van der Aalst, W., van den Bosch, A.: Process mining: Discovering direct successors in process logs. In: Lange, S., Satoh, K., Smith, C.H. (eds.) DS 2002. LNCS, vol. 2534, pp. 364–373. Springer, Heidelberg (2002)
21. Meunier, J.-L.: Optimized xy-cut for determining a page reading order. In: ICDAR 2005: 8th International Conference on Document Analysis and Recognition, pp. 347–351. IEEE Computer Society, Washington (2005)
22. Mitchell, T.: Machine Learning. McGraw-Hill, New York (1997)
23. Muggleton, S.: Inductive Logic Programming. Academic Press, London (1992)
24. Nienhuys-Cheng, S.-W., de Wolf, R.: Foundations of inductive logic programming. Springer, Heidelberg (1997)
25. Taylor, S.L., Dahl, D.A., Lipshutz, M., Weir, C., Norton, L.M., Nilson, R., Linebarger, M.: Integrated text and image understanding for document understanding. In: HLT 1994: Proceedings of the workshop on Human Language Technology, pp. 421–426 (1994)
26. Tsujimoto, S., Asada, H.: Understanding multi-articled documents. In: Proceedings of the 10th International Conference on Pattern Recognition, pp. 551–556 (1990)

Using Semantic Distance in a Content-Based Heterogeneous Information Retrieval System

Ahmad El Sayed, Hakim Hacid, and Djamel Zighed

University of Lyon 2
ERIC Laboratory - 5, avenue Pierre Mendès-France
69676 Bron cedex, France
{asayed,hhacid,dzighed}@eric.univ-lyon2.fr

Abstract. This paper brings two contributions in relation with the semantic heterogeneous (documents composed of texts and images) information retrieval: (1) A new context-based semantic distance measure for textual data, and (2) an IR system providing a conceptual and an automatic indexing of documents by considering their heterogeneous content using a domain specific ontology. The proposed semantic distance measure is used in order to automatically fuzzify our domain ontology. The two proposals are evaluated and very interesting results were obtained. Using our semantic distance measure, we obtained a correlation ratio of 0.89 with human judgments on a set of words pairs which led our measure to outperform all the other measures. Preliminary combination results obtained on a specialized corpus of web pages are also reported.

1 Introduction

An important lack in the current IRS, is that most of them deal with homogeneous data types. We can find those dealing with text content, others with visual content but rarely with both. Let's take the example of a web page. If we compose a web page by its different components, (where each component represents a data type), we'll find that all parts do not necessarily represent the same piece of information. Even on a single document, each component will have a distinct meaning for the user. So if we treat only text for example, we wash out all the information contained in images, and inversely. Let's mention that indexing an image by the text surrounding it, as most search engines do, is not the real solution since text does not necessarily represent the image content.

Any IRS contains mainly two important components: a data representation structure which is generally translated by the use of an index for capturing the semantic of the data and accelerating the access to the low level data, and a querying strategy which enables the end user to express his/her query to the system. To be efficient, these two components necessitate another important element which is a semantic similarity measure. A semantic similarity measure is important to enable capturing the semantic proximity between pieces of information (concepts, words, images, etc...).

Z.W. Raś, S. Tsumoto, and D. Zighed (Eds.): MCD 2007, LNAI 4944, pp. 224–237, 2008.
© Springer-Verlag Berlin Heidelberg 2008

In this paper, we present a novel retrieval system that represent documents by their text and image content and thus, by multiple information sources. So, we bring two contributions in relation with the semantic heterogeneous (documents composed of texts and images) information retrieval: (1) A new context-based semantic distance[1] measure for textual data (since we are dealing with keyword-based information retrieval), and (2) a IR system providing a conceptual and automatic indexing of documents by considering their heterogeneous content using a domain specific ontology. In order to maximize our system's performances, we automatically fuzzify our knowledge unit using our proposed semantic distance measure.

The rest of this paper is organized as follows: In Section 2, we present and evaluate the new semantic distance measure between words. Our heterogeneous information retrieval system is introduced and detailed in Section 3. We conclude and give some future work in Section 4.

2   Semantic Similarity

In text-based applications, beyond managing synonymies and polysemies, one need to measure the degree of semantic similarity between two words/concepts[2]; let's mention: Information retrieval, question answering, automatic text summarization and translation, etc. Many semantic similarity measures[3] have been proposed in the literature. We can distinguish between knowledge-based measures and corpus-based measures.

On the one hand, knowledge-based measures offer reliable results given their hand-crafted 'semantic' logical structure. Taxonomies, like WordNet [15] and Mesh[4], are widely used for that purpose. These measures can be divided into an edge-based measures [20][11][29], a features-based measures [26], or an Information Content (IC) measures [21][10][13].

On the other hand, corpus-based measures are based on a statistical analysis of large text corpora. They have the advantage of being self-independent; they don't need any external knowledge resources, which can overcome the coverage problem in taxonomies. In this category, we can find co-occurrence based measures [5][25] and context-based measures [4][8].

2.1   A MultiSource Context-Dependent Semantic Distance Between Concepts

Our Context-Dependent Measure. A major lack in existing semantic similarity methods is that no one takes into account the context or the considered

---

[1] We consider distance by its dissimilarity which is the inverse of similarity. Then, greater distance values imply greater difference between compared objects.

[2] In the rest of the paper, 'words' is used when dealing with text corpora and 'concepts' is used when dealing with taxonomies where each concept is represented by a list of words holding a sense.

[3] For a more detailed state of art, readers are invited to read our previous paper [22].

[4] http://www.nlm.nih.gov/mesh/

domain. However, two concepts similar in one context may appear completely un-related in another context. Let's take the example of *heart* and *blood*. In a general context, these two concepts can be judged to be very similar. However, if we put ourselves in a medical context, *heart* and *blood* define two largely separated con-cepts. They will be even more distant if the context is more specifically related to *cardiology*. Our context-dependent approach[5] suggest to adapt semantic similar-ities to the target corpus since it's the entity representing the context or the do-main of interest in most text-based applications. This method is inspired by the Information Content (IC) theory of Resnik [21] and by the Jiang [10] measure.

In spite of using a text corpus, the IC proposed measures are unable to capture the target context since they rely uniquely on the probability of encountering a concept in a corpus which is not a sufficiently adaptive measure to reflect its de-pendency to a given context. A concept very frequent in some few documents and absent in many others cannot be considered to be "well" representative for the corpus. Thus, the number of documents where the concept occurs is another im-portant factor that must be considered. In addition to that, it's most likely that a concept c_1 -with a heterogeneous distribution among documents - is more discrimi-native than a concept c_2 with a monotone repartition which can reveal less power of discrimination over the target domain (Experimentations made assess our thesis).

Instead of assigning IC values to concepts, we assign weights for taxonomy's concepts according to a Context-Dependency CD measure for a given corpus C. The goal is to obtain a weighted taxonomy, where 'heavier' subtrees are more context representative than 'lighter' subtrees. This will allow us to calculate se-mantic similarities by considering the actual context. Therefore, lower similarity values will be obtained in 'heavy' subtrees than 'light' subtrees. Thus, in our *heart/blood* example, we tend to give a high similarity for the concept couple in a general context, and a low similarity in a specific context like medicine.

Consequently, we introduce our CD measure which is an adapted version of the standard $tf - idf$. Given a concept c, $CD(c)$ is a function of its total frequency $freq(c)$, the number of documents containing it $d(c)$, and the variance of its frequency distribution $var(c)$ over a corpus C:

$$CD(c) = \frac{log(1 + freq(c))}{log(N)} * \frac{log(1 + d(c))}{log(D)} * (1 + log(1 + var(c))) \qquad (1)$$

Where N denotes the total number of concepts in C and D is the total number of documents in C. The log likelihood seems adaptive to such purpose since it helps to reduce the big margins between values. This formula ensures that if a concept frequency is 0, its CD will equals 0 too. It ensures also that if c have an instance in C, its CD will never be 0 even if $var(c) = 0$.

Note that the CD of a concept c is the sum of its individual CD value with the CD of all its subconcepts in the taxonomy. The weights propagation from the bottom to the top of the hierarchy is a natural way to ensure that a parent even with a low individual CD will be considered as highly context-dependent if its children are well represented in the corpus

---

[5] The approach is presented in more detail in our previous paper [22].

To compare two concepts using the CD values, we assign a Link Strength (LS) to each 'is-a' link in the taxonomy. Assume that c_1 subsumes c_2, the LS between c_1 and c_2 is then calculated as follows: $LS(c_1, c_2) = CD(c_1) - CD(c_2)$. Then our Context-Dependent Semantic Sistance (CDSD) is calculated by summing the log likelihood of LS along the shortest path separating the two concepts in the taxonomy:

$$Dist(c_1, c_2) = \sum_{c \in SPath(c_1, c_2)} log(1 + LS(c, parent(c)))$$

Where $SPath$ denotes the shortest path between c_1 and c_2.

Combinations with other Measures. First, at the corpus level, the promising rates attained by the corpus-based word similarities techniques and especially for the co-occurrence based ones has pushed us to combine them with our context-dependent measure in order to reach the best possible rates. However, two similar words can appear in the same document, paragraph, sentence, or a fixed-size window. It's true that smaller window size can help identifying relations that hold over short ranges with good precisions, larger window size, yet too coarse-grained, allows to detect large-scale relations that could not been detected with smaller windows. We can say that if small windows improve precision, a large windows improve recall.

We have chosen to combine both techniques in order to view relations at different-scales. At the low scale, we use the PMI measure described above with a window size of 10 words. At the large scale, we calculate the Euclidian distance between words vectors where each word is represented by its tf.idf values over the documents.

Secondly, at the taxonomy level, a feature based measure is used. A part of their simple conceptual structure, taxonomies like Wordnet provide users with additional resources which describe most entities. Information in Wordnet is organized around logical groupings called synsets. Each synset is attached to a description set, a list of synonyms, antonyms, etc..In order to take advantage of the full information package in such rich resources, we have chosen to combine our CD measure also with the feature-based measure proposed by Tversky [26] which assumes that the more common characteristics (i.e. synonyms, antonyms, meronyms, etc..) two objects have and the less non common characteristics they have, the more similar the objects are.

2.2 Evaluation and Results

To evaluate our approach, a benchmark composed of a corpus of 30,000 web pages along with the WordNet taxonomy is used. The web pages are crawled from a set of News web sites (reuters.com, cnn.com, nytimes.com...).

The most intuitive way to evaluate a semantic similarity/distance is to compare machine ratings and human ratings on a same data set. A very common set of 30 word pairs is given by Miller and Charles [16]. M&C asked 38 undergraduate students to rate each pair on a scale from 0 (no similarity) to 4 (perfect synonymy). The average rating of each pair represents a good estimate on how similar the two words are. The correlation between individual ratings of human replication was 0.90 which led many researchers to take 0.90 as the upper bound ratio. For our evaluations, we've chosen the M&C subset of 28 words pairs which is the most commonly used subset for that purpose. Note that since our measure calculates distance, the M&C distance will be: $dist = 4 - sim$ where 4 represent the maximum degree of similarity.

When comparing our distance results with the M&C human ratings, the context-dependency CD method alone gave a correlation of 0.83 which seems to be a very promising rate. Then, we have combined our measure with others by trying multiple combination strategies(See the previous section). By doing this, we could increase our correlation ratio to 0.89 which is not too far from human correlations of 0.905. This obtained rate led our approach to outperform the existing approaches for semantic similarity (see Table 1).

Table 1. Comparison between the principal measures and our two-level measure

Similarity method	Type	Correlation with M&C
Human replication	Human	0,901
Rada	Edge-based	0,59
Hirst and St-Onge	Edge-based	0,744
Leacock and Chodorow	Edge-based	0,816
Resnik	Information Content	0,774
Jiang	Information Content	0,848
Lin	Information Content	0,821
CDSD	Context-Dependent	0,830
our multisource measure	Hybrid	0,890

Our method shows an interesting result whether on an individual or on a combination scale. A part of its interesting correlation coefficient of 0.83, our CD method has the advantage to be context-dependent, which means that our results are flexible and can vary from one context to another. We argue that our measure could perform better if we "place" human subjects in our corpus context. In other terms, our actual semantic distance values reflect a specific context that does not necessarily match with the context of the human subjects during the R&C experiments.

We presented in this section our new multisource context dependent semantic distance measure between concepts. In the next section, we detail the architecture as well as he different components of our content-based heterogeneous data retrieval system. We will also demonstrate the use of the proposed similarity measure in the global architecture.

3   Our Content-Based Retrieval System

We propose an approach that enables semantic retrieval on documents containing heterogeneous information sources. The approach we are proposing is based on the translation of all the data forms into a textual form(following an annotation process for images and a normalization process for text documents for example). That's where measuring a distance between two words/concepts takes all its importance in the whole retrieval process.

In the following, we describe the general system architecture of our system and then we detail its three different layers: the structuring layer, the semantic interpretation layer, and the querying layer.

3.1   General Architecture

Figure 1 illustrates the general architecture of our approach. This architecture is composed of three layers: (1) low level layer, (2) high level (semantic) layer, and (3) querying layer.

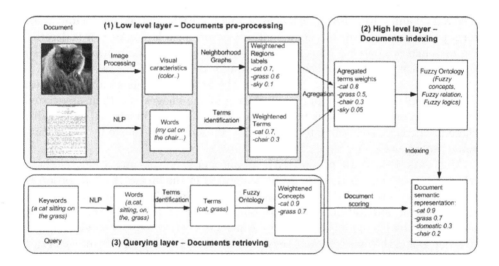

Fig. 1. General Architecture of our semantic based heterogeneous data retrieval framework

First[6], at the low level layer, we take as an input the raw representation of the document. We divide it into multiple parts, where each part corresponds to one data type, in order to process each part separately. The output of this layer is a list of terms representing each document's part content[7]. Second, at the high level layer, we take as an input the obtained lists of keywords which will be combined and then mapped to a fuzzy domain ontology that provides, as an output, one single semantic representation of the whole document. Note that we combine only lists concerning parts of the same document. Finally, at the

---

[6] This operation is made Off-line.

[7] The number of lists depends on the number of document's parts.

querying layer[8], keywords are used to express the user's needs. These keywords are mapped to the same ontology to locate the correspondent concepts. A set of relevant documents is returned from the database according to their scores.

In the following sections, we detail the different steps introduced above.

3.2   Low Level Layer – Document Pre-processing

Image Processing. One of the most important challenges in imagery is semantics association to an image. Indeed, image processing methods associate for each image a features vector (or vectors) calculated on the image. These features are known as "low level features" (color, texture, etc.). The interrogation of an image database is then done by introducing an image query into the system and its comparison to the available ones using similarity measures [27]. Thus, no semantics is associated to images with this process.

The common way for semantics assignment to an image is annotation. Multimedia data annotation is the task of assigning, for each multimedia document, or for a part of the multimedia document, a keyword or a list of keywords describing its semantic contents. This function can be considered as a mapping between the visual aspects of the multimedia data and their high level characteristics.

There does not exist a lot of work on the automatic annotation of images. There are methods which apply a clustering of the images and their associated keywords in order to make it possible to attach a text to images [1][2][3]. With these methods, it is possible to predict the labels of a new image by calculating some probabilities.

Minka and Picard [19] proposed a semi-automatic image annotation system in which the user chooses the area to be annotated in the image. A propagation of the annotations is carried out by considering textures. Maron et al., [14] studied the automatic annotation using only one keyword at the same time. Mori et al., [23] proposed a model based on co-occurrences between images and keywords in order to find the most relevant keywords for an image. The disadvantage of this model is that it requires a large training sample to be effective. Dyugulu et al., [6] proposed another model, called translation model, which is an improvement of the co-occurrence model suggested by Mori et al., [23] by integrating a training algorithm. Probabilistic models such as Cross Media Relevance model [9] and Latent Semantic Analysis [17] were also proposed. Jia and Wang [12] use the two-dimensional hidden markov chains to annotate images.

To handle images semantics in this work, we adopt the proposed method in [7]. This method is based on an interesting geometrical structure, a relative neighborhood graph [24]. This structure combines, at the same time, a distance measure and the topology of the multidimensional space to determine the neighbors of each point (image in this case) in the considered space. The annotation process is performed into two levels: a) the *indexing level* where images are structured as a neighborhood graph using only their low level features (color, texture, etc.), and b) the *Decision making level* where the neighbors of an unlabeled image are

---

[8] This operation is made On-line.

located in the graph and the potential annotations, based on score calculation, are affected to the unknown image. More details about this method are available in [7].

Text Processing. As images, the raw text of the document is treated separately as well. In order to extract terms from text, classic Natural Language Processing (NLP) techniques are applied. A tokenizer is used first to localize words, numbers, punctuations with their different positions in text. A sentence splitter is used next. Then, a morphological and a syntactic analysis are performed in order to identify respectively grammatical Part Of Speech (POS) for each word which will serve for the lemmatisation process. Lemmatisation involves the reduction of words to their basic lexeme. This normalization step is necessary in order to 1)treat the inflected forms of words and to 2) facilitate the matching with ontology concepts that are usually presented in their lemmatised forms.

After that, we apply the trigram approach on words along with a set of grammatical rules in order to identify the candidate terms to be an ontological concepts. Candidate and non candidate terms are then assigned a normalized tf.idf weight (term frequency/inverse document frequency). At the end, the text part of the document is represented simply by a set of terms and weights.

3.3   High Level Layer – Document Indexing

The main goal in this layer is to reach a semantic interpretation of the document. However, keywords or terms extracted at the low level layer are not enough for that purpose. These keywords, are still on an intermediate or object level, and need further treatments to be on a semantic machine understandable level. That's where our knowledge-unit is involved. Extracted terms from text along with deduced labels from images are all redirected to a domain ontology in order to provide a semantic annotation for document content. Let's mention that our system purposes are not for a generic domain. Our system deals with specialized corpora along with domain specific ontologies.

Before the concept mapping, an aggregation step is necessary in order to merge obtained lists from the low level layer into one single list. We use the following formula:

$$w_{ki} = \frac{\sum_{j=1}^{n} w_{kji}}{n} \qquad (2)$$

Where:

- w_{ki} is the weight of term w_k in document D_i;
- n is the total number of parts in document D_i;
- w_{kji} is the weight of term w_k in part P_j of document D_i.

As we've said earlier, each term is associated to a weight representing its importance in the document. Since we treat particular domains, concepts weights should be a function of their document importance and their domain relative importance. Obviously, in a domain ontology, not all concepts represent the

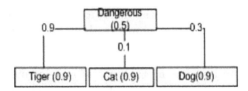

Fig. 2. Illustration Example of a fuzzy ontology

same importance for the target domain. One concept can be more discriminative or more domain-dependent from others, and thus, should be assigned different weights.

That's the reason why we've chosen to use fuzzy ontologies which are an extension of the crisp ontologies [28][18]. Since knowledge can be fuzzy, its representation should be fuzzified. Fuzzification can be integrated to an ontology by using fuzzy concepts, fuzzy relations, and fuzzy logics. It consists of assigning weights to concepts, relation and logical rules. (See figure 2).

Ontology fuzzification is done in an automated manner by making use of our semantic distance method described above. On the one hand, concepts in our domain ontology are assigned weights which correspond to the CD values (described above) that represent concept's dependency to a particular context represented by a text corpus. On the other hand, the 'is-a' relations in the ontology are assigned weights which represent the semantic similarity (as described above) between the two target linked concepts. The semantic similarity is calculated by inversing the semantic distance: $SIM = 1/Dist$.

The aggregated terms with their weights are then sent to the ontology in order to pass from the low level layer to the semantic layer. Certainly, not all terms will be found in the domain ontology. Thus, the result is a set of non-ontological and ontological terms (concepts). The non-ontological terms are then used to semi-automatically enrich the ontology, a process out of the scope of this paper[9]. Terms weight at the low level layer and the concepts weights in the ontology are both used to recalculate each concept weight in the document at the semantic layer using the following formula:

$$fw_{ki} = \frac{(2 \times ow_k) + w_{ki}}{3} \tag{3}$$

Where:

- fw_{ki} is the final weight of concept k in document D_i;
- ow_k is the weight of concept k in the ontology (its CD value);
- w_{ki} is the weight of term k in document D_i (calculated using the above formula);

Note that if a term has been found in the ontology i.e. Ontological term, its ontological weight (ow_k) corresponds to the weight of the concept in that

---

[9] This part will be detailed in our future publications.

ontology. Otherwise, if the term isn't in the ontology i.e. non-ontological term, its ontological weight is set to 0. So, its weight is divided by 3 in order to penalize this term since we consider that it's not a domain close term.

The interest of this formula is that the calculated concepts weights take into account the term importance in the document and its importance for the considered domain.

We've decided not to make any concept expansion at this indexing level. Experimentations have shown that expanding both documents and queries can result to a lot of sense deviations and imprecisions.

3.4   Querying Layer – Document Retrieving

As we mentioned before, we deal only with ontology-based keywords augmenting. The same process done for text indexing is applied for queries. Query keywords are mapped to the ontology in order to extract concepts. Query will then be divided into terms (non-ontological terms) and concepts (ontological terms). These concepts are then expanded to another linked concepts sharing a link weight greater than a fixed threshold ∂. Relations weights (which are semantic similarities) in the ontology are used to calculate the deduced concepts weights. Consider the Figure 2. Assume that the concepts $tiger, cat, dog$ were identified in a query q using a threshold $\partial = 0.2$. The concept $dangerous$ will be used to expand the query q to q' by using the relations weights between $R(dangerous - tiger)$ and $R(dangerous - dog)$ only since $R(dangerous - cat) < \partial$.

Finally, the following formula is used to calculate the weight of a document in the database according to the query:

$$w_{iq} = \sum_{l=1}^{t} (w_{lq} \times fw_{li}) \qquad (4)$$

Where:

- w_{iq} is the weight of the document D_i according to the query q;
- t is the total number of terms within the query q;
- w_{lq} is the weight of term l in a query q;
- fw_{li} is the final weight of concept l in the document D_i.

The weight of term l in a query q is determined according to three situations:

- if the query term l is an ontological term, $w_{lq} = 1$;
- if the query term l is inferred, w_{lq} will be the weight of the relation between the origin query concept and the deducted one;
- if the query term l is a non ontological term, its weight will be the maximum of the weights of the query terms obtained by expansion.

Our objective by setting up these weights query is to create a hierarchy of importance between terms. Thus, the query ontological terms are at the top and the expanded ontological ones are at the bottom.

3.5   System Evaluation

In this section we present some preliminary results of our approach. To perform the experiments, we built a small corpus of 50 web pages. Each web page contains texts and images. All the pages are related to the domain of animals. We have used an animal domain ontology that we fuzzified.

Each web page is then decomposed into two parts, the first part containing images and the second part containing text. Each document is automatically analyzed and annotated by two lists of keywords: a list of keywords describing the image content and another one describing the text content. These lists are merged using the proposed framework described beforehand.

Semantic based systems evaluation is a very hard task. Since in this case the classical evaluation measures (recall and precision for example) are neither efficient nor significant, we make up a user driven evaluation protocol. We considered ten keyword based queries. The user send his query to the system and obtains a list of documents. At each iteration, he selects the pertinent documents to his query. For each selected document, we take into account the part of interest (image, text, image and text) or the manner of obtaining the result (query expansion or not).

Generally speaking, 79% of the returned documents contain interesting information for the user, which seems to be an interesting rate. The graphic of figure 3 illustrates the average of contribution rate of each data type to the global result.

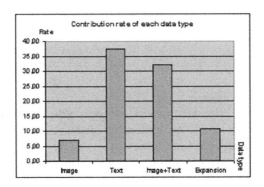

Fig. 3. Contribution rate of each data type to the final results

By considering the graphic, we can note that the different parts of the system: data types (images and texts) and query expansion contribute to the whole result. Text constitutes the most important contribution. We can also note that the combination of image and text gives also interesting results which constitutes a major point.

4   Conclusion and Future Work

Nowadays, retrieving information becomes more and more difficult. This is due especially to the huge volume and the heterogeneity of the modern databases. To interact with these kinds of data, one needs tools which can semantically process them.

In this paper, we have shown the importance of considering the context when calculating semantic distance between words/concepts. We've proposed a context-dependent method that takes the taxonomy as a principal knowledge resource, and a text corpus as a distance adaptation resource for the target context. We've proposed also to combine it with other taxonomy-based and corpus-based methods. The results obtained from the experiments show the effectiveness of our approach which led it to outperform the other approaches. We have also proposed a new framework to handle the heterogeneous data retrieval problem. Each document is then decomposed into different components (text, image, etc.) analyzed separately using appropriate techniques. An indexing level insures the assignment of a significant labels describing the semantic content of each document. The approach supports keywords based querying. Document indexing and query understanding is guided by a domain ontology fuzzified by mean of our semantic distance measure. The obtained results show the effectiveness and the interest of the proposed approach.

As for future work, we aim at evaluating the distance measure and comparing it with others by performing a context-driven human ratings, where human subjects will be asked to rank a same set of words pairs in different contexts. The machine correlation computed next according to each context will be able to show more significantly the added-value of our approach. We plan also to test the retrieval approach on more large databases and compare it to other approaches (text retrieval and image retrieval approaches), and to extend this approach by affecting weights for each part of the document reflecting the relative importance for each data type according to the treated domain.

References

1. Barnard, K., Duygulu, P., Forsyth, D.A.: Clustering art. CVPR (2), 434–441 (2001)
2. Barnard, K., Forsyth, D.A.: Learning the semantics of words and pictures. In: ICCV, pp. 408–415 (2001)
3. Celebi, E., Alpkocak, A.: Semantic image retrieval and auto annotation by covering keyword space to image space. In: MMM, Beijing, China, pp. 153–160 (2006)
4. Manning, C., Schutze, H.: Foundations of statistical natural language processing (1999)
5. Church, K.W., Hanks, P.: Word association norms, mutual information, and lexicography. In: Proceedings of the 27th. Annual Meeting of the Association for Computational Linguistics, pp. 76–83, Vancouver, B.C. (1989)
6. Duygulu, P., Barnard, K., de Freitas, J.F.G., Forsyth, D.A.: Object recognition as machine translation: Learning a lexicon for a fixed image vocabulary. In: Heyden, A., Sparr, G., Nielsen, M., Johansen, P. (eds.) ECCV 2002. LNCS, vol. 2353, pp. 97–112. Springer, Heidelberg (2002)

7. Hacid, H.: Neighborhood graphs for semi-automatic annotation of large image databases. In: Cham, T.-J., Cai, J., Dorai, C., Rajan, D., Chua, T.-S., Chia, L.-T. (eds.) MMM 2007. LNCS, vol. 4351, pp. 586–595. Springer, Heidelberg (2007)
8. Hindle, D.: Noun classification from predicate-argument structures. In: Meeting of the Association for Computational Linguistics, pp. 268–275 (1990)
9. Jeon, J., Lavrenko, V., Manmatha, R.: Automatic image annotation and retrieval using cross-media relevance models. In: SIGIR, pp. 119–126 (2003)
10. Jiang, J.J., Conrath, D.W.: Semantic similarity based on corpus statistics and lexical taxonomy (1997)
11. Leacock, C., Chodorow, M., Miller, G.A.: Using corpus statistics and wordnet relations for sense identification. Computational Linguistics 24(1), 147–165 (1998)
12. Li, J., Wang, J.Z.: Automatic linguistic indexing of pictures by a statistical modeling approach. IEEE Trans. Pattern Anal. Mach. Intell. 25(9), 1075–1088 (2003)
13. Lin, D.: An information-theoretic definition of similarity. In: Proc. 15th International Conf. on Machine Learning, pp. 296–304. Morgan Kaufmann, San Francisco (1998)
14. Maron, O., Ratan, A.L.: Multiple-instance learning for natural scene classification. In: ICML, pp. 341–349 (1998)
15. Miller, G.A.: Wordnet: A lexical database for english. Commun. ACM 38(11), 39–41 (1995)
16. Miller, G.A., Charles, W.: Contextual correlated of semantic similarity. Language and Cognitive Processes 6, 1–28 (1991)
17. Monay, F., Gatica-Perez, D.: On image auto-annotation with latent space models. In: ACM Multimedia, pp. 275–278 (2003)
18. Parry, D.: A fuzzy ontology for medical document retrieval. In: ACSW Frontiers 2004: Proceedings of the second workshop on Australasian information security, Data Mining and Web Intelligence, and Software Internationalisation, Dunedin, New Zealand, pp. 121–126. Australian Computer Society, Inc., Darlinghurst, Australia (2004)
19. Picard, R.W., Minka, T.P.: Vision texture for annotation. Multimedia Syst. 3(1), 3–14 (1995)
20. Rada, R., Mili, H., Bicknell, E., Blettner, M.: Development and application of a metric on semantic nets. IEEE Transactions on Systems, Man, and Cybernetics 19(1), 17–30 (1989)
21. Resnik, P.: Semantic similarity in a taxonomy: An information-based measure and its application to problems of ambiguity in natural language. J. Artif. Intell. Res (JAIR) 11, 95–130 (1999)
22. Sayed, A.E., Hacid, H., Zighed, D.: A multisource context-dependent approach for semantic distance between concepts. In: Wagner, R., Revell, N., Pernul, G. (eds.) DEXA 2007. LNCS, vol. 4653, Springer, Heidelberg (2007)
23. Takahashi, Y.M.H., Oka, R.: Image-to-word transformation based on dividing and vector quantizing images with words. In: Proceedings of the International Workshop on Multimedia Intelligent Storage and Retrieval Management, pp. 341–349 (1999)
24. Toussaint, G.T.: The relative neighborhood graphs in a finite planar set. Pattern recognition 12, 261–268 (1980)
25. Turney, P.D.: Mining the Web for synonyms: PMI–IR versus LSA on TOEFL. In: Flach, P.A., De Raedt, L. (eds.) ECML 2001. LNCS (LNAI), vol. 2167, pp. 491–502. Springer, Heidelberg (2001)
26. Tversky, A.: Features of similarity. Psychological Review 84, 327–352 (1977)

27. Veltkamp, R.C., Tanase, M.: Content-based image retrieval systems : A survey. Technical Report UU-CS-2000-34, Department of Computing Science, Utrecht University (2000)
28. Widyantoro, D.H.: A fuzzy ontology-based abstract search engine and its user studies. FUZZ-IEEE, 1291–1294 (2001)
29. Wu, Z., Palmer, M.: Verb semantics and lexical selection. In: 32nd. Annual Meeting of the Association for Computational Linguistics, pp. 133–138, New Mexico State University, Las Cruces, New Mexico (1994)

Using Secondary Knowledge to Support Decision Tree Classification of Retrospective Clinical Data*

Dympna O'Sullivan[1], William Elazmeh[3], Szymon Wilk[1], Ken Farion[4],
Stan Matwin[1,2], Wojtek Michalowski[1], and Morvarid Sehatkar[1]

[1] University of Ottawa, Ottawa, Canada
{dympna,wilk,wojtek}@telfer.uottawa.ca, mseha092@site.uottawa.ca
[2] Institute of Computer Science, Polish Academy of Sciences, Warsaw, Poland
stan@site.uottawa.ca
[3] University of Bristol, Bristol, United Kingdom
elazmah@cs.bris.ac.uk
[4] Faculty of Medicine, University of Ottawa, Ottawa, Canada
farion@cheo.on.ca

Abstract. Retrospective clinical data presents many challenges for data mining and machine learning. The transcription of patient records from paper charts and subsequent manipulation of data often results in high volumes of noise as well as a loss of other important information. In addition, such datasets often fail to represent expert medical knowledge and reasoning in any explicit manner. In this research we describe applying data mining methods to retrospective clinical data to build a prediction model for asthma exacerbation severity for pediatric patients in the emergency department. Difficulties in building such a model forced us to investigate alternative strategies for analyzing and processing retrospective data. This paper describes this process together with an approach to mining retrospective clinical data by incorporating formalized external expert knowledge (*secondary knowledge sources*) into the classification task. This knowledge is used to partition the data into a number of coherent sets, where each set is explicitly described in terms of the secondary knowledge source. Instances from each set are then classified in a manner appropriate for the characteristics of the particular set. We present our methodology and outline a set of experiential results that demonstrate some advantages and some limitations of our approach.

1 Introduction

In his book [1], Motulsky submits *"the human brain excels at finding patterns and relationships ..."*. Scientists have long exhibited an aptitude to learn and

* The support of the Natural Sciences and Engineering Research Council of Canada, the Canadian Institutes of Health Research and the Ontario Centres of Excellence is gratefully acknowledged.

Z.W. Raś, S. Tsumoto, and D. Zighed (Eds.): MCD 2007, LNAI 4944, pp. 238–251, 2008.
© Springer-Verlag Berlin Heidelberg 2008

generalize from observations leading them to develop refined methods for detecting patterns and identifying coherent conjectures drawn from experience. Since their early days, intelligent computer systems have inspired scientists with their promising potential of supporting such research in medical domains [2]. However, medical data features many difficult domain-specific characteristics and complex properties [3]. Incompleteness (missing data), incorrectness (noise), sparseness (non-representative values), and inexactness (inappropriate parameter selection) make up the short list of challenges faced by any machine learning technique applied in the medical domain [4]. A comprehensive overview of these and other challenges is presented in [5], where medical data is described as often being heterogeneous in source as well as in structure, and that the pervasiveness of missing values for technical and/or social reasons can create problems for automatic methods for classification and prediction. Furthermore, translating physicians' interpretations based on years of clinical experience to formal models represents a serious and complex challenge.

An important requirement of medical problem solving or decision support applications is interpretability for domain users [6]. Such a stipulation dramatically reduces the choice of machine learning models that can be applied to medical problem solving to those that can offer systematic justification and explanation of the prediction process. Such models include classifiers that estimate probabilities (probabilistic), classifiers that identify training examples similar to a test example (case-based), classifiers that produce rules that can be applied to a given test example (rule-based), and classifiers that describe decisions based on a selected set of attributes (tree-based). In this work we have chosen to focus our prediction efforts on tree-based classifiers. Decision tree classification models are especially useful in medical applications as a result of their simple interpretation but also as they are represented in the form typically used for describing clinical algorithms and practice guidelines. As such a tree-based classification model can easily be represented in a comprehensible and transparent format if required, without the need for computer implementation.

In this work, the clinical prediction task is centered on the domain of emergency pediatric asthma where the goal is to develop a classification model that can provide an early prediction of the severity of a child's asthma exacerbation. Asthma is the most common chronic disease in children (10% of Canadian population), and asthma exacerbations are one of the most common reasons for children to be brought to the emergency department [7]. The provision of computer-based decision support to emergency physicians treating asthma patients has been shown to increase the overall effectiveness of health care delivered in emergency departments [8,9]. For a patient suffering from an asthma exacerbation, early identification of severity (*mild*, *moderate*, or *severe*) is a crucial part of the management and treatment process. Patients with a *mild* attack are usually discharged following a brief course of treatment (less than 4 hours) and resolution of symptoms, patients with a *moderate* attack receive more aggressive treatment over an extended observation in the emergency department (up to 12 hours), and patients with a *severe* attack receive maximal therapy before ultimately being transferred to an in-patient hospital bed for ongoing treatment (after about 16 hours in the emergency department).

This paper discusses challenges, issues, and difficulties we face in developing a prediction model for early asthma exacerbation severity using retrospective clinical data. Preliminary analysis of the data without preprocessing resulted in unacceptably low classification accuracy. These results forced a rethink of common methodologies for mining retrospective clinical data. Although not particularly complex, this data set is characterized by a fair amount of missing values such that standard methods of feature extraction and classifier tuning fail to produce acceptable performance. Furthermore, clinically-based "classifiers", such as PRAM (section 3.1) cannot be applied due to the type of data being collected. We employ such a clinical classifier as an external method to evaluate the data which leads to the identification of sets where PRAM can or cannot be readily employed We argue that such partitioning will ultimately improve the classification. Our investigations led us to develop a methodology for classification that involves identification and formalization of expert medical knowledge specific to the clinical domain. This knowledge is referred to as a secondary knowledge source and its incorporation allowed us to exploit implicit domain knowledge in the data for more fine-grained data analysis and processing. This paper demonstrates the usefulness of secondary knowledge to partition medical data and ultimately to to reduce its complexity. Our experimental evaluation demonstrates that with such partitioning a decision tree classifier is capable of overcoming some but not all complexities posed by this dataset. An added benefit is the ability to capture other regularities that should be in asthma data according to PRAM, thus in a sense, we "extend" its interpretation.

This paper is organized as follows. In Section 2 we describe the retrospectively collected asthma data used in this analysis. Section 3 outlines a methodology for identifying, formalizing and applying secondary knowledge sources with the purpose of harnessing and exploiting implicit domain knowledge. An experimental evaluation of this approach is outlined in Section 4, where our results display that the approach can be applied with some degree of success. We conclude with a discussion in Section 5.

2   Retrospective Clinical Dataset

The dataset used in this study was developed as part of a retrospective chart study conducted in 2004 at the Children's Hospital of Eastern Ontario (CHEO), Ottawa, Canada. The study includes patients who visited the hospital's emergency department from 2001 to 2003 for treatment of an asthma exacerbation. To illustrate the underlying structure of the data, we present the workflow by which asthma patients are processed in the emergency department (Figure 1). The workflow shows that a patient is evaluated multiple times by multiple caregivers at variable time intervals. This information is documented on the patient chart with varying degrees of completeness. Furthermore, some aspects of evaluation are objective and therefore reliable measures of the patient's status, however

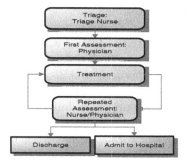

Fig. 1. Asthma Assessment Workflow in the Emergency Department at CHEO

other aspects can be quite subjective and less reliably correlated with the patient's status. In preparing the final dataset, patient information was divided into three subcategories for each record; historical and environmental information, information collected during the triage assessment and information collected at a reassessment approximately 2 hours after triage. The final dataset consisted of 362 records and each record was reviewed by a physician and assigned to one of two classes (*mild* or *moderate/severe*) using predefined criteria related to the duration and extent of treatment required, the final disposition (i.e., discharged or admitted to hospital), and the possible need for additional visits for ongoing symptoms. In this way, the assigned severity group was used as a gold standard for creation and evaluation of a prediction model.

The dynamic nature of asthma exacerbations and the collection of assessments over time would lend itself naturally to a temporal representation for analysis of data. However, inconsistencies in data recording meant it was not possible to incorporate a temporal aspect into the analysis. Further difficulties presented by the data were a significant number of missing values (for some attributes up to 98%), incorrectness, sparseness, and noise due to the variability with which information was recorded, and inexactness due to inappropriate parameter selections as well as the problem of "values as attributes" often encountered in medical data.

3   Secondary Knowledge Sources

Evidence-based medicine is a recent movement that has gained prominence in current clinical practice as a methodology for supporting clinical decision making. The practice of evidence based medicine involves integrating individual clinical expertise with the best available external clinical evidence from systematic research [10]. Individual clinical expertise refers to the proficiency and judgment that individual clinicians acquire through clinical experience and external clinical evidence describes clinically relevant research usually evaluated using randomized control trials. In practice evidence based medicine is applied in a number of ways, including, through the use of clinical practice guidelines, specialty-specific literature and clinical scoring systems.

In this research, we utilize external clinical evidence to support the classification task. The incorporation of a secondary knowledge source into classification leads us to define a three step approach to mining retrospective clinical data. In the first step relevant medical evidence is identified, for example in the form of a clinical practice guideline for the particular clinical domain. The second step is to formalize the medical evidence so it can be applied to available data. The third step involves developing a framework that makes use of the evidence to support the automatic classification task. The advantage of integrating such knowledge is that it allows for more effective and natural organization of information along existing and important data characteristics. As such secondary knowledge can be viewed as a proxy for an expert built classifier and may be incorporated to improve the predictive accuracy of the automatic classification task.

3.1   Secondary Knowledge Sources for Pediatric Asthma

The secondary knowledge source identified as relevant for our retrospective asthma data is the Preschool Respiratory Assessment Measure (PRAM) asthma index [11]. PRAM provides a discriminative index of asthma severity for preschool children. It is based on five clinical attributes commonly recorded for pediatric asthma patients, *suprasternal indrawing, scalene retractions, wheezing, air entry* and *oxygen saturation*. PRAM is based on a 12 point scale (see Table 1) and is calculated using scores of 0, 1, 2, and 3. These scores are assigned to attributes depending on the presence or absence of values as well as observed increasing or decreasing values of attributes. PRAM has been clinically validated as a reliable and responsive measure of the severity of airway obstruction. A patient with a PRAM score of 4 or less is considered to have a *mild* exacerbation,

Table 1. PRAM Scoring System

Signs	0	1	2	3
Suprasternal indrawing	absent		present	
Scalene retractions	absent		present	
Wheezing	absent	expiratory	inspiratory and expiratory	Audible without stethoscope /absent with no air entry
Air entry	normal	decreased bases	widespread decrease	absent/ minimal
Oxygen saturation	≥95%	92-95%	<92%	

a score between 5 and 8 corresponds to a *moderate* exacerbation, and a score of 9 or higher corresponds to a *severe* exacerbation.

In order to identify if the PRAM scoring system was appropriate secondary knowledge, the retrospective asthma dataset was analyzed for the presence of PRAM attributes. It was found that four of the five PRAM attributes were present in our data and values for these attributes may be collected twice for each record, once at triage and again at reassessment. The next step of our approach was to formalize the secondary knowledge source so that it could be applied to the classification task. This process is described in the next subsection.

3.2 Formalizing Secondary Knowledge Sources for Classification

The formalization of the secondary knowledge source involved determining a mapping from the set of attributes outlined by PRAM to a subset of attributes from the retrospective asthma data and an associated assignment of scores for attribute values. This was necessary as not all attributes required to calculate the PRAM score were present in the retrospective asthma data, and for some other attributes a 1:1 mapping did not exist. Specifically, the retrospective data did not contain an attribute corresponding to "Suprasternal Indrawing", and "Wheezing" was captured using two attributes in the retrospective data, inspiratory wheezing and expiratory wheezing. Also, the PRAM scoring system describes "Air Entry" using four values (normal, decreased bases, widespread decrease and absent/minimal), whereas our data defined air entry as either good (i.e., normal) or reduced. Therefore the formalized mapping was developed in conjunction with a domain expert (emergency physician), and the rules devised

Table 2. Mapping PRAM attributes and scores

Attribute(s)	Value(s)	Score
Oxygen Saturation	Greater than 95%	0
Oxygen Saturation	Greater than 92% and Less than 95%	1
Oxygen Saturation	Greater than 88% and Less than 92%	2
Oxygen Saturation	Less than 88%	3
Air Entry	Good	0
Air Entry (class = mild)	Reduced	1
Air Entry (class = other)	Reduced	3
Retractions AND Air Entry	Absent AND Good	0
Retractions AND Air Entry	Absent AND Reduced	1
Retractions AND Air Entry	Absent AND "Missing"	2
Retractions	Present	2
Expiratory AND Inspiratory Wheeze	Absent AND Absent	0
Expiratory AND Inspiratory Wheeze	Present AND Absent	1
Expiratory AND Inspiratory Wheeze	Present AND Present	2
Expiratory AND Inspiratory Wheeze	Absent AND Present	Undefined

for mapping attributes used by the PRAM system to attributes in our data and their corresponding score assignments are shown in Table 2.

3.3   Building a Classifier by a Secondary Knowledge Source

The final step of our approach was to use secondary knowledge to build a model for predicting asthma severity. In the retrospective asthma data a decision (class label) is recorded for each patient along with clinical and historical information. This class is the final outcome for the patient as recorded in the patient chart (not the result of the assessment at the 2-hour point) and indicates whether the patient has suffered a *mild* or *moderate/severe* exacerbation. Using the attributes, values and associated scores mapped from the PRAM scoring system we calculated a PRAM score for each patient in the dataset. This score had possible values between 0 and 12, where a score of less than 5 indicated a *mild* exacerbation and a score of greater than 5 indicated a *moderate/severe* exacerbation. (In our data the *moderate* and *severe* categories outlined by PRAM are collapsed into one group, *moderate/severe*). The score is then compared with the class label for each record in the dataset and the set of patients who comply with the PRAM scoring system are identified. The assignment of PRAM scores allows for the dataset to be partitioned into instances for which all PRAM attributes were present and thus a complete and correct PRAM score could be calculated and instances for which only a partial or no PRAM score could be calculated due to the absence of values for the PRAM attributes. PRAM attributes may be collected at two stages in the asthma workflow (triage and reassessment), however analysis of our data demonstrated that such attributes were more likely to be collected at reassessment (there were many missing values for triage attributes) and as such the dataset was partitioned using the larger set of reassessed values. This resulted in a dataset with 147 instances for which the PRAM score was complete and correct, 206 instances where only a partial or no PRAM score could be calculated due to missing values and 9 instances for which the score calculated by PRAM and the class label completely disagreed. These 9 cases were considered outliers and deleted from the dataset for evaluation.

4   Experimental Evaluation

4.1   Experimental Design

Our evaluation reports results from a number of experiments involving the retrospective asthma dataset where each experiment involved building a decision tree using the J48 decision tree classifier in Weka[12] to classify data. The first experiment involved building a classifier on the entire dataset prior to any application of secondary knowledge. These results serve as a baseline for classifier performance upon which to evaluate all subsequent results. In the next experiment secondary knowledge in the form of the PRAM scoring system was applied to partition the dataset into two sets, one containing PRAM complete and correct instances and one containing PRAM partial or incomplete instances. The purpose of this experiment is to demonstrate that the incorporation of secondary knowledge into the classification tasks allows for enhanced representation of data which results in reducing the complexity of the retrospective clinical dataset for classification. In the final experiment we applied feature selection to the complete original data set twice, once using automatic feature selection (available in Weka), and a second time by manually selecting combinations of expertly selected attributes and removing the remaining attributes. The function of this experiment was to show that neither automatic or expert feature selection can identify and reduce complexities in the data as efficiently as a classifier that incorporates secondary or expert medical knowledge, selects important features and partitions data into sets of similar characteristics.

For each experiment we report classifier performance in terms of percentages of Sensitivity (Sens) and Specificity (Spec), Predictive Accuracy (Acc) and Area Under the Curve (AUC) on the positive class. Sensitivity (the true positive rate) measures how often the classifier finds a set of positive examples. For instance in this research we consider *moderate/severe* to be the critical/positive class, therefore the sensitivity of *moderate/severe* measures how often the classifier correctly identifies patients suffering *moderate/severe* asthma exacerbations. Specificity (1 - false positive rate) measures how often what the classifier finds, is indeed what it was looking for. Therefore the specificity of the positive class (*moderate/severe*) measures how often what the classifier predicts is indeed a patient with a *moderate/severe* asthma exacerbation. Analyzing the trade-off between sensitivity and specificity is common in medical domains and is analogous to Receiver Operating Characteristics (ROC) analysis [13,14] used in machine learning [15].

In addition we report accuracy and AUC where accuracy is the rate at which the classifier classifies patients (in both classes) correctly while AUC represents the probability that the classifier will rank a randomly chosen positive instance higher than a randomly chosen negative instance [16]. Hence AUC measures the classifier's ability to discriminate the positive class from the negative class. In our experiments, we aim to analyze decision tree performance by measuring its ability to discriminate each positive patient with a *moderate/severe* asthma exacerbation. For a given classifier and positive class, an ROC curve [15,16] plots the true positive rate against the rate of false positives produced by the classifier on the test data. The points along the ROC curve are produced by varying the classification threshold from most positive classification to most negative classification and the AUC of a classifier is the area under the ROC curve [16]. For this reason as well as the relatively small size of the dataset we evaluate the classifier for each patient in the dataset using leave-one-out cross-validation.

4.2   Classifying the Entire Dataset

The first experiment involved building a decision tree on the original dataset of 362 instances. The results of this experiment are shown in the first row of Table 3 and demonstrate that the retrospective clinical dataset is complex and that good classification accuracy is difficult to achieve without performing some degree of data preprocessing. We include these results as a baseline by which to measure subsequent classifier performance.

4.3   Classifying PRAM and Non-PRAM Sets

In this experiment the dataset was partitioned by applying the formalized mapping from PRAM scoring system to attributes from the retrospective asthma dataset. This resulted in the dataset being partitioned into those that were PRAM complete and correct (PRAM set) and those that were either PRAM partial or incomplete (non-PRAM set). A decision tree was built for each set and the results are shown in the last two rows of Table 3. Also included for reference purposes are the results for the entire dataset in the first row.

Table 3. Decision Trees built on PRAM and non-PRAM sets

Set	Size	Sens	Spec	Acc	AUC
Entire	362	73	63	69	69
PRAM Set	**147**	**93**	**96**	**95**	**98**
Non-PRAM Set	**206**	**89**	**53**	**74**	**77**

From Table 3 we observe that splitting the datset into different sets based on formalized secondary knowledge increases classification accuracy of the PRAM set. For the non-PRAM set classification improves in terms of Sensitivity, Accuracy and AUC. In particular sensitively on the PRAM set increases by 20% from the baseline. In addition a large gain in AUC from the baseline reflects the increased probability that a positive example is ranked higher than a negative example. In fact, when the decision tree is supplemented with secondary knowledge (the PRAM set) we gain an increase in AUC, and when secondary knowledge cannot be so easily applied (non-PRAM set), the performance only improves marginally on that of the baseline. These results demonstrate that the incorporation of formalized secondary knowledge sources can help with classification in such domains "by exposing" the concept to be learned by the decision tree and thus reducing the overall complexity of the dataset by exploiting domain knowledge implicitly present in the data. The results represent an overall improvement on previous research into classification of clinical data with tree-based classifiers [8,9].

However from the results in Table 3 we also observe a decrease of 10% in specificity between the Non-PRAM set and the baseline. This performance is inadequate in terms of achieving a balance between high sensitivity and high specificity. We note however that in terms of the problem domain high sensitivity and low specificity on the positive class translates to the fact that the classifier is very accurate in identifying *moderate/severe* patients and recommending they are kept for an extended time in the emergency department, however at the same time the classifier is over conservative in recommending that *mild* patients are

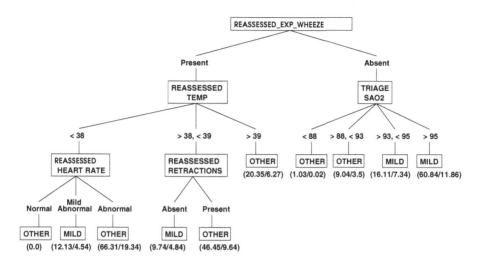

Fig. 2. The resulting decision tree for the classifier trained on the entire data

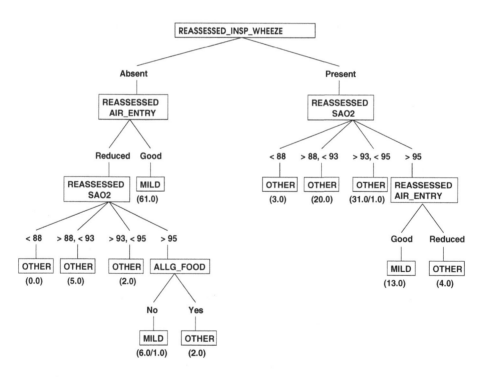

Fig. 3. The resulting decision tree for the classifier trained on PRAM data

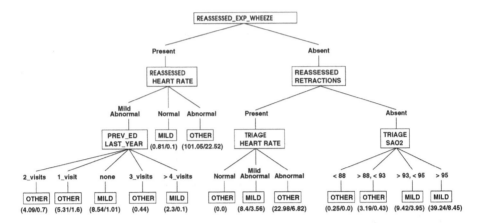

Fig. 4. The resulting decision tree for the classifier trained on Non-PRAM data

kept for longer than usual stays. Such direction of classification a less serious error than one occurring in the opposite situation.

Furthermore, consider figures 2, 3 and 4 which present the resulting decision tree classifier obtained from training with the entire data, PRAM data only, and Non-PRAM data respectively. We clearly see that the classifier trained on the entire data (2) focuses on clinical measures recorded during the TRIAGE and the REASSESSMENT phase of the workflow, yet, this classifier delivers the worst classification performance. On the other hand, building the other two classifiers (figures 3 and 4) allows for the consideration of other historically relevant attributes, such as ALLG_FOOD in figure 3 and PREV_ED_LAST_YEAR in figure 4. Despite the difference in all of these classifiers, lower levels of clinically measured Oxygen saturation remains an indication of severe asthma exacerbation, see REASSESSED_SAO2 and TRIAGE_SAO2 subtrees in all three figures.

An important observation is that most PRAM attributes (data attributes mapped to attributes shown in table 1 on page 242) are being used in all three decision trees. This illustrates the significant relevance of these PRAM attributes to the classification task, thus, providing data-driven evidence to support PRAM. In addition, these decision trees allows us to extend PRAM and present a more fine grained representation of it, including dependencies which could be easily explained and presented to physicians.

4.4   Automatic and Expert-Driven Feature Selection

It is acceptable to state that a reduction in dimensionality of data can reduce the complexity of underlying concepts that it may represent. The purpose of this experiment is to demonstrate that data complexity in this domain requires more than dimensionality reduction to reduce its complexity. We compare results obtained from applying automatic and expert-driven feature selection methods to those obtained by partitioning the data according to PRAM secondary knowledge. Automatic feature selection is based on standard methods used by the data

Table 4. Automatic and Expert Feature Selection

Feature Selection Mode	Mode	Size	Sens	Spec	Acc	AUC
Information Gain	Automatic	362	72	63	68	69
Chi-squared	Automatic	363	72	63	68	69
Combinatorial	Automatic	362	72	65	69	71
Wrapper with Naive Bayes	Automatic	362	71	60	70	77
On All Attributes	Expert	362	72	66	70	73
On Only PRAM Attributes	Expert	362	77	78	70	71

mining community and are available in the Weka software. The expert-driven feature selection methods are based on selecting attributes observed to useful to classification from our repeated experiments and by an expert and those outlined by the PRAM scoring system. 10 methods of feature automatic selection were applied to the dataset where each was used in conjunction with a decision tree for classification. The results for the best four methods are shown in rows 1-4 of Table 4. Comparing the results for automatic feature selection to those for the baseline as outlined in Table 3 we can conclude it is not successful in reducing the complexity of the dataset. In general results do not display any improvement in classification except in the case where a wrapper using a Naive Bayes classifier for optimization is used for feature selection. Here we note an increase in AUC, however this is at the expense of a large decrease in specificity. In applying expert feature selection, we built one classifier using all data records of attributes collected during the reassessment only and another classifier using only the attributes that were mapped from the PRAM scoring system while still using all instances available in the dataset. The results for these two experiments are shown in rows 5-6 of Table 4. Again comparing these results to those outlined for the baseline in Table 3 we observe no significant improvement.

However, by comparing the results from Table 4 to those for classification on the PRAM and non-PRAM sets in Table 3 a number of important conclusions can be drawn. Partitioning data into different sets for classification based on secondary knowledge results in much improved classification that of using either automatic or expert feature selection. Augmenting the developed classifier with external knowledge allows for more effective classification by exploiting underlying domain knowledge in the dataset and by organizing data according to these concepts. Such classification accuracy cannot be captured by a classification model developed on the data alone. The partitioning of data does not reduce the dimensionality of the dataset like traditional methods for classification such as feature selection, however it manages to reduce the complexity of the dataset by using secondary knowledge to identify more coherent sets into which data more naturally fits.

The intention is to use the classification results from the PRAM and non-PRAM sets from Table 3 to implement a prediction model for asthma severity. This can be achieved in a number of ways. One option is to develop a metaclassifier that could learn to direct new instances to either the model built on the

PRAM set or the model built on the non-PRAM set. For such a metaclassifier values of PRAM attributes alone may be sufficient to make the decision or it may be necessary to develop a method by which unseen patients can be related to the sets (PRAM and non-PRAM) we identify in the dataset. Alternatively the predictions from both sets could be combined to perform the prediction task. One option is to use a voting mechanism, another is to build these classifiers in a manner that produce rankings of the severity of the exacerbation. With such a methodology the classifier with the highest ranking provides a better insight into the condition. However, such an approach introduces additional issues in terms of interpretations and calibrations of ranks and probabilities. Such a study remains as part of our future research directions.

5   Discussion

We have introduced an approach to mining complex retrospective clinical data by incorporating secondary knowledge to supplement the classification task by reducing the complexity of the dataset. The methodology involves identification of a secondary knowledge source suitable for the clinical domain, formalization of the knowledge to analyze and organize data according to the underlying principle of the secondary knowledge, and incorporation of the secondary knowledge into the chosen classification model. In this research we concentrated on classifying information using a decision tree to satisfy the requirement that classification should be easily interpreted by domain users. From our experimental results we draw a number of conclusions. Firstly we have demonstrated that domain knowledge is implicit in the data as the dataset partitions naturally into two sets for classification with the application of a formalized mapping from the PRAM scoring system. This is in spite of the fact that the mapping was inexact; our dataset only contained four of the five attributes outlined by PRAM and some attribute values had slightly different representations. In such a way the application of secondary knowledge reduces the complexity of the dataset by allowing for the exploitation of underlying domain knowledge to supplement data analysis, representation and classification. As outlined, this approach is more successful than traditional methods for reducing data complexity such as feature selection which fail to capture a measure of the expert knowledge implicit in the retrospectively collected data. A further advantage of the approach was demonstrated by the ability of the secondary knowledge to help identify outlier examples in the data. However, the results are still somewhat disappointing in terms of achieving a balance acceptable in medical practice between high sensitivity and high specificity in the non-PRAM set. We believe that a high proportion of missing values in this set is causing difficulties for the classification model. This issue remains an open problem for future research. In other future work we are interested in further investigating attributes used by the PRAM system and to test whether all attributes used by PRAM are necessary for enhanced classification.

References

1. Motulsky, H.: Intuitive Biostatistics. Oxford University Press, New York (1995)
2. Ledley, R.S., Lusted, L.B.: Reasoning foundations of medical diagnosis. Science 130, 9–21 (1959)
3. Mullins, I.M., Siadaty, M.S., Lyman, J., Scully, K., Garrett, C.T., Miller, W.G., Muller, R., Robson, B., Apte, C., Weiss, S., Rigoutsos, I., Platt, D., Cohen, S., Knaus, W.A.: Data mining and clinical data repositories: Insights from a 667,000 patient data set. Computers Biology and Medicine 36(12), 1351–1377 (2006)
4. Magoulas, G.D., Prentza, A.: Machine learning in medical applications. In: Paliouras, G., Karkaletsis, V., Spyropoulos, C.D. (eds.) ACAI 1999. LNCS (LNAI), vol. 2049, pp. 300–307. Springer, Heidelberg (2001)
5. Cios, K.J., Moore, G.W.: Uniqueness of medical data mining. A. I. in medicine 26(1-2), 1–24 (2002)
6. Lavrac, N.: Selected techniques for data mining in medicine. Artificial Intelligence in Medicine 16(1), 3–23 (1999)
7. Lozano, P., Sullivan, S., Smith, D., Weiss, K.: The economic burden of asthma in us children: estimates from the national medical expenditure survey. The Journal of allergy and clinical immunology 104(5), 957–963 (1999)
8. Kerem, E., Tibshirani, R., Canny, G., Bentur, L., Reisman, J., Schuh, S., Stein, R., Levison, H.: Predicting the need for hospitalization in children with acute asthma. Chest 98, 1355–1361 (1990)
9. Lieu, T.A., Quesenberry, C.P., Sorel, M.E., Mendoza, G.R., Leong, A.B.: Computer-based models to identify high-risk children with asthma. American Journal of Respiratory and Critical Care Medicine 157(4), 1173–1180 (1998)
10. Sackett, D., Rosenberg, W., Muir Gray, J., Haynes, R., Richardson, W.: Evidence based medicine: what it is and what it isn't. British Medical Journal (1996)
11. Chalut, D.S., Ducharme, F.M., Davis, G.M.: The preschool respiratory assessment measure (pram): A responsive index of acute asthma severity. Pediatrics 137(6), 762–768 (2000)
12. Witten, I.H., Frank, E.: Data mining: Practical machine learning tools and techniques (2005)
13. Sox, H.C.J., Blatt, M.A., Higgins, M.C., Marton, K.I.: Medical Decision Making, Boston (1998)
14. Faraggi, D., Reiser, B.: Computer-based models to identify high-risk children with asthma. American Journal of Respiratory and Critical Care Medicine 157(4), 1173–1180 (1998)
15. Provost, F., Fawcett, T.: Analysis and visualization of classifier performance: comparison under imprecise class and cost distributions. In: The Third International Conference on Knowledge Discovery and Data Mining, pp. 34–48 (1997)
16. Fawcett, T.: Roc graphs: Notes and practical considerations for data mining researchers, Technical Report HPL-2003-4, HP Labs (2003)

POM Centric Multi-aspect Data Analysis for Investigating Human Problem Solving Function

Shinichi Motomura[1], Akinori Hara[1], Ning Zhong[2,3], and Shengfu Lu[3]

[1] Graduate School, Maebashi Institute of Technology, Japan
[2] Department of Life Science and Informatics,
Maebashi Institute of Technology, Japan
[3] The International WIC Institute, Beijing University of Technology, China
motomura@maebashi-it.org

Abstract. In the paper, we propose an approach of POM (peculiarity oriented mining) centric multi-aspect data analysis for investigating human problem solving related functions, in which computation tasks are used as an example. The proposed approach is based on Brain Informatics (BI) methodology, which supports studies of human information processing mechanism systematically from both macro and micro points of view by combining experimental cognitive neuroscience with advanced information technology. We describe how to design systematically cognitive experiments to obtain multi-ERP data and analyze spatiotemporal peculiarity of such data. Preliminary results show the usefulness of our approach.

1 Introduction

Problem-solving is one of main capabilities of human intelligence and has been studied in both cognitive science and AI [9], where it is addressed in conjunction with reasoning centric cognitive functions such as attention, control, memory, language, reasoning, learning, and so on. We need to better understand how human being does complex adaptive, distributed problem solving and reasoning, as well as how intelligence evolves for individuals and societies, over time and place [3,11,12,13,17]. Then, we catch problem solving from the standpoint of Brain Informatics, and address systematically for the solution of a process.

Brain Informatics (BI) is a new interdisciplinary field to study human information processing mechanism systematically from both macro and micro points of view by cooperatively using experimental, theoretical, cognitive neuroscience and advanced information technology [16,17]. It attempts to understand human intelligence in depth, towards a holistic view at a long-term, global vision to understand the principles, models and mechanisms of human information processing system.

Our purpose is to understand activities of human problem solving system by investigating the spatiotemporal features and flow of human problem solving system, based on functional relationships between activated areas of human brain. More specifically, at the current stage, we want to understand:

Z.W. Raś, S. Tsumoto, and D. Zighed (Eds.): MCD 2007, LNAI 4944, pp. 252–264, 2008.

- how a peculiar part (one or more areas) of the brain operates in a specific time;
- how the operated part changes along with time;
- how the activated areas work cooperatively to implement a whole problem solving system;
- how the activated areas are linked, indexed, navigated functionally, and what are individual differences in performance.

Based on this point of view, we propose a way of peculiarity oriented mining (POM) for knowledge discovery in multiple human brain data.

The rest of the paper is organized as follows. Section 2 provides a BI Methodology for multi-aspect human brain data analysis of human problem solving system. Sections 3 explain how to design the experiment of an ERP mental arithmetic task with visual stimuli standing on BI Methodology. Sections 4 describe how to do multi-aspect analysis in the obtained ERP data, respectively, as an example to investigate human problem solving and to show the usefulness of the proposed mining process. Finally, Section 5 gives concluding remarks.

2   Brain Informatics Methodology

Brain informatics pursues a holistic understanding of human intelligence through a systematic approach to brain research. BI regarded the human brain as an information processing system (HIPS) and emphasizes cognitive experiments to understand its mechanisms for analyzing and managing data. Such systematic study includes the following 4 main research issues:

- systematic investigation of human thinking centric mechanisms;
- systematic design of cognitive experiments;
- systematic human brain data management;
- systematic human brain data analysis.

The first issue is based on the observation for Web intelligence research needs and the state-of-the-art cognitive neuroscience. In cognitive neuroscience, although many advanced results with respect to "perception oriented" study have been obtained, only a few of preliminary, separated studies with respect to "thinking oriented" and/or a more whole information process have been reported [1].

The second issue is with respect to how to design the psychological and physiological experiments for obtaining various data from HIPS, in a systematic way. In other words, by systematic design of cognitive experiments in BI methodology, the data obtained from a cognitive experiment and/or a set of cognitive experiments may be used for multi-task/purpose.

The third issue relates to manage human brain data, which is based on a conceptual model of cognitive functions that represents functional relationships among multiple human brain data sources for systematic investigation and understanding of human intelligence.

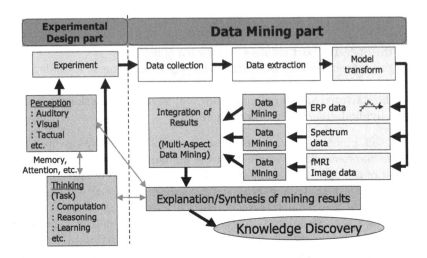

Fig. 1. A flow based on BI methodology

The last issue is concerned with how to extract significant features from multiple brain data measured by using fMRI and EEG in preparation for multiaspect data analysis by combining various data mining methods with reasoning [4,6,10,11,15].

An investigation flow based on BI methodology is shown in Figure 1, in which various tools can be cooperatively used in the multi-step process for experimental design, pre-processing (data extraction, modeling and transformation), multiaspect data mining and post-processing.

3 The Experiment of Mental Arithmetic Task with Visual Stimuli

As mentioned above, based on BI methodology, the data obtained from a cognitive experiment and/or a set of cognitive experiments may be used for multitask/purpose, including for investigating both lower and higher functions of HIPS. For example, it is possible that our experiment can meet the following requirements: investigating the mechanisms of human visual and auditory systems, computation, problem-solving (i.e. the computation process is regarded as an example of problem-solving process), and the spatiotemporal feature and flow of HIPS in general. Figure 2 gives a computation process from the macro viewpoints, with respect to several component functions of human solving problem, such as attention, interpretation, short-term memory, understanding of work, computation, checking.

In this work, the ERP (event-related potential) human brain waves are derived by carrying out a mental arithmetic task with visual stimuli, as an example to investigate human problem solving process. ERP is a light, sound, and

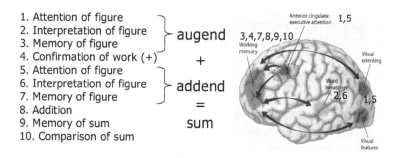

1. Attention of figure
2. Interpretation of figure
3. Memory of figure
4. Confirmation of work (+)
5. Attention of figure
6. Interpretation of figure
7. Memory of figure
8. Addition
9. Memory of sum
10. Comparison of sum

Fig. 2. Computation as an example of problem-solving

brain potential produced with respect to the specific phenomenon of spontaneous movement [2].

3.1 Outline of Experiments

The experiment conducted this time shows a numerical calculation problem to a subject, and asks the subject to solve it in mental arithmetic, and the shown sum has hit, or it pushes a button, and performs a judging of corrigenda. The form of the numerical calculation to be shown are the addition problem of "augend + addend = sum" or "augend, summand 1, summand 2, summand 3, summand 4 = sum". The wrong sum occurs at half the probability, and the distribution is not uniform.

In the experiments, three states (tasks), namely, *on-task*, *off-task*, and *no-task*, exist by the difference in the stimulus given to a human subject. *on-task* is the state which is calculating by looking a number. *off-task* is the state which is looking the number that appears at random. *no-task* is the relaxed state which does not work at all.

Figure 3 gives an example of the screen state transition. We set two presentation types and three levels. Type A is the figure remains on the screen. But, Type B is the figure doesn't remain on the screen. Level 1 is single digit addition (with no carry). Level 2 is double digits addition (with carry), and Level 3 is continuous addition of 5 figures (one digit).

We try to compare and analyze what is the relationship between tasks. By this design, it is possible to analyze the influence, in the different levels of difficulty, with the same *on-task* and *off-task*, and to make a comparison between *on-task* and *off-task* in the same difficulty.

3.2 Trigger Signal and Timing Chart

It is necessary to measure EEG relevant to a certain event to the regular timing in measurement of ERP repeatedly. In this research, since the attention was paid

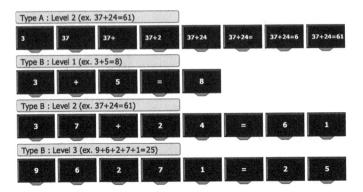

Fig. 3. Experimental design with four difficulty levels

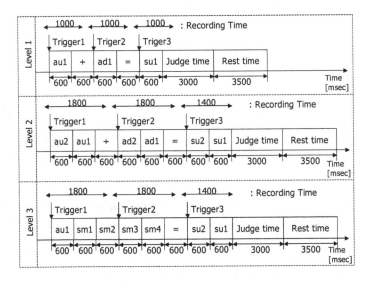

Fig. 4. Timing chart of on-task

to each event of augend, addend, and sum presentation in calculation activities. Pre-trigger was set to 200 [msec], and addition between two digits are recorded in 1800 or 1400 [msec], respectively. Figure 4 gives an example of the time chart. "au" is augend, "ad" is addend, "sm" is summand, and "su" is sum. Therefore "au2" is MSD (last 2-digits) of augend, and "au1" is LSD (last 1-digits) of augend.

3.3   Experimental Device and Measurement Conditions

Electroencephalographic activity was recorded using a 64 channel BrainAmp amplifier (Brain Products, Munich, Germany) with a 64 electrode cap. The electrode cap is based on an extended international 10-20 system. Furthermore, eye movement measurement (2ch) is also used. The sampling frequency is 2500Hz to be processed. The number of experimental subjects is 20.

4   Multi-aspect Data Analysis (MDA)

It is possible that our experiment can meet the following requirements: investigating the mechanisms of human visual and auditory systems, computation, problem-solving, and the spatiotemporal feature and flow of HIPS in general. Furthermore, multi-aspect data analysis (MDA) can mine several kinds of rules and hypotheses from different data sources. In this work, we use the three methods, called the ERP data, potential topography, frequency topography (frequency analysis) for multi-aspect data analysis. Furthermore, a more exact result can be obtained by integration and explanation of MDA results.

The first method is the ERP analysis. In this work, the ERP (event-related potential) human brain waves are derived by carrying out a mental arithmetic task with visual stimuli, as an example to investigate human problem solving process. ERP is a light, sound, and brain potential produced with respect to the specific phenomenon of spontaneous movement [2].

The second method is the potential topography analysis. The advantage of the potential topography is that it can catch electrical fluctuation by the view of spatiotemporal. This means it makes possible for us to recognize distribution and appearance of positive and negative potentials from macroperspective. Thus, we can find the difference of features on the potentials that focus on each step of computation process.

The third method is the frequency topography analysis. The advantage of frequency topography is that it can catch frequency element fluctuation by the view of spatiotemporal. This means it makes possible for us to recognize distribution and appearance of alpha and beta waves from macroperspective. Thus, we can find the difference of features on the frequency elements that focus on each step of computation process. And we can also put medical knowledge to practical use by analyzing frequency element.

4.1   ERP Analysis

For the measured EEG data, a maximum of 40 addition average processing were performed, and the ERPs were derived by using Brain Vision Analyzer (Brain Products, Munich, Germany). Generally speaking, the Wernicke area of a left temporal lobe and the prefrontal area are related to the calculation process [5]. In this study, we pay attention to recognition of the number, short-term memory and ,integrated processing, vision, as well as compare Type A and Type B by focusing on important channels (Fz, FC5, Oz).

Figure 5 shows the ERP. We can see some difference between Type A and Type B, or Level 1, Level 2 and Level 3. We focus on the Fz that is related with memory of numbers and integrated processing. In the Type A which is low on burden of memory, the positive potential is lower than that of Type B. On the

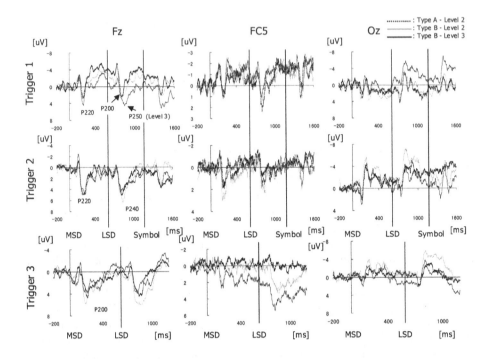

Fig. 5. ERP comparison in Levels 2 and 3

other hand, delaying in the latent time and phenomenon that shows a strong potential fluctuation has occurred in Level 3 which is high on burden and Level 2 of Type B.

4.2 Potential Topography Analysis

Figure 6 shows the potential topography. In this part, the method is different from those of ERP. We depaint to change in potentials from beginning of computation time to end without any trigger signal. In Level 1, a positive potential appears in the frontal cortex just when the subject judges an answer. In Level 3 which is an addition task of continuity, a positive potential appears all of the time. And in the comparison of Type A with Type B in Level 2, a positive potential appears just in the frontal cortex in Type A but it also appears in the visual cortex in Type B. This result seems to be related with the difference of the attention level in the visual scene.

4.3 Frequency Topography Analysis

Figure 7 shows the frequency topography, which compares the strength of theta wave with that of Alfa wave. Both theta and Alfa waves do not change so much in the temporal axis. On the other hand, both Alfa and theta waves are decreasing

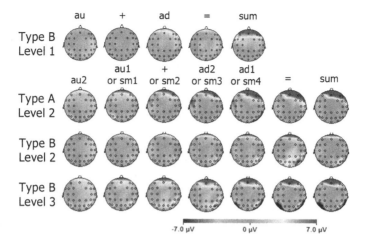

Fig. 6. Potential topography of each event

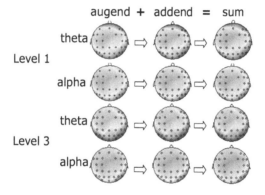

Fig. 7. Frequency topography of theta and Alfa waves

in the frontal cortex of Level 3 when we focus on the difference of levels. This phenomenon can be attributed to workload.

4.4 Peculiarity Oriented Mining

The result of making the best use of each feature was derived from two aspects, the potential change and the frequency element, with different difficulty levels of experiments. Furthermore, POM (Peculiarity Oriented Mining) based methods are used for multi-aspect mining to find interesting time-band and space features in the change of the potential change and the frequency element.

It is clear that a specific part of the brain operates in a specific time and the operations change over time. Although detecting the concavity and convexity (P300 etc.) of ERP data is easy by using the existing tool, it is difficult to find a peculiar one in multiple channels with the concavity and convexity [7,8]. In order to discover new knowledge and models of human information processing activities, it is necessary to pay attention to the peculiar channel and time in ERPs for investigating the spatiotemporal features and flow of a human information processing system.

peculiarity oriented mining (POM) is a proposed knowledge discovery methodology [14,15]. The main task of POM is the identification of peculiar data. The attribute-oriented method of POM, which analyze data from a new view and are different from traditional statistical methods, has been recently proposed by Zhong *et al.* and applied in various real-world problems [14,15]. Unfortunately, such POM is not totally fit for ERP data analysis. The reason is that the useful aspect for ERP data analysis is not amplitude, but the latent time. In order to solve this problem, we extend POM to Peculiarity Vector Oriented Mining (PVOM). After smoothing enough by moving average processing, in the time series, we pay the attention to each potential towards N pole or P pole. Furthermore, the channel with the direction different from a lot of channels is considered to be a peculiar channel at that time. Hence, the distance between the attribute-values is expressed at the angle. And this angle can be obtained from the inner product and the norm in the vector. Let inclination of wave i in a certain time t be x_{it}. The extended PF (Peculiarity Factor) corresponding to ERP can be defined by the following Eq. (1).

$$PF(x_{it}) = \sum_{k=1}^{n} \theta(x_{it}, x_{kt})^{\alpha}. \tag{1}$$

$\alpha = 0.5$ as default. In normally POM, PF is obtained by distance between two attribute values. However, θ in Eq. (1) is an angle which the wave in time t makes. For the θ, we can compute for an angle using Eq. (2).

$$cos\theta = \frac{1 + x_{it} \cdot x_{kt}}{\sqrt{1 + x_{it}^2}\sqrt{1 + x_{kt}^2}}. \tag{2}$$

Based on the peculiarity factor, the selection of peculiar data is simply carried out by using a threshold value. More specifically, an attribute value is peculiar if its peculiarity factor is above minimum peculiarity p, namely, $PF(x_{it}) \geq p$. The threshold value p may be computed by the distribution of PF as follows:

$$threshold = mean\ of\ PF(x_{it}) + \tag{3}$$
$$\beta \times standard\ deviation\ of\ PF(x_{it})$$

where β can be adjusted by a user, and $\beta = 1$ is used as default. The threshold indicates that a data is a peculiar one if its PF value is much larger than the mean of the PF set. In other words, if $PF(x_{it})$ is over the threshold value, x_{it} is a peculiar data. By adjusting the parameter β, a user can control and adjust the threshold value.

In this work, we want to mine four kinds of patterns, which are classified into two types of peculiarity with respect to the temporal and channel axises, respectively, as shown in Figure 8. Mining 1 and Mining 2 are used to find temporal

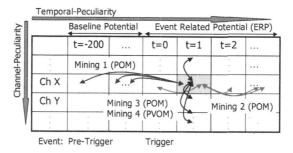

Fig. 8. Views of ERP peculiarity

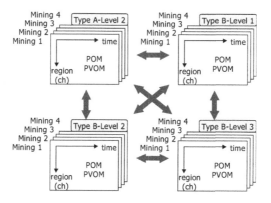

Fig. 9. The integration model of POM based multi-aspect mining

peculiarity, and Mining 3 and Mining 4 are used to find channel peculiarity, respectively.

More specifically, Mining 1 examines whether the potential at arbitrary time is peculiar compared with a baseline potential in channel X. Mining 2 examines whether the potential at arbitrary time is peculiar compared with an ERP in channel X. Mining 3 examines whether the potential of channel X is peculiar compared with the potential on other channels in a specific time. Furthermore, Mining 4 examines whether a potential change of channel X is peculiar compared with a potential change on other channels in a specific time. As shown in Figure 8, the POM method is used for the Mining 1 to Mining 3 and the extended POM method (PVOM) is used for the Mining 4.

4.5 Integration and Explanation of Results

How to explain and integrate the results obtained by POM based multi-aspect mining is a key issue. First of all, we examined an integrated model of the results with MDA in consideration of the spatiotemporal features. Figure 9 provides a global view of the proposed model for such an integration and explanation.

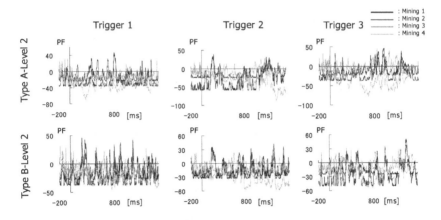

Fig. 10. Multi-POM analysis of FC5

In this figure, the horizontal axis denotes time, and the vertical axis denotes region (channel, Brodmann area etc.). It becomes easy to discover a unique phenomenon and new knowledge because the results obtained by POM based multi-aspect mining are managed in a layered structure. On the other hand, because the difference of the spatiotemporal resolution in MDA, it is important that the representation of the results is based on adequate collation.

We have applied the POM based method to all channels in each difficulty level for finding peculiar channels and time-bands. Some remarkable results were obtained, especially in the FC5 and Iz. FC5 is located a little ahead in the left temporal cortex, which is relevant to the function of language recognition. In contrast, Iz is located a little backward, which is relevant to the function of recognizing the image from eye. Here we discuss the change of peculiar values which focus on the difference of difficulties in these two channels.

First, we discuss the change of PF (Peculiarity Factor) values in FC5 which focuses on the difference between Type A and Type B in Level 2. Figure 10 shows the change of PF values of Type A and Type B of Level 2 in FC5. In this figure, the X axis denotes the time, and the Y axis denotes PF values. A time-band is regarded as a peculiar time-band if its PF value is over 0 since we set y = 0 as a threshold. In Type B, the potential peculiarity is high on both time and space. It is found out, from the result of Minings 1, 3 and 4, that FC5 is with more peculiarity than any other channels, because its potential change is very big. Although Type B is with respect to the two-digit addition task, it displays one-digit at a time. Therefore, we can guess that the subject recognizes a current digit or number strongly every time it is displayed. The hypothesis can been evidenced by the difference between Type A and Type B in Trgger 2, which is taken when addend is displayed. Furthermore, medical knowledge and insight can explain the result well.

Next, we discuss the change of peculiar values in Iz, which focuses on the difference between Type A-Level 2 and Type B-Level 3. Figure 11 shows the change of PF values of Type A-Level 2 and Type B-Level 3 in Iz. From this

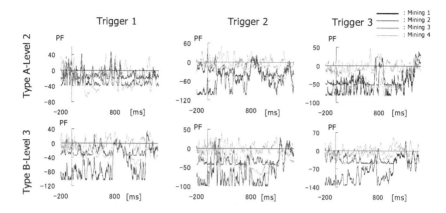

Fig. 11. Multi-POM analysis of Iz

figure, we can see that the potential change in Iz is very smaller than the change in any other channels, from the results of Minings 2, 3 and 4. And it is found out from the result of Mining 1 that the difference of potentials in Iz definitely depends on a time-band. The difference between Type A and Type B is about the visual attention. In Type A, numbers are remained on the screen. In contrast, in Type B the subject must continue to add numbers with watching the change on the screen. The mining result shows that the subjects in Type A is high for the level of the gaze to the number to go in the latter half. We will give a more deep discussion on the rationality of this result in other papers.

5   Conclusion

In this paper, we investigated human problem solving related functions by using computation as an example, which demonstrate what is BI methodology and its usefulness. The proposed POM centric multi-aspect ERP data analysis based on BI methodology shifts the focus of cognitive science from a single type of experimental data analysis towards a deep, holistic understanding of human information processing principles, models and mechanisms.

Our future work includes obtaining and analyzing more subject data, combining with fMRI human brain image data for multi-aspect analysis in various approaches of data mining and reasoning.

Acknowledgments

This work is partially supported by the grant-in-aid for scientific research (No. 18300053) from the Japanese Ministry of Education, Culture, Sports, Science and Technology.

References

1. Gazzaniga, M.S. (ed.): The Cognitive Neurosciences III. The MIT Press, Cambridge (2004)
2. Handy, T.C.: Event-Related Potentials, A Methods Handbook. MIT Press, Cambridge (2004)
3. Liu, J., Jin, X., Tsui, K.C.: Autonomy Oriented Computing: From Problem Solving to Complex Systems Modeling. Springer, Heidelberg (2005)
4. Megalooikonomou, V., Herskovits, E.H.: Mining Structure-Function Associations in a Brain Image Database. In: Cios, K.J. (ed.) Medical Data Mining and Knowledge Discovery, pp. 153–179. Physica-Verlag, Heidelberg (2001)
5. Mizuhara, H., Wang, L., Kobayashi, K., Yamaguchi, Y.: Long-range EEG Phase-synchronization During an Arithmetic Task Indexes a Coherent Cortical Network Simultaneously Measured by fMRI. NeuroImage 27(3), 553–563 (2005)
6. Mitchell, T.M., Hutchinson, R., Niculescu, R.S., Pereira, F., Wang, X., Just, M., Newman, S.: Learning to Decode Cognitive States from Brain Images. Machine Learning 57(1-2), 145–175 (2004)
7. Nittono, H., Nageishi, Y., Nakajima, Y., Ullsperger, P.: Event-related Potential Correlates of Individual Differences in Working Memory Capacity. Psychophysiology 36, 745–754 (1999)
8. Picton, T.W., Bentin, S., Berg, P., Donchin, E., Hillyard, S.A., Johnson, R., et al.: Guidelines for using human event-related potentials to study cognition: Recording standards and publication criteria. Psychophysiology 37, 127–152 (2000)
9. Newell, A., Simon, H.A.: Human Problem Solving. Prentice-Hall, Englewood Cliffs (1972)
10. Sommer, F.T., Wichert, A. (eds.): Exploratory Analysis and Data Modeling in Functional Neuroimaging. MIT Press, Cambridge (2003)
11. Sternberg, R.J., Lautrey, J., Lubart, T.I.: Models of Intelligence. American Psychological Association (2003)
12. Yao, Y.Y.: A Partition Model of Granular Computing. In: Peters, J.F., Skowron, A., Grzymała-Busse, J.W., Kostek, B.z., Świniarski, R.W., Szczuka, M. (eds.) Transactions on Rough Sets I. LNCS, vol. 3100, pp. 232–253. Springer, Heidelberg (2004)
13. Zadeh, L.A.: Precisiated Natural Language (PNL). AI Magazine 25(3), 74–91 (2004)
14. Zhong, N., Yao, Y.Y., Ohshima, M.: Peculiarity Oriented Multi-Database Mining. IEEE Transaction on Knowledge and Data Engineering 15(4), 952–960 (2003)
15. Zhong, N., Wu, J.L., Nakamaru, A., Ohshima, M., Mizuhara, H.: Peculiarity Oriented fMRI Brain Data Analysis for Studying Human Multi-Perception Mechanism. Cognitive Systems Research 5(3), 241–256 (2004)
16. Zhong, N.: Building a Brain-Informatics Portal on the Wisdom Web with a Multi-Layer Grid: A New Challenge for Web Intelligence Research. In: Torra, V., Narukawa, Y., Miyamoto, S. (eds.) MDAI 2005. LNCS (LNAI), vol. 3558, pp. 24–35. Springer, Heidelberg (2005)
17. Zhong, N.: Impending Brain Informatics (BI) Research from Web Intelligence (WI) Perspective. International Journal of Information Technology and Decision Making, World Scientific 5(4), 713–727 (2006)

Author Index

Abe, Akinori 182
Abe, Hidenao 72

Bahri, Emna 131
Basile, T.M.A. 13
Bathoorn, Ronnie 157
Biba, M. 13

Ceci, Michelangelo 209
Ciesielski, Krzysztof 116

d'Amato, Claudia 42
Dardzińska, Agnieszka 104
Delteil, Alexandre 82
Dembczyński, Krzysztof 169
Di Mauro, N. 13

El Sayed, Ahmad 224
Elazmeh, William 238
Esposito, Floriana 13, 42

Fanizzi, Nicola 42
Farion, Ken 238
Ferilli, S. 13
Furutani, Michiko 182
Furutani, Yoshiyuki 182

Grcar, Miha 1
Grobelnik, Marko 1

Hacid, Hakim 224
Hagita, Norihiro 182
Hara, Akinori 252
Hirabayashi, Satoru 72
Hirano, Shoji 27

Kłopotek, Mieczysław A. 116
Kolczyńska, Elżbieta 93
Kontkiewicz, Aleksandra 82
Kotłowski, Wojciech 169
Kryszkiewicz, Marzena 82

Lu, Shengfu 252

Maddouri, Mondher 131
Malerba, Donato 209
Marcinkowska, Katarzyna 82
Matsuoka, Rumiko 182
Matwin, Stan 238
Michalowski, Wojtek 238
Mladenic, Dunja 1
Motomura, Shinichi 252

Nicoloyannis, Nicolas 131

O'Sullivan, Dympna 238
Ohkawa, Takenao 143
Ohsaki, Miho 72
Ozaki, Tomonobu 143

Peters, James F. 57
Protaziuk, Grzegorz 82

Ramanna, Sheela 57
Raś, Zbigniew W. 104, 196
Rybinski, Henryk 82

Sehatkar, Morvarid 238
Siebes, Arno 157
Słowiński, Roman 169

Tsumoto, Shusaku 27

Wasyluk, Hanna 196
Wieczorkowska, Alicja 93
Wierzchoń, Sławomir T. 116
Wilk, Szymon 238
Wyrzykowska, Elżbieta 196

Yamaguchi, Takahira 72
Yamamoto, Tsubasa 143

Zhang, Xin 104
Zhong, Ning 252
Zighed, Djamel 224

Lecture Notes in Artificial Intelligence (LNAI)

Vol. 5012: T. Washio, E. Suzuki, K.M. Ting, A. Inokuchi (Eds.), Advances in Knowledge Discovery and Data Mining. XXIV, 1102 pages. 2008.

Vol. 5009: G. Wang, T. Li, J.W. Grzymala-Busse, D. Miao, A. Skowron, Y. Yao (Eds.), Rough Sets and Knowledge Technology. XVIII, 765 pages. 2008.

Vol. 4994: A. An, S. Matwin, Z.W. Raś, D. Ślęzak (Eds.), Foundations of Intelligent Systems. XVII, 653 pages. 2008.

Vol. 4953: N.T. Nguyen, G.S. Jo, R.J. Howlett, L.C. Jain (Eds.), Agent and Multi-Agent Systems: Technologies and Applications. XX, 909 pages. 2008.

Vol. 4946: I. Rahwan, S. Parsons, C. Reed (Eds.), Argumentation in Multi-Agent Systems. X, 235 pages. 2008.

Vol. 4944: Z.W. Raś, S. Tsumoto, D. Zighed (Eds.), Mining Complex Data. X, 265 pages. 2008.

Vol. 4938: T. Tokunaga, A. Ortega (Eds.), Large-Scale Knowledge Resources. IX, 367 pages. 2008.

Vol. 4933: R. Medina, S. Obiedkov (Eds.), Formal Concept Analysis. XII, 325 pages. 2008.

Vol. 4930: I. Wachsmuth, G. Knoblich (Eds.), Modeling Communication with Robots and Virtual Humans. X, 337 pages. 2008.

Vol. 4929: M. Helmert, Understanding Planning Tasks. XIV, 270 pages. 2008.

Vol. 4924: D. Riaño (Ed.), Knowledge Management for Health Care Procedures. X, 161 pages. 2008.

Vol. 4923: S.B. Yahia, E.M. Nguifo, R. Belohlavek (Eds.), Concept Lattices and Their Applications. XII, 283 pages. 2008.

Vol. 4914: K. Satoh, A. Inokuchi, K. Nagao, T. Kawamura (Eds.), New Frontiers in Artificial Intelligence. X, 404 pages. 2008.

Vol. 4911: L. De Raedt, P. Frasconi, K. Kersting, S. Muggleton (Eds.), Probabilistic Inductive Logic Programming. VIII, 341 pages. 2008.

Vol. 4908: M. Dastani, A. El Fallah Segrouchni, A. Ricci, M. Winikoff (Eds.), Programming Multi-Agent Systems. XII, 267 pages. 2008.

Vol. 4898: M. Kolp, B. Henderson-Sellers, H. Mouratidis, A. Garcia, A.K. Ghose, P. Bresciani (Eds.), Agent-Oriented Information Systems IV. X, 292 pages. 2008.

Vol. 4897: M. Baldoni, T.C. Son, M.B. van Riemsdijk, M. Winikoff (Eds.), Declarative Agent Languages and Technologies V. X, 245 pages. 2008.

Vol. 4894: H. Blockeel, J. Ramon, J. Shavlik, P. Tadepalli (Eds.), Inductive Logic Programming. XI, 307 pages. 2008.

Vol. 4885: M. Chetouani, A. Hussain, B. Gas, M. Milgram, J.-L. Zarader (Eds.), Advances in Nonlinear Speech Processing. XI, 284 pages. 2007.

Vol. 4874: J. Neves, M.F. Santos, J.M. Machado (Eds.), Progress in Artificial Intelligence. XVIII, 704 pages. 2007.

Vol. 4870: J.S. Sichman, J. Padget, S. Ossowski, P. Noriega (Eds.), Coordination, Organizations, Institutions, and Norms in Agent Systems III. XII, 331 pages. 2008.

Vol. 4869: F. Botana, T. Recio (Eds.), Automated Deduction in Geometry. X, 213 pages. 2007.

Vol. 4865: K. Tuyls, A. Nowe, Z. Guessoum, D. Kudenko (Eds.), Adaptive Agents and Multi-Agent Systems III. VIII, 255 pages. 2008.

Vol. 4850: M. Lungarella, F. Iida, J.C. Bongard, R. Pfeifer (Eds.), 50 Years of Artificial Intelligence. X, 399 pages. 2007.

Vol. 4845: N. Zhong, J. Liu, Y. Yao, J. Wu, S. Lu, K. Li (Eds.), Web Intelligence Meets Brain Informatics. XI, 516 pages. 2007.

Vol. 4840: L. Paletta, E. Rome (Eds.), Attention in Cognitive Systems. XI, 497 pages. 2007.

Vol. 4830: M.A. Orgun, J. Thornton (Eds.), AI 2007: Advances in Artificial Intelligence. XIX, 841 pages. 2007.

Vol. 4828: M. Randall, H.A. Abbass, J. Wiles (Eds.), Progress in Artificial Life. XII, 402 pages. 2007.

Vol. 4827: A. Gelbukh, Á.F. Kuri Morales (Eds.), MICAI 2007: Advances in Artificial Intelligence. XXIV, 1234 pages. 2007.

Vol. 4826: P. Perner, O. Salvetti (Eds.), Advances in Mass Data Analysis of Signals and Images in Medicine, Biotechnology and Chemistry. X, 183 pages. 2007.

Vol. 4819: T. Washio, Z.-H. Zhou, J.Z. Huang, X. Hu, J. Li, C. Xie, J. He, D. Zou, K.-C. Li, M.M. Freire (Eds.), Emerging Technologies in Knowledge Discovery and Data Mining. XIV, 675 pages. 2007.

Vol. 4811: O. Nasraoui, M. Spiliopoulou, J. Srivastava, B. Mobasher, B. Masand (Eds.), Advances in Web Mining and Web Usage Analysis. XII, 247 pages. 2007.

Vol. 4798: Z. Zhang, J.H. Siekmann (Eds.), Knowledge Science, Engineering and Management. XVI, 669 pages. 2007.

Vol. 4795: F. Schilder, G. Katz, J. Pustejovsky (Eds.), Annotating, Extracting and Reasoning about Time and Events. VII, 141 pages. 2007.

Vol. 4790: N. Dershowitz, A. Voronkov (Eds.), Logic for Programming, Artificial Intelligence, and Reasoning. XIII, 562 pages. 2007.

Vol. 4788: D. Borrajo, L. Castillo, J.M. Corchado (Eds.), Current Topics in Artificial Intelligence. XI, 280 pages. 2007.

Vol. 4775: A. Esposito, M. Faundez-Zanuy, E. Keller, M. Marinaro (Eds.), Verbal and Nonverbal Communication Behaviours. XII, 325 pages. 2007.

Vol. 4772: H. Prade, V.S. Subrahmanian (Eds.), Scalable Uncertainty Management. X, 277 pages. 2007.

Vol. 4766: N. Maudet, S. Parsons, I. Rahwan (Eds.), Argumentation in Multi-Agent Systems. XII, 211 pages. 2007.

Vol. 4760: E. Rome, J. Hertzberg, G. Dorffner (Eds.), Towards Affordance-Based Robot Control. IX, 211 pages. 2008.

Vol. 4755: V. Corruble, M. Takeda, E. Suzuki (Eds.), Discovery Science. XI, 298 pages. 2007.

Vol. 4754: M. Hutter, R.A. Servedio, E. Takimoto (Eds.), Algorithmic Learning Theory. XI, 403 pages. 2007.

Vol. 4737: B. Berendt, A. Hotho, D. Mladenic, G. Semeraro (Eds.), From Web to Social Web: Discovering and Deploying User and Content Profiles. XI, 161 pages. 2007.

Vol. 4733: R. Basili, M.T. Pazienza (Eds.), AI*IA 2007: Artificial Intelligence and Human-Oriented Computing. XVII, 858 pages. 2007.

Vol. 4724: K. Mellouli (Ed.), Symbolic and Quantitative Approaches to Reasoning with Uncertainty. XV, 914 pages. 2007.

Vol. 4722: C. Pelachaud, J.-C. Martin, E. André, G. Chollet, K. Karpouzis, D. Pelé (Eds.), Intelligent Virtual Agents. XV, 425 pages. 2007.

Vol. 4720: B. Konev, F. Wolter (Eds.), Frontiers of Combining Systems. X, 283 pages. 2007.

Vol. 4702: J.N. Kok, J. Koronacki, R. Lopez de Mantaras, S. Matwin, D. Mladenič, A. Skowron (Eds.), Knowledge Discovery in Databases: PKDD 2007. XXIV, 640 pages. 2007.

Vol. 4701: J.N. Kok, J. Koronacki, R. Lopez de Mantaras, S. Matwin, D. Mladenič, A. Skowron (Eds.), Machine Learning: ECML 2007. XXII, 809 pages. 2007.

Vol. 4696: H.-D. Burkhard, G. Lindemann, R. Verbrugge, L.Z. Varga (Eds.), Multi-Agent Systems and Applications V. XIII, 350 pages. 2007.

Vol. 4694: B. Apolloni, R.J. Howlett, L. Jain (Eds.), Knowledge-Based Intelligent Information and Engineering Systems, Part III. XXIX, 1126 pages. 2007.

Vol. 4693: B. Apolloni, R.J. Howlett, L. Jain (Eds.), Knowledge-Based Intelligent Information and Engineering Systems, Part II. XXXII, 1380 pages. 2007.

Vol. 4692: B. Apolloni, R.J. Howlett, L. Jain (Eds.), Knowledge-Based Intelligent Information and Engineering Systems, Part I. LV, 882 pages. 2007.

Vol. 4687: P. Petta, J.P. Müller, M. Klusch, M. Georgeff (Eds.), Multiagent System Technologies. X, 207 pages. 2007.

Vol. 4682: D.-S. Huang, L. Heutte, M. Loog (Eds.), Advanced Intelligent Computing Theories and Applications. XXVII, 1373 pages. 2007.

Vol. 4676: M. Klusch, K.V. Hindriks, M.P. Papazoglou, L. Sterling (Eds.), Cooperative Information Agents XI. XI, 361 pages. 2007.

Vol. 4667: J. Hertzberg, M. Beetz, R. Englert (Eds.), KI 2007: Advances in Artificial Intelligence. IX, 516 pages. 2007.

Vol. 4660: S. Džeroski, L. Todorovski (Eds.), Computational Discovery of Scientific Knowledge. X, 327 pages. 2007.

Vol. 4659: V. Mařík, V. Vyatkin, A.W. Colombo (Eds.), Holonic and Multi-Agent Systems for Manufacturing. VIII, 456 pages. 2007.

Vol. 4651: F. Azevedo, P. Barahona, F. Fages, F. Rossi (Eds.), Recent Advances in Constraints. VIII, 185 pages. 2007.

Vol. 4648: F. Almeida e Costa, L.M. Rocha, E. Costa, I. Harvey, A. Coutinho (Eds.), Advances in Artificial Life. XVIII, 1215 pages. 2007.

Vol. 4635: B. Kokinov, D.C. Richardson, T.R. Roth-Berghofer, L. Vieu (Eds.), Modeling and Using Context. XIV, 574 pages. 2007.

Vol. 4632: R. Alhajj, H. Gao, X. Li, J. Li, O.R. Zaïane (Eds.), Advanced Data Mining and Applications. XV, 634 pages. 2007.

Vol. 4629: V. Matoušek, P. Mautner (Eds.), Text, Speech and Dialogue. XVII, 663 pages. 2007.

Vol. 4626: R.O. Weber, M.M. Richter (Eds.), Case-Based Reasoning Research and Development. XIII, 534 pages. 2007.

Vol. 4617: V. Torra, Y. Narukawa, Y. Yoshida (Eds.), Modeling Decisions for Artificial Intelligence. XII, 502 pages. 2007.

Vol. 4612: I. Miguel, W. Ruml (Eds.), Abstraction, Reformulation, and Approximation. XI, 418 pages. 2007.

Vol. 4604: U. Priss, S. Polovina, R. Hill (Eds.), Conceptual Structures: Knowledge Architectures for Smart Applications. XII, 514 pages. 2007.

Vol. 4603: F. Pfenning (Ed.), Automated Deduction – CADE-21. XII, 522 pages. 2007.

Vol. 4597: P. Perner (Ed.), Advances in Data Mining. XI, 353 pages. 2007.

Vol. 4594: R. Bellazzi, A. Abu-Hanna, J. Hunter (Eds.), Artificial Intelligence in Medicine. XVI, 509 pages. 2007.

Vol. 4585: M. Kryszkiewicz, J.F. Peters, H. Rybinski, A. Skowron (Eds.), Rough Sets and Intelligent Systems Paradigms. XIX, 836 pages. 2007.

Vol. 4578: F. Masulli, S. Mitra, G. Pasi (Eds.), Applications of Fuzzy Sets Theory. XVIII, 693 pages. 2007.

Vol. 4573: M. Kauers, M. Kerber, R. Miner, W. Windsteiger (Eds.), Towards Mechanized Mathematical Assistants. XIII, 407 pages. 2007.

Vol. 4571: P. Perner (Ed.), Machine Learning and Data Mining in Pattern Recognition. XIV, 913 pages. 2007.

Vol. 4570: H.G. Okuno, M. Ali (Eds.), New Trends in Applied Artificial Intelligence. XXI, 1194 pages. 2007.

Vol. 4565: D.D. Schmorrow, L.M. Reeves (Eds.), Foundations of Augmented Cognition. XIX, 450 pages. 2007.